Laboratory Manual for Exercise Physiology

G. Gregory Haff, PhD, CSCS*D, FNSCA, ASCC

School of Exercise and Health Sciences
Edith Cowan University

Charles Dumke, PhD, FACSM

Health and Human Performance
University of Montana

Human Kinetics

Cataloging-in-Publication Data

Haff, Greg.
    Laboratory manual for exercise physiology / G. Gregory Haff, Charles Dumke.
        p. ; cm.
    Includes bibliographical references and index.
    ISBN 978-0-7360-8413-0 (soft cover) -- ISBN 0-7360-8413-4 (soft cover)
    I. Dumke, Charles, 1966- II. Title.
    [DNLM: 1. Exercise--physiology--Laboratory Manuals. 2. Physical Fitness--physiology--Laboratory Manuals. 3. Exercise Test--Laboratory Manuals. QT 25]

    613.7'1--dc23

                    2011050765

ISBN-10: 0-7360-8413-4
ISBN-13: 978-0-7360-8413-0

The web addresses cited in this text were current as of August 2011, unless otherwise noted.

Acquisitions Editor: Amy Tocco; **Developmental Editor:** Judy Park; **Assistant Editor:** Brendan Shea, PhD; **Copyeditor:** Tom Tiller; **Permissions Manager:** Dalene Reeder; **Graphic Designer:** Joe Buck; **Graphic Artist:** Yvonne Griffith; **Cover Designer:** Keith Blomberg; **Photograph (cover):** © Human Kinetics; **Photographs (interior):** Ryan Gardner, Visual People Design, Inc/© Human Kinetics; **Photo Asset Manager:** Jason Allen; **Visual Production Assistant:** Laura Fitch; **Photo Production Manager**: Joyce Brumfield; **Art Manager:** Kelly Hendren; **Associate Art Manager:** Alan L. Wilborn; **Illustrations:** © Human Kinetics; **Printer:** United Graphics

We thank the University of Montana, in Missoula, Montana, for assistance in providing the location for the photo shoot for this book.

Printed in the United States of America   10   9   8   7   6

The paper in this book is certified under a sustainable forestry program.

Human Kinetics
Website: www.HumanKinetics.com

United States: Human Kinetics
P.O. Box 5076
Champaign, IL 61825-5076
800-747-4457
e-mail: humank@hkusa.com

Canada: Human Kinetics
475 Devonshire Road Unit 100
Windsor, ON N8Y 2L5
800-465-7301 (in Canada only)
e-mail: info@hkcanada.com

Europe: Human Kinetics
107 Bradford Road
Stanningley
Leeds LS28 6AT, United Kingdom
+44 (0) 113 255 5665
e-mail: hk@hkeurope.com

Australia: Human Kinetics
57A Price Avenue
Lower Mitcham, South Australia 5062
08 8372 0999
e-mail: info@hkaustralia.com

New Zealand: Human Kinetics
P.O. Box 80
Torrens Park, South Australia 5062
0800 222 062
e-mail: info@hknewzealand.com

E4876

Contents

Laboratory 13 Pulmonary Function Testing   361

Laboratory 14 Body Composition Assessments   383

Laboratory 15 Electrocardiograph Measurements 413

Acknowledgments

I would like to thank Human Kinetics for the patience with Chuck and I as we slogged through the writing of this text. Specifically I would like to thank Roger Earle for his belief in our abilities and more importantly for his friendship. Additionally, I would like to thank Judy, Brendan, and Amy for working tirelessly to assist us in this process.

While the book was difficult to complete and at times was very stressful I would like to thank Chuck for his patience and dedicated efforts and more importantly for being one of my very best friends and favorite colleagues.

I would like to thank my colleagues Sophia Nimphius, Jeremy Sheppard, Robert Newton, and Michael Newton. Moving to Australia has been a traumatic event for me, but you each have made life a little easier.

To my friends Michael Stone, Bill Sands, and Travis Triplett, I am honored to call you my friends and you each impact my life more than you know. I look forward to enjoying life in your company.

Finally, it would be amiss if I did not acknowledge the most important person in my life—my wife, Erin. You are the rock that supports me in all endeavors that I undertake. While things never seem to go smoothly or appear to be working out, your ability to ground me and to make me laugh and stop and watch the waves is more than anyone could ever want in a life partner. I am blessed to have you in my life.

—*Greg Haff*

Thanks to Greg for lassoing me into this ordeal— no, really it's been great, even through all the deadlines. I wouldn't have wanted to do this by myself or with anybody else. Human Kinetics deserves a shout-out for their experience, professionalism, and patience with our ideas. Thank you for taking this book to product.

I would also like to thank my colleagues in the department of Health and Human Performance at the University of Montana for their patience with me as I worked on this project. I guess I can't use the "book card" anymore to get off of committees!

Acknowledgement should also be made to my parents, Bob and Leah, for their lifetime support and to my wife Shannon and son Carter who make it all worthwhile.

—*Charles Dumke*

Preface

Laboratory Manual for Exercise Physiology is a detailed source of basic tests that you can teach in an undergraduate laboratory course in exercise physiology. The text covers a wide variety of tests typically performed in the exercise physiology laboratory when evaluating athletes, clinical clients, or other generally healthy individuals. The manual's design allows you to choose activities that best suit your individual course needs. Specifically, each chapter offers you a wide variety of laboratory activities that can be mixed and matched depending upon the instrumentation and time allotted for your course. The range of field and laboratory tests presented here gives your students broad exposure to testing that can be applied in a wide variety of professional settings. Organized in a logical progression, the labs build in complexity as students progress through the book and develop their knowledge base. Ultimately, the text serves as a resource for basic testing procedures used in assessing human performance, health, and wellness.

Special Laboratory Features

Each laboratory chapter is a complete lesson beginning with objectives, definitions of key terms, and background information that sets the stage for learning. For each of the laboratory activities, you will find step-by-step instructions, making it easier for those new to the lab setting to complete the procedures. Each laboratory activity has a data sheet to record your individual findings, as well as question sets related to the data collected by students in the class; these questions invite students to put their laboratory experience into context.

Online Resources

A web resource, available at www.HumanKinetics.com/LaboratoryManualForExercisePhysiology, provides additional tools that can assist you in working through the lab activities. Group data sheets found only in the web resource allow you to move beyond collecting individual data. With these group sheets you may compile data from the entire class, calculate values such as mean and range, and compare findings to the normative data discussed in the lab. Here you will also find each of the question sets, providing an easy way to turn in answers after completing a laboratory activity. You will also find practical case study questions for each lab activity. These questions help students begin to critically analyze data collection and synthesize it with material they have learned in lecture and other courses.

Instructors for this course may go online for access to an image bank which features all the photos, illustrations, and tables from the text that can be inserted into tests, quizzes, handouts, and other course materials.

Notes for Instructors

This manual is geared specifically for use in an exercise physiology laboratory course. It is designed to translate the scientific foundation developed in a core exercise physiology lecture course—using, for example, a text such as *Physiology of Sport and Exercise* by Kenney, Wilmore, and Costill (5th edition, Human Kinetics, 2011) into practical applications typically performed in a variety of settings. To accomplish this goal, the manual is divided into 15 laboratories that lead students through a series of activities: the basics of testing, pretesting screening, and methods of evaluating flexibility, muscular strength, anaerobic fitness, aerobic fitness, resting metabolic rate, lactate metabolism, body composition, blood pressure, respiratory fitness, and electrocardiogram assessment.

eBook available at HumanKinetics.com

Each laboratory provides background information and detailed step-by-step procedures for a variety of tests. In addition, since exercise physiology laboratories are equipped in various ways, the labs present multiple methods for introducing the testing concept. For example, laboratory 12 presents multiple methods for assessing vertical jump performance: jump-and-reach, Vertec, and switch mat. Equipment lists at the beginning of each activity make it easier to choose the labs that will work best in the lab facility. This versatility enables you to choose activities that best fit your facilities and best meet the needs of your students.

Resource Finder

Primary Data Collection

Objectives

1. Define basic terminology associated with testing.
2. Learn metric conversions and the units recommended by the International System of Units.
3. Provide a rationale for collecting basic information during testing.
4. Present the methods for evaluating temperature, barometric pressure, and relative humidity.
5. Present basic statistical tests for use in evaluating test results.
6. Describe types of graphics for use in presenting data.

Definitions

accuracy—Degree of a measurement's closeness to the quantity's actual value.

barometric pressure—Pressure exerted by ambient air.

central tendency—Score that best represents all scores collected for a group.

dependent variable—Effect or yield of the independent variable.

displacement—Length that an object moves in a straight line between two points.

distance—Total length that an object travels (may or may not be in a straight line).

energy—Capacity to do work.

field/laboratory test—Test that can be completed in either a field or a laboratory setting.

field test—Test completed in a field setting.

force—Mass multiplied by acceleration.

independent variable—Variable that is manipulated.

laboratory test—Sophisticated test that must be conducted in a laboratory setting.

mass—Measure of matter that constitutes an object; expressed in kg.

mean—Average score of a sample.

median—Middle score of a sample.

mode—Most frequent score in a sample.

normative data—Placement within a population; also referred to as *norms* or *norm data*.

power—Rate at which work can be performed.

precision—Degree to which a test is reproducible with nearly the same value.

range—Distance between end points in the group of scores.

relative humidity—Amount of water or percent saturation in ambient air.

reliability—Repeatability of a measure.

speed—Scalar quantity generally considered to be how fast a body is moving; calculated by dividing distance covered by time.

standard—Desirable or target score.

standard deviation—Measure of variability that shows variation or dispersion from the mean.

validity—Accuracy of a measure.

variable—A characteristic.

variability—Spread of a data set.

velocity—Vector quantity calculated by dividing displacement by time.

wet-bulb globe temperature (WBGT)—Measure of temperature that estimates cooling capacity of the surrounding environment.

work—Force times the distance through which it acts.

Testing human performance under exercise conditions allows us to evaluate the human body's functional ability. This information can give us keen insight into the body's overall health and wellness, as well as the individual's athletic performance capacity. We can also garner information about an individual's ability to tolerate and adapt to exercise by examining his or her post-exercise responses. Numerous tests can be performed in the exercise physiology laboratory in order to evaluate health and wellness (1, 6, 20) or examine athletic performance capacity (13, 26). Many of these tests can be categorized into three basic classifications: field, field/laboratory, and laboratory.

Field tests allow us to assess specific fitness and performance variables in a field setting (16). These tests are generally practical and less expensive than their laboratory-based counterparts (29). Though not often used for research due to difficulty in controlling external variables (e.g., weather, terrain), these tests are extremely useful for screening and monitoring purposes (6). They offer a high degree of validity and are often developed based upon their laboratory counterparts. Examples in exercise physiology include the 1- to 1.5-mile (1.6 to 2.4 km) run test, the 1-mile (1.6 km) jogging test, the 12-minute cycling test, sprints, and body mass index measurements (3, 10, 15). Though typically done in field settings, some of these tests may also be done in laboratory settings (e.g., BMI, 12 min cycling test).

Field/laboratory tests can be conducted in either the field or a laboratory setting. Like field tests, they often require only minimal equipment, but they are subjected to tighter controls (6), and a field/laboratory test performed in the field must be performed with the same tight controls that would be used in the laboratory (3). One example of a field/laboratory test is the step test (6). In the laboratory, this test can be performed using a step box, which limits the number of subjects to one. In the field, the step test can be performed on stadium bleachers with a large number of subjects at the same time. Regardless of location, the step test also requires a metronome and stopwatch. Other examples of field/laboratory tests include the sit-and-reach test, skinfold assessments, vertical jump testing, and blood pressure measurements (6).

Laboratory tests are conducted with the highest level of control and often require expensive equipment that cannot be taken into the field; as a result, they are usually performed on one person at a time (6) and thus tend to be time consuming. In return, they offer a higher degree of accuracy and precision (6). Examples include measurement of maximal oxygen consumption, quantification of resting metabolic rate, exercise electrocardiograms, dual X-ray absorptiometry, underwater weighing, quantification of isometric force-time curves, and anaerobic treadmill testing (4, 5, 6, 11).

Test Variables

One central goal in the exercise physiology laboratory is to quantify specific physiological or performance characteristics. A characteristic is usually termed a **variable**. Variables often quantified in an exercise physiology laboratory include maximal strength, body composition, aerobic power, and flexibility.

Variables are generally categorized as either independent or dependent (27). The **independent variable** is the one that is manipulated, whereas the **dependent variable** is the effect of, response to, or yield of the independent variable (27). In other words, the independent variable is controlled by the person administering the test, and the dependent variable is the physiological or performance response to the independent variable. In a treadmill test, for example, the independent variable is the speed of the treadmill; the dependent variable is the heart rate response or oxygen consumption rate. When graphically representing these variables, we place the independent variable on the *x*-axis (horizontal axis) and the dependent variable on the *y*-axis (vertical axis) (8). In the example of the treadmill test, then, the workload or speed would be the independent variable presented on the *x*-axis, and the heart rate response would be the dependent variable presented on the *y*-axis.

Measurement Terminology

When examining physiological and performance characteristics, an exercise physiologist uses specific measurement terminology to discuss the data. Here are several of those terms:

The most common measurements of **mass** performed in the exercise physiology laboratory are measurements of body mass, lean body mass, and fat mass. Though the terms *body mass* and *body weight* are often used interchangeably in the United States, it is more accurate to use the term *body mass*. The Système International d'Unités (SI) base unit for mass is the kilogram (kg).

Force is a vector quantity characterized by both magnitude and direction. It can be calculated according to Isaac Newton's second law, which states that force is equivalent to mass multiplied by acceleration (14):

$$\text{Force} = \text{mass} \times \text{acceleration}$$

For example, body mass is used often in this laboratory manual when calculating work or work rate, and in such instances body mass is converted to newtons and used in other calculations. Forces are generally expressed as newtons (N), which are calculated by multiplying the object's mass in kg by its downward acceleration due to gravity:

$$\text{Force (N)} = \text{mass} \times \text{acceleration} = \text{mass in kg} \times 9.81 \text{ m} \cdot \text{s}^{-2}$$

When thinking about force, it is not uncommon to consider the expression of strength. Generally, strength is considered to be the maximal force generated by a muscle or group of muscles at a specific velocity (15) or the ability to generate external force (35). Such capability often constitutes a major concern. An individual's force-generating capacity plays an important role in his or her ability to perform in sport (10) and in activities of daily living (20). In addition, the ability to repetitively express submaximal forces is important in endurance activities. Thus strength, or force-generating capacity, exerts a major effect on athletic performance and overall fitness status.

Both **displacement** and **distance** can be considered as lengths. Displacement is measured in a straight line from one point to another, whereas distance is the total length that an object travels and may or may not be limited to a straight line. While both are typically measured in centimeters or meters, the meter is the SI base unit. Quantification of displacement and distance contributes to calculations such as work, velocity of movement, and expression of power.

To calculate the amount of **work** completed, we multiply the amount of force exerted on an object by the distance that the object is moved:

$$\text{Work (J)} = \text{force (N)} \times \text{distance (m)}$$

The results of this equation are reported in joules (J; the SI unit for work), which is equal to N·m. As a whole, work is directly related to the amount of metabolic energy expended—the more work performed, the more kilocalories used (29).

It is important to differentiate between **speed** and **velocity**. Speed is a scalar quantity generally considered to be how fast a body is moving, which is directly related to the total distance covered divided by time. While very similar to speed, velocity is an actual vector that has a magnitude and a direction. Therefore, velocity is based on both speed and direction (25). The difference between speed and velocity centers on the difference between distance and displacement. Mathematically, speed is calculated by dividing distance by time, whereas velocity is calculated by dividing displacement by time:

$$\text{Speed} = \text{distance} / \text{time}$$

$$\text{Velocity} = \text{displacement} / \text{time}$$

For example, if a runner completes a 400 m dash in 49 s, he or she would have an average velocity of 0 m·s^{-1} because his or her displacement would be zero. In this instance, then, it would be better to use average speed to represent how fast the runner is moving, and the result would be 8.16 m·s^{-1}.

Velocity is the linear speed of the object, distance represents how far the object has moved in a given direction, and time represents how long it took to cover that distance (25). For example, if a 100 m sprint was run in 9.58 s, the average velocity of the event would be 10.4 m·s^{-1}.

Power is the rate at which work can be performed (33), and the ability to express high power outputs is one of the most important factors in sport performance (20). Power can be calculated in several ways:

$$Power = work / time$$

$$= force \times velocity$$

$$= force \times (distance / time)$$

Power is generally considered to be proportional to the amount of energy used (33)—thus, the higher the power output, the higher the work rate, which corresponds to a faster expenditure of energy. Because of this relationship, power is often used when discussing the transfer of metabolic energy to physical performance (24, 23). For example, power is used to describe this transformation when talking about aerobic power and anaerobic power.

In a general sense, **energy** should be considered as the capacity to do work (9), and it is often represented in terms of joules (J). When examining metabolic energy release—the result of work done (energy used) and heat released (energy wasted)—the joule is the universally accepted unit (6). In the United States, however, it is more common to represent this release as kilocalories (kcal).

It is possible to estimate caloric or kilojoule (kJ) expenditure by determining oxygen uptake and cost. It can be assumed that 1 liter (L) of oxygen uptake corresponds to approximately 5 kilocalories (kcal) or 21 kilojoules (kJ) of energy expenditure.

Thus caloric expenditure can be calculated with one of the following formulas (3):

$$Energy\ expenditure\ (kcal) = oxygen\ uptake\ (L \cdot min^{-1}) \times 5$$

$$Energy\ expenditure\ (kJ) = oxygen\ uptake\ (L \cdot min^{-1}) \times 21$$

These figures are only approximations, and they can be influenced by the intensity of exercise. Specifically, higher intensities result in more energy being expended per liter of oxygen consumed (6).

Metric Conversions

The use of metric units is standard in most exercise physiology laboratories. Though the metric system is not popular in the United States, it is the preferred system when performing testing and conducting research. Laboratory work in exercise physiology typically involves three categories of conversion: length, weight, and volume.

The standard metric and SI unit for length presented in scientific publications is the meter; it is generally used when reporting an individual's height. The meter is easily converted to other metric units, such as the millimeter, centimeter, and kilometer (e.g., 10 m = 1,000 cm = 10,000 mm). To convert from the American system to the metric system, simply multiply length in inches by 2.54 to figure the length in centimeters. A fuller list of conversions can be found in the accompanying highlight box.

Standard Length Conversions

1 inch	= 0.0254 m	= 2.54 cm	= 25.4 mm
1 foot	= 0.3048 m	= 30.48 cm	= 304.8 mm
1 yard	= 0.914 m	= 91.44 cm	= 914.4 mm
1 mile	= 1.609 km	= 1,609.34 m	= 160,934.4 cm
1 millimeter	= 0.1 cm	= 0.001 m	= 0.000001 km
1 centimeter	= 10 mm	= 0.01 m	= 0.00001 km
1 kilometer	= 1,000,000 mm	= 100,000 cm	= 1,000 m
1 millimeter	= 0.0394 in.	= 0.00328 ft	
1 centimeter	= 0.3937 in.	= 0.03281 ft	
1 kilometer	= 39,370.1 in.	= 3,280.8 ft	= 0.62137 mi

Note: cm = centimeter, ft = foot, in. = inch, km = kilometer, m = meter, mi = mile, mm = millimeter.

When referring to weight, the base metric and SI unit is the kilogram (kg), which is the preferred method for representing mass in scientific literature (33, 36). In the exercise physiology laboratory, it is commonplace to represent body mass, lean body mass, and fat mass in terms of kilograms (3). The conversion of pounds to kg is easily accomplished by dividing the pound weight by 2.2046. A summary of basic conversions for measurements of mass can be found in the accompanying highlight box.

Standard Mass Conversions

1 kg	= 1,000 g	= 100,000 cg
	= 1,000,000 mg	
1 g	= 100 cg	= 1,000 mg
1 kg	= 2.2046 lb	= 35.274 oz
1 g	= 0.0022046 lb	= 0.035274 oz
1 pound	= 16 oz	
	= 0.4536 kg	= 453.6 g
1 oz	= 0.0283495 kg	= 28.3495 g

Note: cg = centigram, g = gram, kg = kilogram, lb = pound, mg = milligram, oz = ounce.

The basic metric unit for volume is the liter (L), also known as a cubic decimeter (dm^3) (35). The SI unit for volume is the cubic meter (m^3) because it can be used to express the volumes of solids, liquids, and gases (33, 36). For example, 1,500 cubic centimeters is equivalent to 1,500 milliliters or 1.5 liters. The liter is commonly used when quantifying lung volume, oxygen consumption, cardiac output, stroke volume, and sweat loss. Length and weight measures can also be converted easily to volume measures. For example, if an athlete loses 1 kg of body mass during a training session, this is equivalent to 1 L of sweat loss and thus would require 1 L of water to restore fluid balance (6). A summary of volume conversions can be found in the accompanying highlight box.

Common Volume Conversions

1 L	= 10 dl	= 100 cl	= 1,000 ml
1 dl	= 0.1 L	= 10 cl	= 100 ml
1 cl	= 0.01 L	= 0.1 dl	= 10 ml
1 ml	= 0.001 L	= 0.01 dl	= 0.1 cl
1 cm³	= 1 ml	= 0.01 dl	= 0.001 L
1 L	= 1.0567 qt		
1 L	= 1 kg		

Note: cl = centiliter, cm^3 = cubic centimeter, dl = deciliter, kg = kilogram, L = liter, ml = milliliter, qt = quart.

Background and Environmental Information

Collecting basic information is an important organizational part of the testing process (6). In planning a testing session, you should consider several distinct items as part of the basic information collected—for example, the subject's name or identification number, age, and sex. It is also important to note the date, time of day, and who conducted the testing session.

• Name or identification number: Typically, the subject's surname is noted first, followed by a comma and the subject's first name; however, if the data are being used for research, a subject number should be noted instead of the subject's name in order to ensure confidentiality and compliance with human subject research procedures (7). This information is generally placed at the top of each data sheet and on all forms related to the test.

• Age and sex: It is crucial to note the subject's age, especially when comparing the subject's data with those presented in normative data tables. The subject's age is typically recorded to the nearest year, though some instances may warrant reporting the year to the closest tenth of a year (6). For example, if a subject is 18 years and 6 months old, his or her age should be recorded as 18.5 years. It is also necessary to document the subject's sex, generally by recording either an M for male or an F for female on the data sheet.

• Date: The date should be recorded as either month/day/year or day/month/year. For example, the date of August 5, 2012, could be noted as 8/5/2012 in the appropriate location on the laboratory data sheet. This information should be clearly noted on any data sheet used or generated in the testing process (6).

• Time: It is especially important to note time in the data collection process when performing longitudinal testing because some biological and performance measures exhibit diurnal or circadian variations (10, 31, 32). As a result, longitudinal testing sessions should generally be conducted at the same time each day in order to minimize the possibility of diurnal- or circadian-induced variations in performance.

• Tester's initials: The tester should initial the data sheet in order to create a record of who conducted the testing session (6). This information identifies a contact person to whom one can direct questions about the session. It also enables you to match the subject and tester if differences arise between various testers in the facility.

In testing procedures in the exercise physiology laboratory, the two basic variables most often assessed are height and weight. These measures can serve simply as descriptors or as integral parts of a testing program.

Height, technically referred to as *stature*, is routinely measured in most exercise physiology laboratories. Stature is generally measured with a physician's scale, a stadiometer, or a metric scale attached to a wall (6). The measurement should be made to the nearest tenth of a centimeter (0.1 cm) or to the nearest quarter or half of an inch. Measurements made in inches should be converted to centimeters. For example, if stature is determined to be 5 feet 11 inches, the following conversion would be made:

$$cm = in. \times 2.54$$

$$stature\ in\ cm = 71\ in. \times 2.54 = 180.34\ cm$$

Rounding to the nearest tenth of a centimeter, 180.34 cm would be reported as a value of 180.3 cm. Similarly, if the height had been calculated to be 182.88 cm, the reported value would be rounded to 182.9 cm.

The assessment of weight is probably the most common test done in the exercise physiology laboratory because it plays a role in many calculations performed in the laboratory. Body weight, which is equivalent to body mass under normal gravitational forces (6), is represented as a kilogram value in the scientific literature. Most Americans are familiar with pound measurements, but this representation of weight does not meet SI standards. To convert body mass from pounds to kg, use the following equation:

$$mass\ (kg) = mass\ (lb)\ /\ 2.2046$$

Every kilogram is equal to 2.2046 pounds, a figure that is more commonly expressed as 2.2 pounds based upon rounding. If a person weighs 225 pounds, his or her body mass would be calculated as follows:

$$mass\ in\ kg = 225\ lb\ /\ 2.2046 = 102.1\ kg$$

If this equation were used with 2.2 pounds in the denominator, then the individual's body mass would be recorded as 102.3 kg. Body mass

results should be rounded to the nearest tenth of a kilogram.

Once you have recorded the background information, the next step is to measure and record meteorological information about the testing environment. Typically, you will assess temperature, barometric pressure, and relative humidity (6) because they can profoundly affect results achieved in some physiological and performance tests (12, 27, 30). It is well documented that high temperature can exert a significant physiological effect on the results of testing (12, 30). In fact, it appears that heart rate increases 1 beat per minute for every increase of 1 degree Celsius above 24 degrees (30). In contrast, cold environments can increase the respiratory rate, which can negatively affect performance by increasing the risk of dehydration.

Temperature is commonly represented in units of Fahrenheit, Celsius, or Kelvin (33). Most Americans are familiar with the Fahrenheit temperature scale, in which 32 degrees represents the melting point of ice (6), but this measure is not recommended by the international system (SI) that sets the standard in research settings (35).

Another common method for representing temperature is the Celsius scale, in which 0 degrees represents the freezing point of water and 100 degrees represents the boiling point (3). To convert from Fahrenheit to Celsius, use either of the following formulas:

$$°C = (°F - 32)\ /\ 1.8$$

$$°C = 0.56 \times (°F - 32)$$

This measure of temperature is commonly seen in the scientific literature, but, like Fahrenheit, it is not the recommended SI unit (33). Instead, the SI thermal unit is the kelvin, which contains no negative or below-zero temperatures. The conversion from Celsius to the Kelvin scale is accomplished by means of the following formula:

$$K = 273.15 + °C$$

This system represents the coldest possible temperature as 0 K (6).

The pressure of ambient air is represented as **barometric pressure** and can fluctuate with changes in altitude (22) and weather pattern (6). As barometric pressure changes, so do the partial pressures of the gases that make up ambient air (oxygen, carbon dioxide, and nitrogen). Regardless of the change in overall barometric pressure, the percentage of the gases contained in the ambient air remain constant. Ambient air contains 79.04% nitrogen (N_2), 20.93% oxygen (O_2), and 0.03% carbon dioxide (CO_2). In order to determine the partial pressures of these gases, simply multiply the total barometric pressure by the gas' percent contribution.

Barometric pressure is measured with the use of an aneroid or mercury barometer in units of millimeters of mercury (mmHg). For example, at sea level, barometric pressure is about 760 mmHg, and the partial pressure of oxygen is around 159 mmHg. At an elevation of 2,000 m above sea level, the barometric pressure would fall to about 596 mmHg, whereas the partial pressure of oxygen would decrease to 125 mmHg. This reduction in barometric pressure and concomitant decrease in the partial pressure of oxygen can result in a significant decrease in an individual's ability to perform aerobic exercise (18, 27). Because of the impact of changes in barometric pressure on both pulmonary and cardiovascular function, both respiratory ventilation and metabolic volumes are often corrected for these changes (6).

In the scientific literature, barometric pressure is typically reported in the following units: mmHg, torr, hectopascal (hPa), or kilopascal (kPa). Generally, the following formulas can be used to convert between the various units:

$$1 \text{ torr} = 1 \text{ mmHg}$$

$$\text{kPa} = \text{torr} \times 0.1333 = \text{torr} / 7.50$$

$$\text{hPa} = \text{torr} \times 1.333 = \text{torr} / 0.750$$

Therefore, a barometric pressure of 674 mmHg or 674 torr would be converted as follows:

$$674 \text{ torr} \times 0.1333 = 89.8 \text{ kPa}$$

$$674 \text{ torr} \times 1.333 = 898.4 \text{ hPa}$$

The **relative humidity** of the environment is the amount of water or percent saturation in the ambient air (6, 22). To quantify relative humidity in the exercise physiology laboratory, use an instrument called a *hygrometer,* which yields a percent relative humidity value (e.g., 60% relative humidity). If, for example, the ambient air is completely saturated with water vapor, then the relative humidity is represented as 100% at that temperature. Generally, the amount of water that can be contained in ambient air increases with temperature.

Relative humidity values between 20% and 60% generally do not affect exercise, but values above or below this range can influence physical performance (2). Specifically, high humidity limits the evaporative capacity of sweat, which can significantly reduce blood plasma volume and thus increase cardiovascular stress. High relative humidity can also affect thermoregulation, which can increase the effects of temperature on cardiovascular function (27). Instead of the 1-beat-per-minute increase in heart rate typically associated with a 1 °C increase in temperature above 24 °C, an increase in relative humidity can result in a 2- to 4-beat-per-minute increase in heart rate (6, 30). Because of these potential effects on physical performance, relative humidity is commonly assessed in the exercise physiology laboratory.

Heat stress can be estimated in terms of what has been called the **wet-bulb globe temperature (WBGT)**. This measure simultaneously accounts for three thermometer readings and provides a single temperature reading in order to estimate the cooling capacity of the surrounding environment (22). The first thermometer measurement is the dry-bulb temperature (T_{db}), which is taken with a standard thermometer and evaluates the actual air temperature. The second thermometer measurement is taken with a wet-bulb thermometer, which reflects the effect of sweat evaporating from the skin (22). With this measure, water evaporates from the bulb, which effectively lowers the temperature below that represented by the dry-bulb, thus yielding what is termed the wet-bulb temperature (T_{wb}). The difference between the wet- and dry-bulb temperatures is used to represent the environment's capacity for cooling. The third thermometer is placed in a black globe and generally has a higher temperature than that indicated by the dry bulb because the black globe absorbs heat. This globe temperature (T_g) is used to estimate the environment's radiant heat load (22).

Once the temperature is measured with the three thermometers, the three results can be combined to estimate the overall atmospheric challenge to body temperature in outdoor environments by means of the following equation:

$$\text{Wet-bulb globe temperature} = 0.1T_{db} + 0.7T_{wb} + 0.2T_g$$

Careful examination of this equation reveals that the T_{wb} reflects the importance of sweat evaporation in the physiology of heat exchange. If the relative humidity is high, this measure reflects impairment in the ability to evaporate sweat, which in turn increases the heat load encountered by the body (22). As a general rule, if the WBGT is >28 °C (>82–83 °F), then modifications to exercise or practice should be considered—including canceling practice, moving to indoor facilities, or reducing the intensity of training.

Descriptive Statistics

Statistics are a mathematical method for describing and analyzing numerical data (34). Exercise physiology laboratories typically involve calculating descriptive statistics such as measures of central tendency and **variability**.

Central Tendency

Measures of **central tendency** are commonly used to present a score that best represents all of the scores collected for a group (34). The most common statistics calculated when representing central tendency are the mean, median, and mode. The **mean** is calculated by summing all the scores, then dividing the result by the number of scores. The following formula represents the calculation of the mean:

$$\text{Mean} = \frac{1}{N}\sum X$$

In this equation, X is the individual scores, and N equals the number of scores. For example, if you measured the body mass of 5 people as 53, 55, 65, 48, and 60 kg, the mean would be calculated as follows:

$$\text{Mean} = \frac{53+55+65+48+60}{5} = \frac{281}{5} = 56.2 \text{ kg}$$

Therefore, for this example, the mean weight for the four individual tests is 56.2 kg.

The **median** represents the middle score in a series of data. It is generally calculated by means of the following equation when the sample is placed in order:

$$\text{Median} = [(N + 1) / 2]\text{th score}$$

In this equation, N represents the total number of scores in the sample. For the preceding example involving five body weights—now placed in order as 48, 53, 55, 60, and 65 kg—the median would be calculated as follows:

$$\text{Median} = (5 + 1) / 2 = 3\text{rd score}$$

You would then count 3 places from the first score, thus revealing the median as 55 kg. If trying to calculate the median with an even number of values, you need to first find the middle pair of numbers. For example, if the six weights are 48, 53, 55, 58, 60, and 65 kg, the preceding equation would give (6 + 1) / 2 = 3.5. You would then move in 3.5 positions and find the two middle values, 55 and 58. These would then be added together, 55 + 58 = 113, and then divided by two, which yields a median value of 56.5.

The **mode** is defined as the most frequent score in a sample, and it is possible to have more than one mode in a group of scores. If the body masses of 10 people were measured as 48, 49, 53, 53, 55, 59, 60, 60, 60, and 62 kg, the most frequently occurring value would be 60 kg, which would thus be identified as the mode of this sample population. If each score in a sample appears with the same frequency, then the mode is undefined for that sample.

Variability

The variability of a data set allows the spread of the data to be depicted. In examining a group of data in the exercise physiology laboratory, variability is commonly examined with the use of the standard deviation and the range.

The **standard deviation** of a data set is easily calculated with the use of many spreadsheet programs, but it can be just as easily be calculated by hand with the following equation:

$$\text{Standard deviation} = \sqrt{\frac{1}{(N-1)}\sum (X - M)^2}$$

In this equation, the mean (M) is subtracted from each score (X) in the group. The result for each score is then squared, and these figures are summed to provide a total, which is then divided by the number (N) of scores in the group minus 1. The square root of this is the standard deviation. An example of how one might calculate the standard deviation is presented in the accompanying highlight box; the example involves body mass measurements of five subjects.

Sample Calculation of Standard Deviation

Sample Laboratory Data Table

Sample	X	X – M	(X – M)²
1	53	–3.2	10.2
2	65	8.8	77.4
3	48	–8.2	67.2
4	60	3.8	14.4
5	55	–1.2	1.4
Σ=	281	0.0	170.6

$$\text{Mean} = \frac{1}{N}\Sigma X = \frac{281}{5} = 56.2$$

$$\text{SD} = \sqrt{\frac{1}{(N-1)}\Sigma(X-M)^2}$$

$$= \sqrt{\frac{170.6}{4}} = \sqrt{42.7} = 6.53$$

Note: SD = standard deviation, X = sample score, M = mean, N = number of samples.

Thus the mean body mass was 56.2 kg, and the standard deviation was 6.53 kg. In a results section of a manuscript or laboratory report, we would then represent these data as 56.2 ± 6.5 kg.

The second measure of variability often used in the exercise physiology laboratory is **range**—the distance between the end points in a group of scores (7). The range is easily calculated with the following formula:

Range = (high score – low score) + 1

Thus, for our previous example of body mass scores, the range would be calculated as follows:

Range = (65 – 48) + 1 = 17 + 1 = 18

The high score can also be considered as the *maximum*, and the low score would be considered as the *minimum*. These variables are also sometimes reported in the results generated by performance testing in the exercise physiology laboratory.

Reliability and Validity

Whether you are conducting basic testing or using test results for research purposes, it is essential that you determine the reliability and validity of any testing process. If the tests chosen do not meet both of these criteria, you may produce false information about the individual's physiological or physical performance capacity. Therefore, before using any testing procedure—whether in the field or in a laboratory—you must determine its reliability and validity. Typically, you can do so by using established testing protocols that have been previously validated and determined to be reliable.

Reliability refers to the repeatability or consistency of a measure (7, 19, 34). The reliability of a measure is often affected by experimental or biological errors (6). Experimental errors may include technical errors such as miscalibration of instruments or changes to the testing environment. Biological errors may include natural shifts in performance that occur in response to the time of day at which testing is undertaken (31, 32) or in response to accumulated fatigue. The subject's familiarity with the testing process can also affect a test's reliability (6).

Highly reliable field, field/laboratory, and laboratory tests generate high intraclass correlation coefficients (ICCs) when repeated trial data are compared (6, 19). Acceptable ICCs range between 0 and 1 or between 0 and –1. Perfect correlations are either 1 or –1 (4); they rarely occur with the variables analyzed in the exercise physiology laboratory. Generally, correlations greater than 0.90 are considered to demonstrate high reliability, whereas correlations less than 0.70 are considered to exhibit poor reliability (6).

Even if a measure or test is found to be reliable, it may not be valid; in other words, rather than providing a suitable assessment of what you are trying to measure, your measures might be dependable but consistently wrong. Therefore, you must determine that a test is reliable *and* valid before using it in either a clinical or sport testing program.

Validity is defined as the degree to which a test or instrument measures what it is supposed to measure (34). In the exercise physiology laboratory, the main form of validity is criterion validity—the ability of a test to be related to a recognized standard or criterion measure (34). For example, when examining body composition, underwater weighing is considered to be the gold standard or criterion measure. When considering another body composition assessment method—say, skinfolds—validity is determined by correlating the results obtained from the skinfold estimate to the body fat as determined by underwater weighing. Typically, this correlation is performed with the use of a Pearson product-moment correlation coefficient, and a high correlation (r) between the criterion variable and the test is generally held to represent high validity.

A measurement is considered valid if it is both accurate and precise. The closer a measurement is to an accepted value, the higher the **accuracy** of measurement. If, for example, multiple arrows are shot at a target, the closeness of the arrows to the bull's-eye indicates the accuracy of the measure. If the arrows are clustered close to one another, then there would be a high degree of **precision**. Thus precision is considered to be the reproducibility of the measurement. Ultimately, a measure can be highly accurate but not precise, highly precise but not accurate, highly accurate and highly precise, or neither highly precise nor highly accurate (see figure 1.1).

Presentation of Results

Data collected on a group of subjects can be presented in many ways in a publication or laboratory report. One of the first steps in determining how to present data is to decide whether or not it should be placed in a table, a figure, or the text of the document. Consider whether you can best represent the information in the form of the actual numbers collected or as a picture of the results. Once you make this decision, you can create a table or figure to best represent the data.

Though it is relatively easy to create a table with modern computer software, it remains important to consider some basic rules. First, tables provide a means for communicating information about data—not storing them (34). One example of good use of tables is that of creating normative data tables for evaluating testing. You can use these tables for communicating information about typical results for any given field or laboratory test. When creating a table, follow these basic rules suggested by Thomas, Nelsen, and Silverman (34):

1. Table headings should be clearly labeled and easy to follow. To improve readability of headings, avoid excessive abbreviation and make formatting easy to follow.

2. Like characteristics should be presented vertically. The columns and rows contained in the table should make sense. For example, columns could contain variables, means, and standard deviations, and each row could represent a specific variable.

3. Readers should not have to refer to the text to understand the content of a table. As a rule, a table should stand alone and not require the reader to search for abbreviations or text to explain what the table is attempting to present.

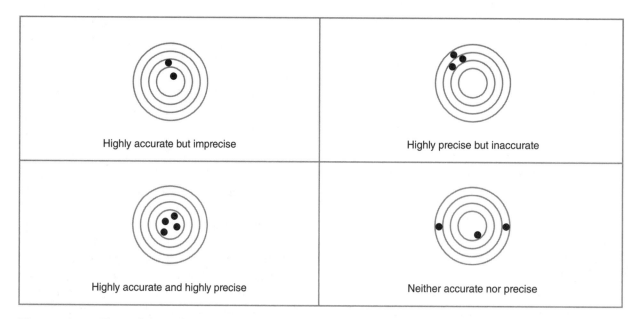

Figure 1.1 The relationship between accuracy and precision.

Table 1.1   Sample Laboratory Table

Characteristics	Mean	SD
Age (y)	22.2	± 1.0
Height (cm)	172.3	± 2.3
Weight (kg)	83.5	± 5.6
Percent body fat (%)	8.2	± 3.1
$\dot{V}O_2$max (ml·kg^{-1}·min^{-1})	45.2	± 2.2
1RM back squat (kg)	110.2	± 10.5
1RM bench press (kg)	85.4	± 4.6

1RM = 1-repetition max; SD = standard deviation.

Table 1.1 shows a sample that meets these rules and presents some basic data collected in an exercise physiology laboratory. This basic tabular format can be modified to fit many different data sets.

Another way to present data is by using a figure. Deciding whether to place data in table form or figure form depends largely on whether a picture of the results (i.e., a figure) will work better than the actual numbers (presented in a table). If the data are better represented as a figure, consider the following basic rules (34):

1. Ensure that the figure is clearly labeled and easy to read.

2. Present important information so that it is easily evaluated.

3. Create a figure that is free from visual distractions.

Determine what type of figure will best represent the data. Types include bar and column charts, line graphs, scatter plots, flow charts, and pie charts.

 Bar and column charts: These figures are useful when comparing single responses—often a mean or a single most important time point—between treatment groups. While not the best method for determining trends over time, column charts can be used for comparing amounts over time so that trends can be seen or for comparing variables between groups. You can often use shading or coloring to distinguish between columns and thus help your audience easily interpret the data. You can also depict standard deviations and standard errors of the mean in order to make the chart more descriptive.

 Line graphs: This type of graphical representation is often used to present longitudinal data and show how things change over time. When creating charts of this type, time is often placed on the *x*-axis and the variable being measured or quantified on the *y*-axis. When multiple lines are placed on these charts, it is prudent to use shading, differing symbols, and broken or colored lines to allow the reader to easily distinguish between variables.

 Scatter plots: In this type of chart, each point on the plot represents a data point on the *x*-axis and the *y*-axis. As a result, this type of graphical representation makes it easy to see the basic patterns of individual scores. Since scatter plots illustrate individual data points, researchers can use them to gain a sense of the spatial distribution of the results and to determine linearity, outliers, or clumps of data. Researchers often facilitate interpretation of the data by using multiple colors or symbol shapes to distinguish between groups.

 Flow charts: These charts are useful in depicting procedures or steps within a process. This type of representation can often be used in methods sections, especially when depicting decisions and concomitant steps.

Pie charts: This type of graphic representation can be used to show proportions within a whole. The entire circle represents the whole, and each segment represents a percentage of the whole. For example, a pie chart might be used to show the leading causes of death from cardiovascular disease; each segment would represent a percentage of the overall death rate.

Pie charts should generally contain a maximum of about five segments or variables. Each segment should be shaded so that the various segments can be easily distinguished from one another.

Interpretation of Data

Once you have collected the data, you will need to interpret them in order to develop meaning from them. In clinical settings, the data collected in the exercise physiology laboratory are generally compared with normative data or standards (6).

One method of interpreting data is to compare the subject's results with those of the general population. When a population is normally distributed, the mean, median, and mode are all the same, and the data will look similar to figure 1.2 when graphed. **Normative data** are used to derive percentile ranks and standard deviations of the data, which are then associated with a descriptive category such as poor, average, or excellent. For example, in figure 1.2, normative data are presented as a basic normal curve, wherein the middle portion contains the majority of the population (68%), which would correspond to the 50th percentile; thus subjects whose results fell within this area of the curve would be considered average.

Standards and normative data are often wrongly used interchangeably (6). A **standard** is better used to describe a desirable or target score (3), whereas normative data represent a subject's placement within a population (6). Standards are often set for items such as the recommended quantity and quality of exercise for maintaining a minimal level of physical fitness (3). In many cases, both standards and normative data are presented for a given population in order to facilitate interpretation of the subject's results (6).

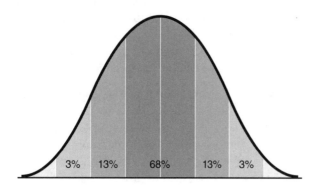

Figure 1.2 The normal curve.

References

1. Acevedo EO and Starks MA. *Exercise Testing and Prescription Lab Manual.* Champaign, IL: Human Kinetics, 2003.

2. Adams V, Jiang H, Yu J, Mobius-Winkler S, Fiehn E, Linke A, Weigl C, Schuler G, and Hambrecht R. Apoptosis in Skeletal Myocytes of Patients With Chronic Heart Failure Is Associated With Exercise Intolerance. *J Am Coll Cardiol* 33: 959–965, 1999.

3. American College of Sports Medicine. ACSM Position Stand: The Recommended Quantity and Quality of Exercise for Developing and Maintaining Cardiorespiratory and Muscular Fitness, and Flexibility in Healthy Adults. *Med Sci Sports Exerc* 30: 975–991, 1998.

4. American College of Sports Medicine. *ACSM's Guidelines for Exercise Testing and Prescription.* 8th ed. Philadelphia, PA: Lippincott Williams & Wilkins, 2010.

5. American College of Sports Medicine. *ACSM's Health-Related Physical Fitness Assessment Manual,* 3rd ed. Baltimore: Lippincott Williams & Wilkins, 2010.

6. Beam WC and Adams GM. *Exercise Physiology Laboratory Manual.* 5th ed. Boston: McGraw-Hill, 2007.

7. Berg KE and Latin RW. *Essentials of Research Methods in Health, Physical Education, Exercise Science, and Recreation.* Baltimore, MD: Lippincott Williams & Wilkins, 2004, p. 292.

8. Bompa TO and Haff GG. *Periodization: Theory and Methodology of Training.* 5th ed. Champaign, IL: Human Kinetics, 2009.

9. Brooks GA, Fahey TD, White TP, and Baldwin KM. *Exercise Physiology: Human Bioenergetics and Its Applications.* 4th ed. Mountain View, CA: Mayfield, 2004.

10. Brown FM, Neft EE, and LaJambe CM. Collegiate Rowing Crew Performance Varies by Morningness-Eveningness. *J Strength Cond Res* 22: 1894–1900, 2008.

11. Dolezal BA, Thompson CJ, Schroeder CA, Haub MD, Haff GG, Comeau MJ, and Potteiger JA. Laboratory Testing to Improve Athletic Performance. *Strength and Cond J* 19: 20–24, 1997.

12. Gonzalez-Alonso J, Teller C, Andersen SL, Jensen FB, Hyldig T, and Nielsen B. Influence of Body Temperature on the Development of Fatigue During Prolonged Exercise in the Heat. *J Appl Physiol* 86: 1032–1039, 1999.

13. Gore CJ, editor. *Physiological Tests for Elite Athletes.* Champaign, IL: Human Kinetics, 2000, p. 464.

14. Harman E. Biomechanics of Resistance Exercise. In: Baechle TR and Earle RW, eds., *Essentials of Strength Training and Conditioning*. 3rd ed. National Strength and Conditioning Association. Champaign, IL: Human Kinetics, 2008, p. 66–91.

15. Harman E and Garhammer J. Administration, Scoring, and Interpretation of Selected Tests. In: Baechle TR and Earle RW, eds., *Essentials of Strength Training and Conditioning*. 3rd ed. National Strength and Conditioning Association. Champaign, IL: Human Kinetics, 2008, pp. 249–292.

16. Heyward VH. *Advanced Fitness Assessment and Exercise Prescription*. 6th ed. Champaign, IL: Human Kinetics, 2010.

17. Hoffman JR. *Norms for Fitness, Performance, and Health*. Champaign, IL: Human Kinetics, 2006.

18. Hoffman JR. *Physiological Aspects of Sport Training and Performance*. Champiagn, IL: Human Kinetics, 2002, p. 342.

19. Hopkins WG. Measures of reliability in sports medicine and science. *Sports Med* 30: 1–15, 2000.

20. Housh TJ, Cramer JT, Weir JP, Beck TW, and Johnson GO. *Physical Fitness Laboratories on a Budget*. Scottsdale, AZ: Holcomb-Hathaway, 2009, p. 234.

21. Hruda KV, Hicks AL, and McCartney N. Training for Muscle Power in Older Adults: Effects on Functional Abilities. *Can J Appl Physiol* 28: 178–189, 2003.

22. Kenney WL, Wilmore JH, and Costill DL. *Physiology of Sport and Exercise*. 5th ed. Champaign, IL: Human Kinetics, 2012.

23. Komi PV. *Strength and Power in Sport*. Oxford, UK: Blackwell Scientific Publications, 1991.

24. Komi PV. *Strength and Power in Sport*. 2nd ed. Malden, MA: Blackwell Scientific, 2003, p. 523.

25. Kreighbaum E and Barthels KM. *Biomechanics: A Qualitative Approach for Studying Human Movement*. New York: MacMillan, 1990, p. 749.

26. Maud PJ and Foster C, editors. *Physiological Assessment of Human Fitness*. 2nd ed. Champaign, IL: Human Kinetics, 2006.

27. Mcardle WD, Katch FI, and Katch VL. *Exercise Physiology: Energy, Nutrition, and Human Performance*. Baltimore: Lippincott Williams & Wilkins, 2007, p. 1067.

28. National Institute of Diabetes and Digestive and Kidney Diseases. *Understanding Adult Obesity*. Bethesda, MD: National Institutes of Health, U.S. Department of Health and Human Services, 2008.

29. Nieman DC. *Exercise Testing and Prescription: A Health-Related Approach*. New York: McGraw-Hill, 2003, p. 774.

30. Pandolf KB, Cafarelli E, Noble BJ, and Metz KF. Hyperthermia: Effect on Exercise Prescription. *Arch Phys Med Rehabil* 56: 524–526, 1975.

31. Souissi N, Bessot N, Chamari K, Gauthier A, Sesboue B, and Davenne D. Effect of Time of Day on Aerobic Contribution to the 30-s Wingate Test Performance. *Chronobiol Int* 24: 739–748, 2007.

32. Souissi N, Gauthier A, Sesboue B, Larue J, and Davenne D. Circadian Rhythms in Two Types of Anaerobic Cycle Leg Exercise: Force-Velocity and 30-s Wingate Tests. *Int J Sports Med* 25: 14–19, 2004.

33. Stone MH, Stone ME, and Sands WA. *Principles and Practice of Resistance Training*. Champaign, IL: Human Kinetics, 2007, p. 376.

34. Thomas JR, Nelson JK, and Silverman SJ. *Research Methods in Physical Activity*. 5th ed. Champaign, IL: Human Kinetics, 2005.

35. Thompson A and Taylor BN. NIST Special Publication 811, 2008 edition: Guide for the Use of International System of Units (SI). Gaithersburg, MD: United States Department of Commerce, National Institute of Standards and Technology, 2008.

36. Young DS. Implementation of SI Units for Clinical Laboratory Data. Style Specifications and Conversion Tables. *Ann Intern Med* 106 (1): 114–129, 1987. [Published errata appear in *Ann Intern Med* 107 (2): 265 (1987, Aug); *Ann Intern Med* 110 (4): 328 (1989, Feb 15); *Ann Intern Med* 114 (2):172 (1991, Jan 15), and *Ann Intern Med* 148 (9): 715 (2008, May 6)].

BASIC DATA

EQUIPMENT

- Platform scale (or digital, Chatillon, or other scale)
- Certified weights for calibrating scale
- A stadiometer (wall or freestanding) or physician's scale with attached anthropometer
- Mercury or aneroid barometer
- Hygrometer and hygrometer chart
- Laboratory thermometer
- Sliding caliper
- Screwdriver
- Individual and group data sheets

Find the group data sheets for this laboratory online at www.HumanKinetics.com/ LaboratoryManualForExercisePhysiology.

BACKGROUND INFORMATION

In most testing situations, you can follow these steps when collecting basic informational data:

Step 1: Note the subject's name or identification number, as well as the time and date of the testing session, in the appropriate locations on the individual data sheet.

Step 2: Indicate the sex and age of the subject on the individual data sheet.

Step 3: Initial the data sheet in the appropriate location.

HEIGHT

Step 1: Verify the accuracy of the stadiometer by measuring the distance from the floor to the base to the horizontal headboard. If using a physician's scale, measure the distance from the scale platform to the hinged horizontal level.

Step 2: Have the subject stand with his or her back to the stadiometer and heels placed evenly apart. If using a wall-mounted stadiometer, have the subject stand with his or her buttocks, scapulae, and heels in contact with the wall.

Step 3: Instruct the subject to stand as tall as possible, inhale, and hold his or her breath while looking straight ahead.

Step 4: Lower the headboard so that it rests atop the subject's head and is perpendicular to the measuring scale. If using a physician's scale, move the hinged lever so that it is perpendicular to the graduated vertical rod while resting evenly on top of the subject's head.

Step 5: Measure the subject's height to the nearest tenth of a centimeter (0.1 cm) or to the nearest eighth or quarter of an inch and record this value on the individual data sheet.

Step 6: Repeat steps 2 through 5 twice for a total of three measurements.

Step 7: To begin interpreting the data, convert (if necessary) the measured height from inches to centimeters, then calculate the average of the three trial measures.

Step 8: Compare the results with the normative data presented in table 1.2 for average height of Americans and record the result on the data sheet.

Step 9: Calculate a difference score by subtracting the average height from the measured height. Use the following equation:

$$\text{Percent difference} = \frac{\text{actual height (cm)} - \text{average height (cm)}}{\text{average height (cm)}} \times 100$$

WEIGHT SCALE CALIBRATION

Scales come in several forms, including the following: digital scales, which contain a load cell; spring resistance scales, such as a Chatillon scale, which measures weight by means of pressure resistance on an internal spring; and balance beam scales, typically referred to as *physician's scales* (1). Calibration procedures depend largely on the type of scale. Every digital scale is slightly different, but in most cases calibration is initiated by turning the device on and off, by pressing the reset button, or by pressing the calibration button. For a Chatillon scale, to tare or set the zero point, a screw or knob is turned until the unweighted device displays a zero. Once the scale is zeroed, the tester can verify that it is calibrated by using the procedures described in the following paragraph.

Simply zeroing a scale, however, is not enough to ensure that the scale is appropriately calibrated. It is also important to verify the scale's accuracy at three points—zero, midpoint, and high point—in order to ensure its accuracy across a variety of weights. The basic steps for calibrating a scale are as follows (4, 6):

Step 1: To set the zero point, move the beam weights to the zero points.

Step 2: Observe the position of the pointer. If the zero point is set, the pointer at the end of the beam should sit midway between the top and bottom of the pointer window. If the pointer is centered, then the tester should perform the steps listed in steps 5 through 7. If the pointer is not centered, follow the next calibration step.

Step 3: Adjust the pointer by manipulating the tare screw weight with a screwdriver. If the pointer is below the middle position, turn the tare screw clockwise to move the pointer upward; if the pointer is above the middle position, turn the tare screw counterclockwise to move the

Table 1.2 Average Height of Americans

Age (y)	Males		Females	
	in.	cm	in.	cm
Young children				
2	35.9	91.2	35.5	90.2
3	38.8	98.6	38.4	97.5
4	41.9	106.4	41.7	105.9
5	44.5	113.0	44.3	112.5
6	46.9	119.1	46.1	117.1
7	49.7	126.2	49.0	124.5
8	52.2	132.6	51.5	130.8
Preadolescents				
9	54.4	138.2	53.9	136.9
10	55.7	141.5	56.4	143.3
11	58.5	148.6	59.6	151.4
12	60.9	154.7	61.4	156.0
Adolescents				
13	63.1	160.3	62.6	159.0
14	66.3	168.5	63.7	161.8
15	68.4	173.8	63.8	162.1
16	69.0	175.3	63.8	162.0
17	69.0	175.3	64.2	163.1
Adults				
18	69.5	176.5	64.2	163.1
19	69.6	176.8	64.2	163.1
20–29	69.6	176.8	64.2	162.8
30–39	69.5	176.5	64.2	163.1
40–49	69.7	177.0	64.3	163.3
50–59	69.2	175.8	63.9	162.3
60–74	68.6	174.2	63.0	160.0
≥75	67.4	171.2	62.0	157.5

Reprinted, by permission, from J. Hoffman, 2006, *Norms for fitness, performance, and health* (Champaign, IL: Human Kinetics), 82. Adapted from C.L. Ogden et al., 2004, *Mean body weight, height, and body mass index, United States 1960-2002,* Advance data from vital and health statistics, no. 347. (Hyattsville, MD: National Center for Health Statistics).

pointer downward. Fine-tune the position of the pointer by adjusting the tare screw until the pointer rests in the middle of the pointer window.

Step 4: Repeat steps 1 through 3 twice for a total of three times, then perform the verification-of-calibration procedure described in steps 5 through 7.

Step 5: To perform the high-point calibration, place the heaviest certified weight onto the scale (with some scales, you may have to hang the weights). In most instances, this weight should correspond to the heaviest person you might weigh.

Laboratory Activity 1.1

Step 6: Record the actual weight of the certified weight and the weight measured by the scale in the appropriate locations on the individual data sheet.

Step 7: If the actual weight and the weight measured are the same, the scale is accurate at the high-point range. If not, adjust the scale so that it reads correctly.

Step 8: Calculate a difference score between the actual and measured weights and use this score to create a correction factor. For example, if the scale consistently weighs 2 kg heavier at weights over about 120 kg, note that anyone who weighs more than 120 kg should have 2 kg subtracted from his or her weight.

Step 9: To perform the midpoint calibration, estimate the average weight of the subjects you will measure, then place this amount of certified weight on the scale.

Step 10: Record the actual weight of the certified weight and the weight measured by the scale in the appropriate locations on the individual data sheet.

Step 11: If the actual weight and the weight measured are the same, the scale is accurate at the midpoint range. If not, adjust the scale so that it reads correctly.

Step 12: Record the actual weight of the certified weight and the weight measured by the scale in the appropriate locations on the individual data sheet.

Step 13: Calculate a difference score between the actual and measured weights and use it to create a correction factor.

Step 14: Repeat steps 5 through 13 twice for a total of three times.

BODY WEIGHT

Step 1: Determine whether the scale is calibrated; if it is not, follow the steps listed in the Weight Scale Calibration section.

Step 2: After completing the calibration, have the subject remove as much clothing as possible, including shoes and jewelry, in order to enable the most accurate assessment. It is best to perform a nude weight or have the subject wear a paper gown, which is considered to be weightless. If this is not possible, one other option is to weigh the subject's weigh-in clothing to the nearest quarter pound or 0.02 kg, then subtract this value from the total weight measured when the subject is wearing the items. If it is not possible to weigh the subject's weigh-in clothing, then the subject should be weighed while wearing as little clothing as possible.

Step 3: Have the subject stand on the scale either nude, wearing a minimal amount of clothing, or wearing his or her weigh-in clothing. Stand behind the platform scale in order to adjust the beam weights and accurately read the measurement without touching the subject.

Step 4: If using a beam scale, adjust the beam weights until they are balanced and an accurate reading can be made of the subject's body weight.

Record this weight in the appropriate location on the individual data sheet.

Step 5: Repeat steps 2 through 4 twice for a total of three measurements.

Step 6: Determine the subject's frame size by using the sliding caliper to measure elbow breadth. Have the subject extend the right arm forward, perpendicular to the body, and bend the arm to 90° so that the fingers point upward and the hand points away from the body. Determine the breadth of the elbow by taking a measurement of the distance between the two prominent bones on either side of the elbow (i.e., measuring the widest point). Record the inch value for the elbow breadth in the appropriate location on the individual data sheet.

Step 7: To begin interpreting the body mass data, determine the subject's frame size by comparing the elbow breadth measurement with the values presented in table 1.3. Indicate the appropriate size on the individual data sheet.

Step 8: Compare the actual weight measured with the weight recommendations from the National Institutes of Health shown in table 1.4. Record the range and determine whether the subject falls into, below, or above the recommended range.

Step 9: Using the frame size estimation, compare the measured weight to the ranges presented in table 1.5 (for men) or 1.6 (for women) and record the corresponding range on the individual data sheet. Indicate whether the subject falls into, below, or above the recommended range.

Table 1.3 Metropolitan Life Insurance Frame Size Chart

| Sex | Height | | Elbow breadth | | | | | |
| | | | Small frame | | Medium frame | | Large frame | |
	in.	cm	in.	cm	in.	cm	in.	cm
Men	61–62	154.9–157.5	<2.5	<6.4	2.5–2.88	6.4–7.3	>2.88	>7.3
	63–66	160.0–167.6	<2.63	<6.7	2.63–2.88	6.7–7.3	>2.88	>7.3
	67–70	170.2–177.8	<2.75	<7.0	2.75–3	7.0–7.6	>3	>7.6
	71–74	180.3–188.0	<2.75	<7.0	2.75–3.13	7.0–8.0	>3.13	>8.0
	≥75	≥190.5	<2.88	<7.3	2.88–3.25	7.3–8.3	>3.25	>8.3
Women	57–58	144.8–147.3	<2.25	<5.7	2.25–2.5	5.7–6.4	>2.5	>6.4
	59–62	149.9–157.5	<2.25	<5.7	2.25–2.5	5.7–6.4	>2.5	>6.4
	63–66	160.0–167.6	<2.38	<6.0	2.38–2.63	6.0–6.7	>2.63	>6.7
	67–70	170.2–177.8	<2.38	<6.0	2.38–2.63	6.0–6.7	>2.63	>6.7
	≥71	≥180.3	<2.5	<6.4	2.5–2.75	6.4–7.0	>2.75	>7.0

Height is measured without shoes. Elbow breadth is measured by extending the right arm forward, perpendicular to the body, and bend the arm to 90° so that the fingers point upward and the hand points away from the body. The breadth of the elbow is determined by taking a measurement of the distance between the two prominent bones on either side of the elbow (i.e., measuring the widest point).

Adapted from the Metropolitan Life Insurance Company 1983.

Laboratory Activity 1.1

Table 1.4 National Institutes of Health Weight Recommendations

Height			Age 19 years to 34 years		Age 35 years or older	
ft, in.	in.	cm	lb	kg	lb	kg
5'0"	60	152	97–128	44–58	108–138	49–63
5'1"	61	155	101–132	46–60	111–143	51–65
5'2"	62	158	104–137	47–62	115–148	52–67
5'3"	63	160	107–141	49–64	119–152	54–69
5'4"	64	163	111–146	50–66	122–157	56–71
5'5"	65	165	114–150	52–69	126–162	57–74
5'6"	66	168	118–155	54–71	130–167	59–76
5'7"	67	170	121–160	55–73	134–172	61–78
5'8"	68	173	125–164	57–75	138–178	63–81
5'9"	69	175	129–169	59–77	142–183	65–83
5'10"	70	178	132–174	60–79	146–188	66–86
5'11"	71	180	136–179	62–81	151–194	69–88
6'0"	72	183	140–184	64–84	155–199	71–91
6'1"	73	185	144–189	66–86	159–205	72–93
6'2"	74	188	148–195	67–89	164–210	75–96
6'3"	75	191	152–200	69–91	168–216	76–98
6'4"	76	193	156–205	71–93	173–222	79–101
6'5"	77	196	160–211	73–96	177–228	81–104
6'6"	78	198	164–216	75–98	182–234	83–106

Note: Height and weight are measured without shoes.

Adapted from American College of Sports Medicine (5) and the U.S. Department of Health and Human Services (28).

Table 1.5 Metropolitan Life Insurance Height and Weight Table for Men

Height			Small frame		Medium frame		Large frame	
ft, in.	in.	cm	lb	kg	lb	kg	lb	kg
5'2"	62	158	128–134	58–61	131–141	60–64	138–150	63–68
5'3"	63	160	130–136	59–62	133–143	61–65	140–153	64–69
5'4"	64	163	132–138	60.0–63	135–145	61–66	142–156	65–71
5'5"	65	165	134–140	61–64	137–148	62–67	144–160	66–73
5'6"	66	168	136–142	62–65	139–151	63–69	146–164	66–75
5'7"	67	170	138–145	63–66	142–154	65–70	149–168	68–76
5'8"	68	173	140–148	64–67	145–157	66–71	152–172	69–78
5'9"	69	175	142–151	65–69	148–160	67–73	155–176	71–80
5'10"	70	178	144–154	66–70	151–163	68–74	158–180	72–82
5'11"	71	180	146–157	66–71	154–166	70–76	161–184	73–84
6'0"	72	183	149–160	68–73	157–170	71–77	164–188	75–86
6'1"	73	185	152–164	69–75	160–174	73–79	168–192	76–87
6'2"	74	188	155–168	71–76	164–178	75–81	172–197	78–90
6'3"	75	191	158–172	72–78	167–182	76–83	176–202	80–92
6'4"	76	193	162–176	74–80	171–187	78–85	181–207	82–94

Weights are based on height and frame size for men wearing indoor clothing of 5 lb (2.3 kg) and 2 in. (5.1 cm) heels.

Adapted from the Metropolitan Life Insurance Company 1983.

Laboratory Activity 1.1

Table 1.6 Metropolitan Life Insurance Height and Weight Table for Women

Height			Small frame		Medium frame		Large frame	
ft, in.	in.	cm	lb	kg	lb	kg	lb	kg
4'10"	58	147	102–111	46–51	109–121	50–55	118–131	54–60
4'11"	59	150	103–113	47–51	111–123	51–56	120–134	55–61
5'0"	60	152	104–115	47–52	113–126	51–57	122–137	56–62
5'1"	61	155	106–118	48–54	115–129	52–59	125–140	57–64
5'2"	62	158	108–121	49–55	118–132	54–60	128–143	58–65
5'3"	63	160	111–124	51–56	121–135	55–61	131–147	60–67
5'4"	64	163	114–127	52–58	124–138	56–63	134–151	61–69
5'5"	65	165	117–130	53–59	127–141	58–64	137–155	62–71
5'6"	66	168	120–133	55–61	130–144	59–66	140–159	64–72
5'7"	67	170	123–136	56–62	133–147	61–67	143–163	65–75
5'8"	68	173	126–139	57–63	136–150	62–68	146–167	66–76
5'9"	69	175	129–142	59–65	139–153	63–70	149–170	68–77
5'10"	70	178	132–145	60–66	142–156	65–71	152–173	69–79
5'11"	71	180	135–148	61–67	145–159	66–72	155–176	71–80
6'0"	72	183	138–151	62–69	148–162	67–74	158–179	72–81

Weights are based upon height and frame size for women wearing indoor clothing of 3 lb (1.4 kg) and 2 in. (5.1 cm) heels.

Adapted from the Metropolitan Life Insurance Company 1983.

ENVIRONMENTAL INFORMATION

One of the first things that can be done in the testing environment is to collect basic meteorological data.

Temperature

Step 1: Use a thermometer to measure the temperature of the laboratory in degrees Fahrenheit and record this value on the individual data sheet.

Step 2: Convert the temperature from degrees Fahrenheit to degrees Celsius and note the result on the laboratory data sheet. Next, convert the Celsius value to Kelvin and note the result in the appropriate location on the individual data sheet.

Barometric Pressure

Step 1: Use a mercury or aneroid barometer to determine the barometric pressure in mmHg, then record this value on the individual data sheet.

Step 2: Convert the barometric pressure in mmHg to torr and hPa values.

Relative Humidity

Step 1: Use a hygrometer to measure the relative humidity of the laboratory. To use this device, fill the water bottle and attach it to the device, then hang it on the wall and wait 5 to 10 min.

Step 2: Read and record both the wet- and dry-bulb temperatures.

Laboratory Activity 1.1

Step 3: Subtract the wet-bulb temperature from the dry-bulb temperature to get the wet-bulb depression value. Then determine the relative humidity from the hygrometer chart by finding the intersection between the wet-bulb depression and the dry-bulb temperature.

Step 4: Record the relative humidity on the individual data sheet.

QUESTION SET 1.1

1. Explain the calibration process and discuss the results achieved during this process.

2. What are the SI units for length, mass, and volume? How does one convert to these values from the American system?

3. Explain the process by which body mass is assessed. How can the method used affect the accuracy of the measurement?

4. How does the average height of the group compare with the normative data tables presented in this text?

5. What does the percent difference score indicate?

6. How do the average body weight results for the males and females you measured in this laboratory activity compare with the Metropolitan Life Insurance ranges? With the National Institutes of Health recommendations?

Find the case studies for this laboratory online at www.HumanKinetics.com/ LaboratoryManualForExercisePhysiology.

Laboratory Activity 1.1 Individual Data Sheet

Basic Information

Name or ID number: _____ Date: _____

Tester: _____ Time: _____

Sex: M / F (circle one)   Age: _____ y Height: _____ in. _____ cm

Weight: _____ lb _____ kg

Temperature, Barometric Pressure, and Relative Humidity

Temperature: _____ °F

(_____ °F − 32) / 1.8 = _____ °C   (_____ °C) + 273.15 = _____ K

Barometric pressure: _____ mmHg = _____ torr × 1.333 = _____ hPa = _____ kPa

Relative humidity: _____ − _____ = _____ = _____ % humidity

| | Dry-bulb temp °C | Wet-bulb temp °C | Wet-bulb depression °C |

Scale Calibration

Zero Point Calibration Verification

Trial 1: Known weight: _____ lb or _____ kg   Measured weight: _____

Trial 2: Known weight: _____ lb or _____ kg   Measured weight: _____

Trial 3: Known weight: _____ lb or _____ kg   Measured weight: _____

Average weight: _____

Difference score: _____ = _____ − _____
Known weight   Average weight

Midpoint Calibration Verification

Trial 1: Known weight: _____ lb or _____ kg   Measured weight: _____

Trial 2: Known weight: _____ lb or _____ kg   Measured weight: _____

Trial 3: Known weight: _____ lb or _____ kg   Measured weight: _____

Average weight: _____

Difference score: _____ = _____ − _____
Known weight   Average weight

High Point Calibration Verification

Trial 1: Known weight: _____ lb or _____ kg   Measured weight: _____

Trial 2: Known weight: _____ lb or _____ kg   Measured weight: _____

Trial 3: Known weight: _____ lb or _____ kg   Measured weight: _____

Average weight: _____

Difference score: _____ = _____ − _____
Known weight   Average weight

Height Assessment

Trial 1: _____ in. × 2.54 = _____ cm

Trial 2: _____ in. × 2.54 = _____ cm

Trial 3: _____ in. × 2.54 = _____ cm

Average: _____ in.   _____ cm

Average height from table 1.2 for _____ y old = _____ in. =_____ cm

Difference score calculation: _____ − _____ = _____
                                    Actual height (cm)   Average height (cm)   Difference score

Percent difference = (_____ / _____) × 100 = _____
                              Difference score   Average height

Weight Assessment

Trial 1: (_____ lb − _____ lb) / 2.2 = _____ kg
                                        Weigh-in clothes*

Trial 2: (_____ lb − _____ lb) / 2.2 = _____ kg
                                        Weigh-in clothes*

Trial 3: (_____ lb − _____ lb) / 2.2 = _____ kg
                                        Weigh-in clothes*

Average: _____ lb   _____ kg

Note: *If weigh-in clothes are not measured, simply use a zero here.

Elbow breadth: _____ in. × 2.54 = _____ cm

Frame size: _____

Metropolitan Life Insurance height and weight range: _____ (actual weight) is _____ range.

NIH weight recommendations range: _____ (actual weight) is _____ range.

STATISTICAL PROCEDURES

EQUIPMENT

- Data collected in laboratory activity 1.1
- Calculator
- Individual and group data sheets
- Microsoft Excel or equivalent spreadsheet software

 Find the group data sheets for this laboratory online at www.HumanKinetics.com/ LaboratoryManualForExercisePhysiology.

CALCULATING THE MEAN

Once the basic data have been collected in laboratory 1.1, they can be analyzed.

Hand Calculation

Step 1: Note the number of subjects for whom data was collected in the appropriate location on the data sheet for laboratory 1.2.

Step 2: Sum all scores recorded in laboratory 1.1 and note the result as variable *X* on the laboratory 1.2 data sheet.

Step 3: Calculate the mean for these scores and mark the result on the individual data sheets for laboratory 1.2. The following formula should be used:

$$\text{Mean} = \frac{1}{N}\sum X$$

Step 4: Transfer the mean score to the group data sheet.

Excel Calculation

Step 1: Enter the data presented in table 1.1 into an Excel spreadsheet. The rows should represent subjects, and the columns should represent variables. Place the Subject label in cell A2, then highlight cells B1 and C1 and select the Merge Cells function located on the toolbar under the Alignment tab. Use the same process to merge cells D1 and E1. Now, label cells B2 and D2 as Male, C2 and E2 as Female, B1 and C1 as Height, and D1 and E1 as Weight. See table 1.7 (rows 1–11) for an example of what your data table should look like. (The other rows are described in the steps that follow.)

Step 2: In a cell under the last subject's variables, enter the following formula:

$$=\text{AVERAGE}(\text{number1:number2})$$

Number1 indicates the first number of the range, and number2 indicates the last number of the range. In the sample shown in table 1.7, the formula appears in cells B11, C11, D11, and E11.

Step 3: Transfer the mean score to the group data sheet.

Table 1.7 Sample Format and Formula Placements in a Spreadsheet

	A	B	C	D	E
		Height (cm)		Weight (kg)	
1					
2	Subject	Male	Female	Male	Female
3	1				
4	2				
5	3				
6	4				
7	5				
8	6				
9	7				
10	8				
11	Mean	=AVERAGE(B3:B10)	=AVERAGE(C3:C10)	=AVERAGE(D3:D10)	=AVERAGE(E3:E10)
12	Median	=MEDIAN(B3:B10)	=MEDIAN(C3:C10)	=MEDIAN(D3:D10)	=MEDIAN(E3:E10)
13	Mode	=MODE(B3:B10)	=MODE(C3:C10)	=MODE(D3:D10)	=MODE(E3:E10)
14	SD	=STDEV(B3:B10)	=STDEV(C3:C10)	=STDEV(D3:D10)	=STDEV(E3:E10)

Once a formula is entered into a cell, it will disappear, and the calculated values will appear in its place.

CALCULATING THE MEDIAN AND THE MODE

The median and the mode offer two other methods for examining the central tendency of a data set.

Hand Calculation

Step 1: Place all of the scores for height and weight in order from lowest to highest for each group of data (male and female).

Step 2: Calculate the location of the median score for each group by using the following formula:

$$\text{Median} = [(N + 1) / 2]\text{th term}$$

Step 3: Determine the median score for males and the median score for females and record them in the appropriate locations on the laboratory 1.2 data sheets.

Step 4: Calculate the mode by determining which value occurs most frequently. Note the result on the laboratory 1.2 data sheets.

Excel Calculation

Step 1: Use the following equations:

$$=MEDIAN(number1:number2)$$

$$=MODE(number1:number2)$$

Number1 indicates the first number in the range, and number2 indicates the last number. Place the median formula in cells B12, C12, D12, and E12. Place the mode formula in cells B13, C13, D13, and E13. Thus your spreadsheet will be formatted as shown in table 1.7 (rows 13 and 14).

Step 2: Transfer the results to the laboratory 1.2 data sheets.

CALCULATING THE STANDARD DEVIATION

The standard deviation is one of the most commonly reported indicators of variability in a data set.

Hand Calculation

Step 1: Sum all the results for the variable being evaluated (X).

Step 2: Subtract the mean, calculated previously, from each variable collected ($X - M$). Hint: This must be done for each subject. If summed, the value should be 0.

Step 3: Square the value calculated in step 2 [$(X - M)^2$], then sum all the values.

Step 4: Use the following equation to calculate the standard deviation:

$$Standard\ deviation = \sqrt{\frac{1}{(N-1)} \sum (X-M)^2}$$

Step 5: Record the standard deviation results in the appropriate location in the laboratory 1.2 data sheets.

Excel Calculation

Place the following formula in the cell underneath the data column:

$$=STDEV(number1:number2)$$

See row 14 of the sample spreadsheet shown in table 1.7.

CALCULATING THE RANGE

The range of scores in a data set can also be used to depict the variability of the set.

Step 1: Determine the high and low scores within the data set for all the height and weight data: height for men, weight for men, height for women, and weight for women.

Step 2: For each data set, subtract the low score from the high score.

Step 3: Add one to each of the scores calculated in step 2.

Step 4: Record the range for each set in the appropriate location on the laboratory 1.2 data sheets.

QUESTION SET 1.2

1. Are the height and weight data collected in this laboratory normally distributed? How can you tell?

2. When working with Excel, what formulas are used for mean, median, mode, and standard deviation?

3. Do the men or the women in the class have higher body mass? Was this expected? How do their heights compare? Explain the possible reasons for these results.

Find the case studies for this laboratory online at www.HumanKinetics.com/ LaboratoryManualForExercisePhysiology.

Laboratory Activity 1.2 Individual Data Sheet

Basic Information

Name or ID number: _____ Date: _____

Tester: _____ Time: _____

Sex: M / F (circle one)   Age: _____ y   Height: _____ in. _____ cm

Weight: _____ lb _____ kg

Hand Calculation of Mean

Male Data

Number of subjects: _____ = N

Trial 1: Height (cm): _____ = X Mean = _____ / _____ = _____
 Sum of trials X N

Trial 2: Height (cm): _____ = X Mean = _____ / _____ = _____
 Sum of trials X N

Trial 3: Height (cm): _____ = X Mean = _____ / _____ = _____
 Sum of trials X N

Trial 1: Weight (kg): _____ = X Mean = _____ / _____ = _____
 Sum of trials X N

Trial 2: Weight (kg): _____ = X Mean = _____ / _____ = _____
 Sum of trials X N

Trial 3: Weight (kg): _____ = X Mean = _____ / _____ = _____
 Sum of trials X N

Female Data

Number of subjects: _____ = N

Trial 1: Height (cm): _____ = X Mean = _____ / _____ = _____
 Sum of trials X N

Trial 2: Height (cm): _____ = X Mean = _____ / _____ = _____
 Sum of trials X N

Trial 3: Height (cm): _____ = X Mean = _____ / _____ = _____
 Sum of trials X N

Trial 1: Weight (kg): _____ = X Mean = _____ / _____ = _____
 Sum of trials X N

Trial 2: Weight (kg): _____ = X Mean = _____ / _____ = _____
 Sum of trials X N

Trial 3: Weight (kg): _____ = X Mean = _____ / _____ = _____
 Sum of trials X N

Excel Spreadsheet Mean

Males

Trial 1: Height: _____ cm   Weight: _____ kg

Trial 2: Height: _____ cm   Weight: _____ kg

Trial 3: Height: _____ cm   Weight: _____ kg

Females

Trial 1: Height: _____ cm   Weight: _____ kg

Trial 2: Height: _____ cm   Weight: _____ kg

Trial 3: Height: _____ cm   Weight: _____ kg

Hand Calculation of the Median and Mode

Male Data

Trial 1: Median height: (_____ + 1) / 2 = _____ = _____ cm
                           Number of samples

Trial 2: Median height: (_____ + 1) / 2 = _____ = _____ cm
                           Number of samples

Trial 3: Median height: (_____ + 1) / 2 = _____ = _____ cm
                           Number of samples

Trial 1: Median weight: (_____ + 1) / 2 = _____ = _____ kg
                           Number of samples

Trial 2: Median weight: (_____ + 1) / 2 = _____ = _____ kg
                           Number of samples

Trial 3: Median weight: (_____ + 1) / 2 = _____ = _____ kg
                           Number of samples

Trial 1: Mode: Height: _____ cm   Weight: _____ kg

Trial 2: Mode: Height: _____ cm   Weight: _____ kg

Trial 3: Mode: Height: _____ cm   Weight: _____ kg

Female Data

Trial 1: Median height: (_____ + 1) / 2 = _____ = _____ cm
                           Number of samples

Trial 2: Median height: (_____ + 1) / 2 = _____ = _____ cm
                           Number of samples

Trial 3: Median height: (_____ + 1) / 2 = _____ = _____ cm
                           Number of samples

Trial 1: Median weight: (_____ + 1) / 2 = _____ = _____ kg
                           Number of samples

Trial 2: Median weight: (_____ + 1) / 2 = _____ = _____ kg
                           Number of samples

Trial 3: Median weight: (_____ + 1) / 2 = _____ = _____ kg
                           Number of samples

Trial 1: Mode: Height: _____ cm   Weight: _____ kg

Trial 2: Mode: Height: _____ cm   Weight: _____ kg

Trial 3: Mode: Height: _____ cm   Weight: _____ kg

Excel Spreadsheet Median and Mode

Male Median

Trial 1: Height: _____ cm   Weight: _____ kg

Trial 2: Height: _____ cm   Weight: _____ kg

Trial 3: Height: _____ cm   Weight: _____ kg

Male Mode

Trial 1: Height: _____ cm   Weight: _____ kg

Trial 2: Height: _____ cm   Weight: _____ kg

Trial 3: Height: _____ cm   Weight: _____ kg

                                                 Laboratory Activity 1.2

Female Median

Trial 1: Height: _____ cm   Weight: _____ kg

Trial 2: Height: _____ cm   Weight: _____ kg

Trial 3: Height: _____ cm   Weight: _____ kg

Female Mode

Trial 1: Height: _____ cm   Weight: _____ kg

Trial 2: Height: _____ cm   Weight: _____ kg

Trial 3: Height: _____ cm   Weight: _____ kg

Calculation of the Standard Deviation

Male Excel Calculation of the SD

Height: Trial 1: _____   Trial 2: _____   Trial 3: _____

Weight: Trial 1: _____   Trial 2: _____   Trial 3: _____

Female Excel Calculation of the SD

Height: Trial 1: _____   Trial 2: _____   Trial 3: _____

Weight: Trial 1: _____   Trial 2: _____   Trial 3: _____

Males	Trial 1 height			Trial 2 height			Trial 3 height		
Subject	X	$X - M$	$(X - M)^2$	X	$X - M$	$(X - M)^2$	X	$X - M$	$(X - M)^2$
1									
2									
3									
4									
5									
6									
7									
8									
9									
10									
11									
12									
13									
14									
15									
16									
17									
18									
19									
20									
Sum									
SD =		SD=			SD=				

Laboratory Activity 1.2

Females	Trial 1 height			Trial 2 height			Trial 3 height		
Subject	X	$X - M$	$(X - M)^2$	X	$X - M$	$(X - M)^2$	X	$X - M$	$(X - M)^2$
1									
2									
3									
4									
5									
6									
7									
8									
9									
10									
11									
12									
13									
14									
15									
16									
17									
18									
19									
20									
Sum									
SD =		SD=			SD=				

Males	Trial 1 weight			Trial 2 weight			Trial 3 weight		
Subject	X	$X - M$	$(X - M)^2$	X	$X - M$	$(X - M)^2$	X	$X - M$	$(X - M)^2$
1									
2									
3									
4									
5									
6									
7									
8									
9									
10									
11									
12									
13									
14									
15									
16									
17									
18									
19									
20									
Sum									
SD =		SD=			SD=				

Laboratory Activity 1.2

Females	Trial 1 weight			Trial 2 weight			Trial 3 weight		
Subject	X	$X - M$	$(X - M)^2$	X	$X - M$	$(X - M)^2$	X	$X - M$	$(X - M)^2$
1									
2									
3									
4									
5									
6									
7									
8									
9									
10									
11									
12									
13									
14									
15									
16									
17									
18									
19									
20									
SD =		SD =			SD =				

Calculation of the Range

Hand calculation of the range (note that these are the high and low scores for the group data):

Male Data

Trial 1 height: (_____ − _____) + 1 = _____ = Range
               High score   Low score

Trial 2 height: (_____ − _____) + 1 = _____ = Range
               High score   Low score

Trial 3 height: (_____ − _____) + 1 = _____ = Range
               High score   Low score

Trial 1 weight: (_____ − _____) + 1 = _____ = Range
               High score   Low score

Trial 2 weight: (_____ − _____) + 1 = _____ = Range
               High score   Low score

Trial 3 weight: (_____ − _____) + 1 = _____ = Range
               High score   Low score

Female Data

Trial 1 height: (_____ − _____) + 1 = _____ = Range
       High score Low score

Trial 2 height: (_____ − _____) + 1 = _____ = Range
       High score Low score

Trial 3 height: (_____ − _____) + 1 = _____ = Range
       High score Low score

Trial 1 weight: (_____ − _____) + 1 = _____ = Range
       High score Low score

Trial 2 weight: (_____ − _____) + 1 = _____ = Range
       High score Low score

Trial 3 weight: (_____ − _____) + 1 = _____ = Range
       High score Low score

Laboratory Activity 1.2

TABLES AND GRAPHS

EQUIPMENT

- Data from laboratories 1.1 and 1.2
- Individual and group data sheets
- Microsoft Word or equivalent word-processing software
- Microsoft Excel or equivalent spreadsheet software

 Find the group data sheets for this laboratory online at www.HumanKinetics.com/ LaboratoryManualForExercisePhysiology.

CREATING TABLES

Step 1: Use either Microsoft Word or Microsoft Excel. If using Word, first select the Insert Table Function command, then select the number of columns and rows the table should have. For this lab, choose five rows and seven columns (see the sample table on the next page for an example).

Step 2: Insert labels into the table. The first column should contain the variables Age, Height, and Weight.

Step 3: Merge the second, third, and fourth columns. To merge the columns, highlight the top cell of each column, then select the Merge Cells function. Label the resulting merged column as Male (click on the Center option to center the label). Repeat this process for columns five through seven but use the label Female.

Step 4: Below the Male label and the Female label, insert the mean, ±, and SD labels for each grouping. To insert the ± symbol, find the Insert Symbol link; if necessary (i.e., if the ± symbol is not included in the frequently used list), select the More Symbols option. To find the ± symbol, change the font to Symbol and scroll through the available symbols.

Step 5: Center the ± symbol by highlighting the columns in which it appears and selecting the Center option located under the Home tab. Right-justify the Mean and SD labels by highlighting the columns in which they appear and selecting the Right Justify option under the Home tab.

Step 6: Resize the columns containing the ± symbol in order to make the table smaller and thus easier to interpret. Do so by double-clicking on the right column line to prompt the software package to automatically resize the column. In addition, insert the ± symbol in each row under the label for ±.

Step 7: Now remove the lines on the table. To do so, highlight the entire table, select the Borders option, and then choose No Border.

Step 8: Insert a border line at the top and at the bottom of the table and below the row containing the column labels. If done correctly, the table should be formatted as shown in the table below.

Step 9: Enter into the table the pertinent data collected in laboratory activities 1.1 and 1.2. Once the data are entered, you can adjust the formatting of the table to make it easier to read.

Variables	Males			Females		
	Mean	±	SD	Mean	±	SD
Age (y)		±			±	
Height (cm)		±			±	
Weight (kg)		±			±	

CREATING COLUMN GRAPHS

Bar graphs offer a great way to represent data and are relatively easy to create with software such as Microsoft Excel. Here, you will create a column graph based on data collected in laboratory activity 1.1 and analyzed in laboratory activity 1.2.

Step 1: Locate the mean and standard deviation data recorded on the group data sheet for laboratory 1.2. If you analyzed these data by using Excel, use the data set created in Excel to do this new analysis.

Step 2: If you did not use Excel for your analysis in laboratory activity 1.2, enter the mean and standard deviation data into Excel now.

Step 3: To create a bar graph for the height data, highlight cells B11 and C11, then find the Insert Column Graph link.

Step 4: A basic column graph should be created. To clean up the graph, click on the horizontal gridlines to effectively highlight them, then delete them.

Step 5: To add a y-axis label, find your software's option for editing the axis titles, then find the option for editing the primary vertical axis. Rotate the title so that it stretches along the y-axis, then type in the units that correspond to the y-axis. In this case, use height (cm).

Step 6: Delete the Series 1 legend entry by clicking on it and then pressing the delete key.

Step 7: Next, edit the x-axis. Find the option for editing the horizontal axis, then highlight the range. Find the function to label the axis so that you can highlight the Male and Female labels located in cells B2 and C2, respectively. At this point, the columns should be labeled according to their contents.

Step 8: To add standard deviation bars, click on the column chart, find your software's option for inserting error bars, and then find the option for formatting error bars. Now find and select the custom option and insert a custom positive and negative range. In this example, highlight cells B14 and C14, then click the box with the red arrow pointing downward to return to the custom error bar selection screen. Select the negative error value range option and then highlight B14 and C14. After both negative and positive error values are selected, click OK and then close, which will place the appropriate error bars in the column graph.

Step 9: Add a title above the chart.

Step 10: Perform these steps again to create a column graph for the weight variable.

QUESTION SET 1.3

1. If you were creating a graphical depiction of a series of anthropometric data, what important variables would you need to present?

2. Is it important to include the standard deviation of the data in the graphical depiction of the data set? Why or why not?

3. Describe how you might set up your spreadsheet to create a graphical depiction of your data. Be very specific.

4. How would you determine whether a graphical depiction is necessary for your data set?

Find the case studies for this laboratory online at www.HumanKinetics.com/ LaboratoryManualForExercisePhysiology.

LABORATORY (2)

Pretest Screening

Objectives

1. Introduce the basic components of a pretesting screening process.
2. Explain the informed consent process and outline what is contained in an appropriately structured informed consent form.
3. Present the basic components of a medical and health history questionnaire.
4. Practice the performance of pretesting screening.
5. Describe the American College of Sports Medicine's current risk stratifications and guidelines for physician involvement in the testing process.

Definitions

arrhythmia—Abnormal heart rhythm.

arteriosclerosis—Abnormal thickening and hardening of arterial walls resulting in a loss of elasticity.

atherosclerosis—Progressive narrowing of the arteries due to changes in the arterial lining and buildup of plaque; a form of arteriosclerosis.

coronary heart disease—Condition in which blood flow through the coronary arteries is reduced, typically as a result of atherosclerosis; also known as *coronary artery disease*.

high coronary heart disease risk—Presence of one or more signs or symptoms of cardiovascular or pulmonary disease or presence of known risk factors for cardiovascular, pulmonary, or metabolic disease.

hypertension—Chronically elevated blood pressure; also referred to as *high blood pressure*.

informed consent—Process in which the subject acknowledges having been informed of the purposes, methods, risks, and benefits of a study or test.

low coronary heart disease risk—Absence of relevant symptoms along with the presence of no more than one coronary heart disease risk factor; characteristic of young individuals (men under age 45, women under age 55).

moderate coronary heart disease risk—Status typically assigned to older individuals (men older than 45, women older than 55), as well as individuals of any age who have two or more coronary heart disease risk factors).

myocardial infarction—Heart attack.

palpitation—Racing or pounding of the heart.

PAR-Q—Physical Activity Readiness Questionnaire, which is used to establish an individual's fitness for physical activity.

Prior to performing any testing, it is essential to use a prescreening process in order to detect possible diseases and initially classify the subject's disease risk (1, 3, 4). More specifically, the American College of Sports Medicine (ACSM) suggests that a prescreening program be used to accomplish the following goals (2):

1. Identify individuals with medical contraindications that exclude them from performing certain health-related fitness assessments.

2. Determine which individuals should have a medical evaluation and consult a physician before undergoing a health-related fitness assessment.

3. Establish which subjects should undergo a medically supervised fitness assessment.

4. Determine whether an individual has other health or medical concerns (e.g., diabetes mellitus, orthopedic injuries).

At minimum, a pretesting screening should include these three components (4):

1. Health history questionnaire

2. Physical activity readiness questionnaire (PAR-Q)

3. Medical or health exam

Generally, however, it is probably warranted to perform a more comprehensive, seven-step screening process in order to decrease the subject's potential risk for an adverse effect during the testing process (3). See table 2.1 for a description of this process.

Informed Consent

Before performing any part of a testing process, you must obtain full informed consent from any subject or individual who is going to be tested (2, 3). Doing so documents that the individual has been informed about the procedures, benefits, and risks associated with the assessment process that he or she will undergo. At this time, you should also present any available alternative tests (2).

Your informed consent forms should clearly document the benefits and risks associated with the assessment so that the subject can decide whether the scientific or clinical benefits clearly outweigh the personal risks (2). It is also essential that you inform the subject that (a) he or she is volunteering to participate in the assessment process, (b) he or she should inform the testers of any problems experienced during the assessment, and (c) he or she has the right to withdraw from the testing process without any consequences at any time. The accompanying highlight box presents eight items generally included in an **informed consent form;** for a sample, see form 2.1 (p. 57).

Table 2.1 Seven-Step Comprehensive Screening Process

Questionnaire/screening form	Purpose
Informed consent	Explains the purpose, risks, and benefits to the individual of participating in the testing process.
PAR-Q	Helps determine the individual's readiness to perform physical activity.
Medical/health history	Gives information about the client's past and present personal health history and insight into the individual's family history. Taken collectively, this information focuses on conditions that require medical referrals and clearance.
Signs and symptoms of disease and medical clearance	Identifies individuals who need a physician's approval, via a medical referral, to perform exercise testing.
Coronary risk factor analysis	Establishes the individual's risk by quantifying his or her number of coronary heart disease risk factors.
Lifestyle evaluation	Gives insight into the individual's living habits.
Disease risk classification	Categorizes the individual as high, medium, or low risk.

Adapted, by permission, from V. Heyward, 2010, *Advanced fitness assessment and exercise prescription,* 6th ed. (Champaign, IL: Human Kinetics), 20.

Items Needed in an Informed Consent Form

- General statement of the testing process and objectives
- Easily understood description of the testing procedures
- Description of all risks associated with the testing process
- Explanation of the benefits associated with the testing procedures
- Statement informing the subject that the testers will answer any questions
- Statement informing the subject that he or she is free to withdraw consent and cease the testing process at any time without consequence
- Statement informing the subject that he or she is free to refuse to answer any questions or respond to specific items contained in any questionnaires
- General statement of the procedures that will be used to maintain confidentiality of the data obtained from the subject's screening and testing

Adapted from Nieman (4) and American College of Sports Medicine (2).

Physical Activity Readiness Questionnaire

The Physical Activity Readiness Questionnaire (**PAR-Q**) contains seven questions designed to identify individuals who need medical clearance before either being tested or entering an exercise regime (form 2.2) (3). The PAR-Q is widely recognized as a safe prescreening questionnaire for those who plan to undertake low- to moderate-intensity exercise (2, 4). The PAR-Q contains seven yes-or-no questions that are generally very easy for a subject to understand and answer (2). If a subject answers *yes* to any of the questions, he or she should be referred to a physician for a more comprehensive health screening and for medical clearance for any testing procedure or exercise regime (3).

Health History Questionnaire

During prescreening, you should also use a comprehensive health history questionnaire to establish the subject's medical risk (2, 3). You can use the information gathered from this questionnaire to stratify the subject's risk and determine whether he or she needs a medical referral prior to undergoing any testing procedures. This process will also help you determine whether the subject needs medical supervision during any of the testing procedures.

A health history questionnaire generally includes 10 items essential to risk assessment (6): medical diagnoses; previous physical examination results; history of symptoms; recent illness, hospitalization, new medical diagnosis, or surgical procedures; orthopedic issues; medications taken and drug allergies; lifestyle habits; exercise habits; work history; and family history of disease. For sample specifics in these categories, see table 2.2.

A sample comprehensive health history questionnaire is presented in form 2.3 on page 61 (4). This health history contains all major items recommended as part of a comprehensive questionnaire. When you have a subject or client fill out this type of document, make someone available to answer any questions. Once the document has been completed, the tester should examine it closely for contraindications to exercise testing (see the highlight box on p. 43). Individuals who have absolute contraindications should not be tested until the condition is stabilized or resolved (2). For those with relative contraindications, testing may be performed only after the risk-to-benefit ratio has been carefully evaluated (1). In general, you should direct all subjects who present either absolute or relative contraindications to a physician for a complete examination and to receive clearance for the testing (3).

Table 2.2   Components of a Health History Questionnaire

Component	Items of concern (not a comprehensive list)
Any medical diagnosis	• Cardiovascular disease, including myocardial infarction • Percutaneous coronary artery procedures, including angioplasty • Valvular dysfunction and surgeries • Symptoms of ischemic coronary syndrome • Peripheral vascular disease • Hypertension • Diabetes • Obesity • Pulmonary disease • Anemia • Stroke • Cancer
Previous physical examination findings	• Heart murmur • Abnormal heart sounds • Abnormal pulmonary or cardiovascular findings • Abnormal blood sugar or heart rate, high blood pressure, high cholesterol, or other lab abnormalities
History of symptoms	• Discomfort in the chest, jaw, neck, arms, or back • Light-headedness, dizziness, or fainting • Unilateral weakness • Shortness of breath • Palpitation
Recent illness, hospitalization, new medical diagnosis, or surgical procedure	Any recent changes in the subject's medical status
Orthopedic problems	• Arthritis • Joint swelling • Other issues that impede ability to exercise
Medications used and drug allergies	Any medications that the subject takes and any to which he or she may be allergic
Lifestyle habits	• Dietary habits, including alcohol use • Drug use • Tobacco use
Exercise habits	List of types, duration, and intensity of exercise currently undertaken
Work history	Description of the subject's current work atmosphere and types of work activities
Family history of disease	• Cardiovascular disease • Pulmonary disease • Metabolic disease • Stroke • Sudden death

Adapted from Nieman (4) and American College of Sports Medicine (1, 2, 6).

Contraindications to Exercise Testing

Absolute Contraindications

Acute myocardial infarction (within 2 days)

Changes in resting ECG suggesting significant ischemia (within 2 days)

Unstable angina

Uncontrolled cardiac arrhythmias causing symptoms of hemodynamic compromise

Acute aortic dissection

Symptomatic severe aortic stenosis

Suspected or known dissecting aneurysm

Acute pericarditis or myocarditis

Acute pulmonary embolus or pulmonary infarction

Uncontrolled symptomatic heart failure

Acute systemic infection

Relative Contraindications

Left main artery stenosis

Moderate stenotic valvular heart disease

Known electrolyte abnormalities (hypokalemia, hypomagnesemia)

Severe arterial hypertension (resting diastolic blood pressure >110 mmHg or systolic blood pressure >200 mmHg)

Tachyarrhythmias or bradyarrhythmias

Hypertrophic cardiomyopathy and other forms of outflow tract obstruction

High-degree atrioventricular block

Mental or physical impairment leading to inability to exercise adequately

Chronic infection (e.g., mononucleosis, hepatitis, AIDS)

Uncontrolled metabolic disease (e.g., diabetes, thyrotoxicosis, myxedema)

Adapted from American College of Sports Medicine (1) and Gibbons, R.J., et al. 2002. *ACC/AHA 2002 Guideline update for exercise testing. A report of the American College of Cardiology/American Heart Association Task Force on Practice Guidelines (Committee on Exercise Testing).*

Signs and Symptoms of Disease and Medical Clearance

It is important to screen all subjects or clients for signs and symptoms of disease prior to testing (3). This can be done in several ways:

1. In the health history questionnaire (form 2.2)
2. In a separate checklist you create to address signs and symptoms of disease (form 2.4)
3. In the Physical Activity Readiness Medical Examination (PARmed-X) form (form 2.5) (3, 4)

Form 2.4 (p. 67) presents a sample checklist for signs and symptoms of disease (3). In this example, subjects would be referred to their physician if they have any of the listed conditions or more than two of the listed risk factors. The PARmed-X form (form 2.5, p. 69) can be used by the subject's physician to evaluate and give medical clearance for subjects who have answered yes to any questions on the PAR-Q form; that is, the PARmed-X form is used to either clear a subject for testing or exercise or to refer him or her to a medically supervised test or exercise program (3).

Coronary Risk Factor Analysis

In determining whether it is safe to conduct exercise testing with a given subject, it is important to establish his or her coronary risk profile. Form 2.6 (p. 73) presents a series of items positively associated with **coronary heart disease** risk (1, 2). Though these items make up part of the American College of Sports Medicine's risk stratification, the list presented in form 2.6 should not be considered all inclusive (6). The items should, however, be viewed as "clinically relevant thresholds" to consider along with other screening items when making decisions about a subject's risk for an adverse coronary event and about the level of medical clearance needed prior to testing (1, 2).

As a screening tool, form 2.6 can be used to assess the individual's risk by summing the subject's risk factors. If the subject has a high-density lipoprotein-C (HDL-C) level equal to or greater than 60 mg · dl⁻¹, the tester should subtract 1 from the total number of risk factors (1, 2)

Lifestyle Evaluation

Another way to gain insight into an individual's overall risk profile is to develop an understanding of his or her lifestyle (3). A lifestyle questionnaire addresses questions such as smoking, physical activity, diet, and alcohol consumption habits. These items can easily be folded into the medical or health history questionnaire, as is done in form 2.3 (4); they can also be handled in a separate questionnaire as shown in form 2.7 (p. 75), which presents a tool called the Fantastic Lifestyle Checklist for assessing a subject's current health-related behaviors.

Disease Risk Stratification

Once you have collected all of the prescreening data, you can establish an overall risk profile in order to stratify the individual's risk and decide

about the need for a physician's referral or presence during the testing process.

The ACSM suggests using a three-tier risk stratification (low, moderate, high) when performing an initial evaluation of a subject's or client's risk profile (1). For example, taking the data collected in forms 2.3 and 2.4, the tester can stratify the subject's overall risk level by using the guidelines presented in table 2.3. The basic stratification presented by the ACSM suggests that individuals who have no more than one risk factor (from form 2.6) are in the low-risk category, while individuals who have two or more risk factors fall into the moderate-risk category. Individuals with known diseases (see table 2.3) are classified as high risk and require further medical screening before undergoing exercise testing.

Once the subject's health status has been classified into one of the three tiers, you can make decisions about testing conditions (table 2.4) (1, 2). If the subject is classified as low risk, it is not necessary for him or her to have a pretest medical examination, and physician supervision of the test is unnecessary. However, if the subject's risk is either moderate or high, then he or she should undergo a medical screening, and a physician should be in close proximity and readily available during the test (1, 2). When the risk is high, it is

Table 2.3 ACSM Initial Risk Stratification

Low risk	Asymptomatic men and women who have ≤1 CVD risk factor from form 2.6
Moderate risk	Asymptomatic men and women who have ≥2 risk factors from form 2.6

High risk	Individuals with
	• known cardiac, peripheral vascular, or cerebrovascular disease; chronic obstructive pulmonary disease, asthma, interstitial lung disease, or cystic fibrosis; or diabetes mellitus (type 1 or type 2), thyroid disorders, renal or liver disease; or
	• one or more of the following signs or symptoms:
	  – heart murmur;
	  – unexplained fatigue;
	  – dizziness or syncope;
	  – swelling of the ankles;
	  – tachycardia or irregular heartbeat;
	  – unexplained shortness of breath;
	  – intermittent claudication;
	  – breathing discomfort when not in upright position, or interrupted breathing at night; or
	  – pain or discomfort in the jaw, neck, chest, arms, or elsewhere that could be caused by ischemia

Adapted from American College of Sports Medicine (1).

Table 2.4 ACSM Recommendations for Physician Involvement in Exercise Testing

	Low risk	Moderate risk	High risk
Current (within past year) medical examination and exercise testing before participation in exercise[1]			
Moderate exercise[3]	Not necessary[2]	Not necessary	Recommended
Vigorous exercise[3]	Not necessary	Recommended	Recommended
Physician supervision of exercise tests			
Submaximal test	Not necessary	Not necessary	Recommended
Maximal test	Not necessary	Recommended	Recommended

[1] When looking at exercise, the classification "not necessary" suggests that it is not essential to employ medical examination, exercise testing, or physician supervision as part of the preparticipation screening process. However, it could still be done based on the judgment of the exercise physiologist. The "recommended" classification suggests that prior to beginning an exercise program the subject should undergo a medical examination and exercise testing conducted under the supervision of a physician.

[2] When applied to exercise testing, the "not necessary" classification suggests that exercise testing could be done without a pretest medical screening or physician supervision. The "recommended" classification suggests that the subject should undergo a medical examination prior to undergoing any testing and that the exercise testing should be conducted under the supervision of a physician.

[3] *Moderate exercise* is generally defined as 3 to 6 METS (5), whereas *vigorous exercise* is defined as >6 METS or >60 percent of maximal oxygen consumption (1, 2).

Adapted from American College of Sports Medicine (1).

preferable to have the physician present in order to maximize the safety of the testing procedures.

References

1. American College of Sports Medicine. *ACSM'S Guidelines for Exercise Testing and Prescription*. 8th ed. Baltimore: Lippincott Williams & Wilkins, 2010.

2. American College of Sports Medicine. *ACSM's Health-Related Physical Fitness Assessment Manual*. 3rd ed. Baltimore: Lippincott Williams & Wilkins, 2009.

3. Heyward VH. *Advanced Fitness Assessment and Exercise Prescription*. 6th ed. Champaign, IL: Human Kinetics, 2010.

4. Nieman DC. *Exercise Testing and Prescription: A Health-Related Approach*. 5th ed. New York: McGraw-Hill, 2003, p. 774.

5. Pate RR, Pratt M, Blair SN, et al. Physical activity and public health. A recommendation from the Centers for Disease Control and Prevention and the American College of Sports Medicine. *JAMA* 273: 402-407, 199.

6. Whaley, MH, Brubacker PH, and Otto RM, eds. *ACSM's Guidelines for Exercise Testing and Prescription*. 6th ed. Baltimore: Lippincott Williams & Wilkins, 2000.

BASIC SCREENING PROCEDURES

EQUIPMENT

- Informed consent form (form 2.1)
- PAR-Q form (form 2.2)
- Health history questionnaire (form 2.3)
- Checklist for signs and symptoms of disease (form 2.4)
- Coronary heart disease risk form (form 2.6)
- Fantastic Lifestyle Checklist (form 2.7)
- Simulated subject data cards (available online)

 The simulated subject data cards for this laboratory are available online at www.HumanKinetics.com/ LaboratoryManualForExercisePhysiology.

PRETEST SCREENING

Pair up and administer the five questionnaires with your partner. One student per pair serves as the tester, and the other uses the simulated client cards to act as the subject. The tester should administer the pretesting screening process in order to simulate what would be undertaken in a laboratory testing scenario. Once screening is completed, the tester evaluates the subject's overall risk and stratifies him or her according to ACSM guidelines. Then reverse roles with your partner and repeat the activity.

When initiating a testing scenario, it is important for the tester to follow a basic step-by-step procedure (3). A multitude of possible procedural steps are available. Here is a basic model.

Step 1: Greet the subject.

Step 2: Explain the basic process for the pretesting screening. Give a basic overview of all the data that will be collected during the prescreening process.

Step 3: Give the subject the informed consent form (form 2.1). Answer any questions while the subject reads, fills out, and signs the document. Be aware that form 2.1 is merely an example and that a more detailed or specific version may be required for your testing scenario. This example meets the basic criteria for informed consent, but it is important for you to realize that each institutional review board has more specific requirements.

Step 4: Administer the PAR-Q screening form (form 2.2). Encourage the subject to answer all questions truthfully. Have the subject sign and date the form. Evaluate the form according to the criteria presented on the form.

Step 5: Give the subject a health history questionnaire (form 2.3). Encourage the subject to read the document carefully, to answer all questions, and to ask questions when needed. After the form has been completed, read it over and look for items that may need clarification. Note any contraindications to exercise and testing (this will aid in the risk stratification process).

Step 6: Give the subject the checklist for signs and symptoms of disease (form 2.4). Ask the subject to fill out the checklist to the best of his or her knowledge and note any signs or symptoms of disease.

Step 7: Administer the coronary heart disease risk factors form (form 2.6). Ask the subject to answer *yes* to any questions that correspond to his or her health status. After completing the form, add up all the points to determine the subject's overall risk.

Step 8: Give the subject the Fantastic Lifestyle Checklist (form 2.7) and instruct him or her to fill out the form as truthfully as possible. Evaluate the subject's answers as follows:

  a. Total the number of Xs contained in each column.

  b. Multiply the totals by the numbers indicated on the form.

  c. Total the scores to yield a grand score.

Step 9: Evaluate the subject's overall disease risk by comparing the grand total score with the normative data presented in form 2.7. A grand total between 85 and 100 constitutes an excellent score, whereas a grand total between 0 and 34 indicates a need for improvement. In the middle, grand total scores between 70 and 84 are classified as very good, scores between 55 and 69 as good, and scores between 35 and 54 as fair.

Step 10: Determine the subject's initial risk stratification according to the ACSM stratifications presented in table 2.3. Then refer to table 2.4 to determine whether physician's supervision is needed for any testing procedures.

QUESTION SET 2.1

  1. What 10 components should be included in a health history questionnaire?
  2. What are the three tiers of ACSM's initial risk stratification? What factors determine a subject's risk level?
  3. What eight items should be included in an informed consent form?
  4. What is a PAR-Q, and what is it used for?
  5. What factors should you concentrate on when evaluating a client's or subject's pretest screening?
  6. Do all clients need medical clearance or supervision during testing? Why or why not?
  7. How does one use the information collected in pretesting screening to evaluate a client's or subject's risk?

Laboratory Activity 2.1

PRETEST RESULTS

EQUIPMENT

- Clinical case study health history questionnaire for sample client (figure 2.1)
- Checklist for signs and symptoms of disease (form 2.4)
- Coronary heart disease risk factors form (form 2.6)

PRETEST SCREENING DATA

Imagine that a new client has come into your health and wellness facility. She has already completed an informed consent form and filled out a comprehensive health history questionnaire. Now it is time to use her answers to make decisions about her overall health and wellness. Form groups of two or three students each and go over the sample client's health history questionnaire (figure 2.1).

When reviewing the results of a pretesting screening, you should perform a comprehensive and systematic analysis of the information collected (3). You can choose from many methods in conducting this evaluation; what follows should serve as a basic model.

Step 1: Examine the sample client's health history questionnaire for any contraindications to exercise testing (see the highlight box on page 43 for the full list) and note them on a copy of the checklist for signs and symptoms of disease (form 2.4). Make sure to comment specifically on any items for which the answer was *yes*. Use this form to make a preliminary decision about the need for physician's clearance for testing.

Step 2: Use the coronary heart disease risk factors form (form 2.6) to evaluate the client's risk in this area. Pay particular attention to the seven major risk factor categories. Note any risk factors presented by the client and tabulate the point values.

Step 3: After evaluating the client's health history questionnaire, refer to table 2.3 to determine her initial risk stratification.

Step 4: Once you have determined the client's risk stratification, use table 2.4 to determine whether she needs a current medical examination and exercise testing and whether a physician needs to supervise the test.

QUESTION SET 2.2

1. What general risk factors does this client exhibit? Would these risk factors affect the client's ability to perform physical activity or participate in testing?
2. Does this client present any coronary heart disease risk factors? If so, which ones?
3. Does this client exhibit any contraindications to exercise? If so, which ones?
4. Under which risk stratification would this client be classified?
5. Based upon the client's health history questionnaire, does she need to be screened by a physician prior to testing?

Figure 2.1   Clinical Case Study Health History Questionnaire for Sample Client

Contact Information

Jones	*Jamie*	*J*
Last Name	First Name	Middle Initial

9/30/1978	*31*	Sex: ☐ *Male* ☑ *Female*
Date of Birth (mm/dd/yy)	Age	

(304)-293- 6318	*372 Main Street*	*Morgantown, WV*	*26505*
Home Phone Number	Street Address	City, State	Zip Code

John Jones	*(304)-293-0647*	*Husband*
Emergency Contact	Emergency Contact Number	Relationship to Emergency Contact

Matt Lively	*(304)-293-8828*
Primary Physician	Physician's Contact Number

Background

Please circle the highest grade in school you have completed:

Elementary school:	1	2	3	4	5	6	7	8
High school:	9	10	11	(12)				
College/postgrad	13	14	15	16	17	18	19	20+

What is your marital status?   ☐ Single   ☑ Married   ☐ Widowed   ☐ Divorced/separated

What is your race or ethnic background?

☑ White, not of Hispanic origin   ☐ American Indian/Alaskan native   ☐ Asian

☐ Black, not of Hispanic origin   ☐ Pacific Islander   ☐ Hispanic

What is your occupation?

☐ Health professional   ☐ Disabled, unable to work   ☐ Service   ☐ Manager, educator, professional

☐ Skilled crafts person   ☐ Operator, fabricator, laborer   ☑ Homemaker   ☐ Technical, sales, support

☐ Retired   ☐ Unemployed   ☐ Student   ☐ Other

Symptoms or Signs Suggestive of Disease

Chronic Disease Risk Factors

Do you know your blood pressure?   ☑ = yes: *130/80* systolic/diastolic   ☐ = no

Do you know your cholesterol levels?   ☑ = yes: *200* total cholesterol; _____ = HDL; _____ = LDL

☐ = no

Do you know your fasting glucose levels?   ☐ = yes: _____ mg/dl   ☑ = no

Please check the appropriate box.

YES	NO	
☐	☑	Have you experienced any unusual pain or discomfort in your chest, neck, jaw, arms, or other areas that may be due to heart problems?
☑	☐	Have you experienced unusual fatigue or shortness of breath at rest, during usual activities (e.g., climbing stairs, carrying groceries, brisk walking), or during mild to moderate exercise?
☐	☑	Have you ever had any problems with dizziness or fainting?
☐	☑	When you stand up, do you have difficulty breathing?
☐	☑	Do you have difficulty breathing while sleeping?
☐	☑	Do your ankles swell (ankle edema)?
☑	☐	Have you ever experienced an unusual and rapid heartbeat or fluttering of the heart?
☐	☑	Have you experienced severe pain in your legs while walking?
☐	☑	Have you ever been told by a doctor that you have a heart murmur?

Chronic Disease Risk Factors

Please check the appropriate box.

YES	NO	
☐	☑	Are you a male over age 45 or a female over age 55?
☐	☑	If you are a female, have you experienced premature menopause and are *not* on estrogen replacement therapy?
☑	☐	Has your father or brother had a heart attack or died suddenly from heart disease before age 55?
☐	☑	Has your mother or sister had a heart attack or died suddenly from heart disease before age 65?
☐	☑	Has anyone in your family died before age 40 (excluding accidental death)?
☑	☐	Are you a current cigarette smoker?
☑	☐	Has a doctor told you that you have high blood pressure (more than 130/80 mmHg)?
☑	☐	Are you on medication to control high blood pressure?
☑	☐	Has a doctor ever told you that you have high cholesterol?
☑	☐	Is your serum cholesterol greater than 200 mg/dl?
☑	☐	Do you have diabetes mellitus?
☑	☐	Are you physically inactive or sedentary (i.e., do you perform little physical activity on the job or during leisure time)?
☑	☐	During the past year, have you experienced levels of stress, strain, and pressure that might affect your health?
☑	☐	Do you eat foods that are high in fat and cholesterol (e.g., fatty meats, cheese, fried food, butter, whole milk, or eggs) on a daily basis?
☑	☐	Do you tend to avoid foods that are high in fiber (e.g., whole grain breads and cereals, fresh fruits, vegetables)?
☑	☐	Do you weigh 30 pounds (14 kg) more than you should?
☑	☐	Do you average more than two alcoholic beverages a day?

(continued)

Figure 2.1   *(continued)*

Medical History

Please check all conditions that you or someone in your family have had or now have.

YOU	YOUR FAMILY	
☐	☑	Coronary heart disease, heart attack, coronary artery surgery
☐	☑	Angina
☐	☐	Peripheral vascular disease
☐	☐	Phlebitis or emboli
☐	☐	Other heart problems (specify: _____)
☐	☐	Lung cancer
☐	☑	Breast cancer
☐	☐	Prostate cancer
☐	☐	Colorectal cancer (bowel cancer)
☐	☐	Skin cancer
☐	☐	Other cancer (specify: _____)
☐	☐	Stroke
☐	☐	Chronic obstructive pulmonary disease (emphysema)
☐	☐	Pneumonia
☐	☑	Asthma
☑	☐	Bronchitis
☑	☑	Diabetes mellitus
☐	☐	Thyroid problems
☐	☐	Kidney disease
☐	☐	Liver disease (cirrhosis of the liver)
☐	☐	Hepatitis (A, B, C, D, or E)
☐	☐	Gallstone or gallbladder disease
☐	☐	Osteoporosis
☐	☑	Arthritis
☐	☐	Gout
☐	☐	Anemia (low iron)
☐	☑	Bone fracture
☐	☐	Major injury to foot, leg, knee, hip, or shoulder
☑	☐	Major injury to back or neck
☐	☐	Stomach or duodenal ulcer
☐	☐	Rectal growth or bleeding
☐	☐	Cataracts
☐	☐	Glaucoma

❑	❑	Hearing loss
❑	❑	Depression
❑	❑	High anxiety, phobias
❑	❑	Substance abuse problems (e.g., alcohol, drugs)
❑	❑	Eating disorders (anorexia, bulimia)
❑	❑	Problems with menstruation
❑	❑	Hysterectomy
❑	❑	Sleeping problems
❑	❑	Allergies
❑	❑	HIV/AIDS
❑	❑	Any other problems (please be specific and include information on any recent illnesses, hospitalizations, or surgical procedures): _____

_____

_____

Medications

Please check any of the following types of medication that you currently take regularly. Also give the name of the medication.

MEDICATION		NAME OF MEDICATION
❑	Heart medicine	
☑	Blood pressure medicine	*Captopril*
☑	Blood cholesterol medicine	*Vytorin*
❑	Hormones	
☑	Birth control pills	*Seasonale*
❑	Medicine for breathing or lungs	
☑	Insulin	*Humulin N*
❑	Other medicines for diabetes	
❑	Arthritis medicine	
❑	Medicine for depression	
❑	Medicine for anxiety	
❑	Thyroid medicine	
❑	Medicine for ulcers	
❑	Painkiller medicine	
❑	Allergy medicine	
❑	HIV/AIDS medicine	
❑	Hepatitis medicine	
❑	Other (please specify)	

(continued)

Figure 2.1 *(continued)*

Physical Fitness, Physical Activity, Exercise

In general, as compared with other persons your age, rate how physically fit you are:

1	2	3	4	5	6	7	8	9	10
❏	❏	❏	☑	❏	❏	❏	❏	❏	❏

Not at all fit   Somewhat fit   Extremely fit

Outside your normal work or daily responsibilities, how often do you engage in exercise that increases your breathing and heart rate at least moderately, and makes you sweat, for at least 20 minutes (e.g., brisk walking, cycling, swimming, aerobic dance, stair climbing, rowing, basketball, racquetball, vigorous yard work).

❏ 5 or more times per week   ❏ 3–4 times per week   ❏ 1–2 times per week

❏ Less than once a week   ☑ Seldom or never

How much hard physical work is required in your job?

❏ A great deal   ❏ A moderate amount   ❏ A little   ☑ None

How long have you exercised or played sports regularly?

☑ Do not exercise regularly   ❏ Less than 1 year   ❏ 1–2 years

❏ 2–5 years   ❏ 5–10 years   ❏ More than 10 years

Diet

On average, how many servings of fruit do you eat per day?
(One serving = 1 medium apple, banana, orange; 1/2 cup chopped, cooked, or canned fruit; 3/4 cup fruit juice.)

☑ None   ❏ 1   ❏ 2   ❏ 3   ❏ 4 or more

On average, how many servings of vegetables do you eat per day?
(One serving = 1/2 cup cooked or chopped raw, 1 cup raw leafy, 3/4 cup vegetable juice.)

❏ None   ☑ 1   ❏ 2   ❏ 3   ❏ 4 or more

On average, how many servings of bread, cereal, rice, or pasta do you eat per day?

❏ None   ❏ 1–3   ☑ 4–6   ❏ 7–9   ❏ 10 or more

When you use grain and cereal products, do you emphasize:

❏ Whole grain, high fiber   ❏ Mixture of whole grain and refined   ☑ Refined, low fiber

On average, how many servings of fish, poultry, lean meat, cooked dry beans, peanut butter, or nuts do you eat per day?
(One serving = 2–3 ounces meat, 1/2 cup cooked dry beans, 2 tablespoons peanut butter, or 1/3 cup nuts)

❏ None   ☑ 1   ❏ 2   ❏ 3   ❏ 4 or more

On average, how many servings of dairy products do you eat per day?
(One serving = 1 cup milk or yogurt, 1.5 ounces natural cheese, 2 ounces processed cheese)

❏ None   ❏ 1   ☑ 2   ❏ 3   ❏ 4 or more

When you use dairy products, do you emphasize:

    ☑ Regular   ❑ Low-fat   ❑ Nonfat

How would you characterize your intake of fats and oils (e.g., regular salad dressings, butter or margarine, mayonnaise, vegetable oils)

    ☑ High   ❑ Moderate   ❑ Low

Body Weight

How tall are you (without shoes)?   _5_ feet _5_ inches (1.7 m)

How much do you weigh (with minimal clothing and without shoes)?   _190_ pounds (86 kg)

What is the most you have ever weighed?   _210_ pounds (95 kg)

Are you *now* trying to:

    ☑ Lose weight   ❑ Gain weight   ❑ Stay about the same   ❑ Not trying to do anything

Psychological Health

How have you been feeling in general during the past month?

    ❑ In excellent spirits   ❑ In very good spirits   ❑ In good spirits
    ☑ Up and down in spirits a lot   ❑ In low spirits   ❑ In very low spirits

During the past month, what level of stress would you say that you experienced?

    ❑ A lot   ☑ Moderate   ❑ Relatively little   ❑ Almost none

In the past year, how much has stress affected your health?

    ❑ A lot   ☑ Some   ❑ Hardly or none

On average, how many hours of sleep do you get in a 24-hour period?

    ❑ Fewer than 5 hours   ☑ 5–6 hours   ❑ 7–9 hours   ❑ More than 9 hours

Substance Use

Have you smoked at least 100 cigarettes in your life?   ☑ Yes   ❑ No

How would you describe your cigarette smoking habit?

❑ Never smoked

❑ Used to smoke—how many years since you smoked?   _____ years

☑ Currently smoke—how many cigarettes do you smoke on average?   _10_ cigarettes/day

How many alcoholic drinks do you consume?
(A "drink" is a glass of wine, a wine cooler, a bottle or can of beer, a shot glass of liquor, or a mixed drink.)

❑ Never use alcohol   ❑ Less than 1 drink per week   ☑ 1–6 drinks per week
❑ 1 drink per day   ❑ 2–3 drinks per day   ❑ More than 3 drinks per day

(continued)

Figure 2.1   *(continued)*

Occupational Health

Please explain your main job duties: *I am a homemaker. Mostly do things around the house such as*
cleaning and taking care of the kids.

After a day's work, do you often have pain or stiffness that lasts more than 3 hours?

    ❑ All the time   ❑ Most of the time   ☑ Some of the time   ❑ Rarely or never

How often does your work entail repetitive pushing and pulling movements, or lifting while bending or twisting, leading to back pain?

    ❑ All the time   ❑ Most of the time   ❑ Some of the time   ☑ Rarely or never

I hereby state that, to the best of my knowledge, my answers to the preceding questions are complete and correct.

Jane Jones	*Jane Jones*	*10/2/2009*
Printed Name of Respondent	Signature of Respondent	Date (mm/dd/yy)

Printed Name of Parent of Guardian	Signature of Parent of Guardian	Date (mm/dd/yy)

James Thomas	*James Thomas*	*10/2/2009*
Printed Name of Witness	Signature of Witness	Date (mm/dd/yy)

Adapted, by permission, from D.C. Nieman, 2003, *Exercise testing and prescription: A health-related approach,* 5th ed. (New York, NY: McGraw-Hill), 774.

Form 2.1 Informed Consent Form

I, _____, have been informed that I will perform a series of tests in order to determine my physical fitness status as well as enhance my understanding of my own health and physical fitness status. I understand that I can voluntarily withdraw from these tests at any time without any penalty. Additionally, I understand that I can ask questions about the tests at any time and will have those questions answered to my satisfaction. In the event of any side effects or injuries related to these tests, I understand that I may contact _____ _____ at any time with my concerns.

Explanation of Tests: I, _____, understand that I will fill out a series of questionnaires including a health history questionnaire and a PAR-Q in order to ensure my safety during the testing process. I understand that if these questionnaires reveal that I am at significant risk of an adverse event then no further tests will be conducted. I understand that if the questionnaires reveal little risk to my safety, I will perform the remainder of the assessments. I will have my blood pressure, body weight, and height assessed with methods typically used in a physician's office. After completing these assessments, I will have my body composition measured by _____, which will evaluate how much fat and fat-free mass my body contains. I will then perform an assessment of my muscular strength and endurance, which will require me to lift weights for a number of repetitions using a _____. After completing this assessment, I will perform a graded exercise test on a _____ in which the workload increases every few minutes until exhaustion or until the test is terminated. I understand that this assessment will give insight into my cardiorespiratory fitness.

Risks and Discomforts: It has been explained to me that during a graded exercise test, certain physical changes can occur, including abnormal blood pressure responses, fainting, irregular heartbeat, and in some instances fatal heart attacks. In order to minimize these risks, the personnel conducting the test are trained to handle such adverse effects. Additionally, they are trained to recognize potential warning signs and will stop any testing if these arise. I understand that the measurement of my height, weight, and body composition are of minimal risk to me. I have also been informed that the assessment of muscular strength and endurance involves some risk of pulling or spraining a muscle and that these risks will be minimized by employing a proper warm-up period and by means of technical monitoring by the testing staff. I also understand that after the completion of the testing bouts, I may experience some local muscle soreness that may last for 24 to 48 hours.

If muscle soreness does I occur, I understand that I can perform a series of stretches that have been demonstrated to me by the testing staff. If these symptoms persist, I will report them to _____ _____.

Benefits From Testing: I understand that the results of these tests will give insight into my overall physical health and wellness. Additionally, I have been informed that this information will reveal any potential health hazards and can be used to better individualize my exercise program.

Inquiries: I understand that if I have any questions, I can ask them of the testing staff. These questions will be answered to my satisfaction by the testing staff.

Confidentiality: I understand that all of my personal health and physical fitness data will be kept confidential.

(continued)

Form 2.1 *(continued)*

I have read the information contained in this document and understand it. All questions pertaining to the procedures that I am volunteering to undergo have been answered to my satisfaction. I understand that I am free to decline answering any questions and to withdraw from this testing at any time without penalty. Additionally, I have been informed that all of the information gathered about me and the tests undertaken by me are confidential and will not be disclosed to anyone but me or others in my care or used for exercise prescription without my written permission. Finally, I am aware of all risks associated with this testing and voluntarily give my consent to participate in this testing.

_____	_____
Date	Signature of patient/client
_____	_____
Date	Signature of witness
_____	_____
Date	Signature of supervisor

Physical Activity Readiness
Questionnaire - PAR-Q
(revised 2002)

PAR-Q & YOU

(A Questionnaire for People Aged 15 to 69)

Regular physical activity is fun and healthy, and increasingly more people are starting to become more active every day. Being more active is very safe for most people. However, some people should check with their doctor before they start becoming much more physically active.

If you are planning to become much more physically active than you are now, start by answering the seven questions in the box below. If you are between the ages of 15 and 69, the PAR-Q will tell you if you should check with your doctor before you start. If you are over 69 years of age, and you are not used to being very active, check with your doctor.

Common sense is your best guide when you answer these questions. Please read the questions carefully and answer each one honestly: check YES or NO.

YES	NO		
☐	☐	1.	**Has your doctor ever said that you have a heart condition <u>and</u> that you should only do physical activity recommended by a doctor?**
☐	☐	2.	**Do you feel pain in your chest when you do physical activity?**
☐	☐	3.	**In the past month, have you had chest pain when you were not doing physical activity?**
☐	☐	4.	**Do you lose your balance because of dizziness or do you ever lose consciousness?**
☐	☐	5.	**Do you have a bone or joint problem (for example, back, knee or hip) that could be made worse by a change in your physical activity?**
☐	☐	6.	**Is your doctor currently prescribing drugs (for example, water pills) for your blood pressure or heart condition?**
☐	☐	7.	**Do you know of <u>any other reason</u> why you should not do physical activity?**

If you answered

YES to one or more questions

Talk with your doctor by phone or in person BEFORE you start becoming much more physically active or BEFORE you have a fitness appraisal. Tell your doctor about the PAR-Q and which questions you answered YES.

- You may be able to do any activity you want — as long as you start slowly and build up gradually. Or, you may need to restrict your activities to those which are safe for you. Talk with your doctor about the kinds of activities you wish to participate in and follow his/her advice.
- Find out which community programs are safe and helpful for you.

NO to all questions

If you answered NO honestly to <u>all</u> PAR-Q questions, you can be reasonably sure that you can:
- start becoming much more physically active — begin slowly and build up gradually. This is the safest and easiest way to go.
- take part in a fitness appraisal — this is an excellent way to determine your basic fitness so that you can plan the best way for you to live actively. It is also highly recommended that you have your blood pressure evaluated. If your reading is over 144/94, talk with your doctor before you start becoming much more physically active.

→ **DELAY BECOMING MUCH MORE ACTIVE:**
- if you are not feeling well because of a temporary illness such as a cold or a fever — wait until you feel better; or
- if you are or may be pregnant — talk to your doctor before you start becoming more active.

PLEASE NOTE: If your health changes so that you then answer YES to any of the above questions, tell your fitness or health professional. Ask whether you should change your physical activity plan.

<u>Informed Use of the PAR-Q</u>: The Canadian Society for Exercise Physiology, Health Canada, and their agents assume no liability for persons who undertake physical activity, and if in doubt after completing this questionnaire, consult your doctor prior to physical activity.

No changes permitted. You are encouraged to photocopy the PAR-Q but only if you use the entire form.

NOTE: If the PAR-Q is being given to a person before he or she participates in a physical activity program or a fitness appraisal, this section may be used for legal or administrative purposes.

"I have read, understood and completed this questionnaire. Any questions I had were answered to my full satisfaction."

NAME _____

SIGNATURE _____ DATE_____

SIGNATURE OF PARENT _____ WITNESS _____
or GUARDIAN (for participants under the age of majority)

Note: This physical activity clearance is valid for a maximum of 12 months from the date it is completed and becomes invalid if your condition changes so that you would answer YES to any of the seven questions.

CSEP|SCPE
THE HEALTH STANDARD FOR EXERCISE SCIENCE AND PERSONAL TRAINING

© Canadian Society for Exercise Physiology www.csep.ca/forms

Source: Physical Activity Readiness Questionnaire (PAR-Q) © 2002. Reprinted with permission from the Canadian Society for Exercise Physiology. http://www.csep.ca/forms.asp.

Form 2.3   Health History Questionnaire

Contact Information

_____   _____   _____
Last name   First name   Middle initial

_____   _____   Sex: ❑ Male ❑ Female
Date of birth (mm/dd/yy)   Age

(___) -_____-_____   _____   _____   _____
Home phone number   Street address   City, State   Zip code

_____   (___)-_____-_____   _____
Emergency contact   Emergency contact number   Relationship to emergency contact

_____   (___)-_____-_____
Primary physician   Physician's contact number

Background

Please circle the highest school grade you have completed:

Primary:	1	2	3	4	5	6	7	8
Secondary:	9	10	11	12				
College/postgrad:	13	14	15	16	17	18	19	20+

What is your marital status?   ❑ Single   ❑ Married   ❑ Widowed   ❑ Divorced or separated

What is your race or ethnic background?

❑ White, not of Hispanic origin   ❑ American Indian/Alaskan native   ❑ Asian

❑ Black, not of Hispanic origin   ❑ Pacific Islander   ❑ Hispanic

What is your occupation?

❑ Health professional   ❑ Disabled, unable to work   ❑ Service   ❑ Manager, educator, professional

❑ Skilled crafts   ❑ Operator, fabricator, laborer   ❑ Homemaker   ❑ Technical, sales, support

❑ Retired   ❑ Unemployed   ❑ Student   ❑ Other

Symptoms or Signs Suggestive of Disease

Please place a check the appropriate box:

Yes	No	
❑	❑	Have you experienced any unusual pain or discomfort in your chest, neck, jaw, arms, or other areas that may be due to heart problems?
❑	❑	Have you experienced unusual fatigue and/or shortness of breath at rest, during usual activities (e.g., climbing stairs, carrying groceries, walking briskly) or during mild or moderate exercise?
❑	❑	Have you ever had any problems with dizziness or fainting?
❑	❑	When you stand up, do you have difficulty breathing?
❑	❑	Do you have difficulty breathing while sleeping?
❑	❑	Do your ankles swell (ankle edema)?

(continued)

Form 2.3 *(continued)*

Yes	No	
☐	☐	Have your ever experienced an unusual or rapid heartbeat or fluttering of the heart?
☐	☐	Have you experienced severe pain in your legs while walking?
☐	☐	Have you ever been told by a doctor that you have a heart murmur?

Chronic Disease Risk Factors

Do you know your blood pressure?   ☐ = Yes: _____ / _____ systolic/diastolic   ☐ = No

Do you know your cholesterol levels?   ☐ = Yes: _____ total cholesterol; _____ HDL; _____ LDL   ☐ = No

Do you know your fasting glucose levels?   ☐ = Yes: _____ mg/dl   ☐ = No

Please place a check in the appropriate box:

Yes	No	
☐	☐	Are you a male over the age of 45 or a female over the age of 55?
☐	☐	If you are a female, have you experienced premature menopause and are not on estrogen replacement therapy?
☐	☐	Has your father or brother had a heart attack or died suddenly from heart disease before the age of 55?
☐	☐	Has your mother or sister had a heart attack or died suddenly from heart disease before the age of 65?
☐	☐	Has anyone in your family died before the age of 40 (excluding accidental death)?
☐	☐	Are you a current cigarette smoker?
☐	☐	Has a doctor told you that you have high blood pressure (more than 130/80)?
☐	☐	Are you on medication to control high blood pressure?
☐	☐	Has a doctor ever told you that you have high cholesterol?
☐	☐	Is your serum cholesterol greater than 200 mg/dl?
☐	☐	Do you have diabetes mellitus?
☐	☐	Are you physically inactive or sedentary (i.e., do you perform little physical activity on the job or during leisure time)?
☐	☐	During the past year, have you experienced levels of stress, strain, and pressure that might affect your health?
☐	☐	Do you eat foods that are high in fat and cholesterol, such as fatty meats, cheese, fried food, butter, whole milk, or eggs on a daily basis?
☐	☐	Do you tend to avoid foods that are high in fiber, such as whole-grain breads and cereals, fresh fruits, and vegetables?
☐	☐	Do you weigh 30 pounds more than you should?
☐	☐	Do you average more than two alcoholic beverages a day?

Medical History

Please check all conditions that you or your family have had or now have.

You	Your family		You	Your family	
☐	☐	Coronary heart disease, heart attack, coronary artery surgery	☐	☐	Peripheral vascular disease
☐	☐	Angina	☐	☐	Phlebitis or emboli

You	Your family		You	Your family	
☐	☐	Other heart problems (Specify: _____)	☐	☐	Major injury to foot, leg, knee, hip, or shoulder
☐	☐	Lung cancer	☐	☐	Major injury to back or neck
☐	☐	Breast cancer	☐	☐	Stomach or duodenal ulcer
☐	☐	Prostate cancer	☐	☐	Rectal growth or bleeding
☐	☐	Colorectal cancer (bowel cancer)	☐	☐	Cataracts
☐	☐	Skin cancer	☐	☐	Glaucoma
☐	☐	Other cancer (Specify: _____)	☐	☐	Hearing loss
			☐	☐	Depression
☐	☐	Stroke	☐	☐	High anxiety, phobia
☐	☐	Chronic obstructive pulmonary disease (emphysema)	☐	☐	Substance abuse problems (e.g., alcohol, drugs)
☐	☐	Pneumonia	☐	☐	Eating disorders (anorexia, bulimia)
☐	☐	Asthma			
☐	☐	Bronchitis	☐	☐	Problems with menstruation
☐	☐	Diabetes mellitus	☐	☐	Hysterectomy
☐	☐	Thyroid problems	☐	☐	Sleeping problems
☐	☐	Kidney disease	☐	☐	Allergies
☐	☐	Liver disease (cirrhosis of the liver)	☐	☐	HIV/AIDS
☐	☐	Hepatitis (A, B, C, D, or E)	☐	☐	Any other problems (please be specific and include information on any recent illnesses, hospitalizations, or surgical procedures): _____
☐	☐	Gallstone/gallbladder disease			
☐	☐	Osteoporosis			
☐	☐	Arthritis			_____
☐	☐	Gout			_____
☐	☐	Anemia (low iron)			_____
☐	☐	Bone fracture			_____

Medications

Please check any of the following types of medication that you currently take regularly. Also give the name of the medication.

Type of medication	Name of medication
☐ Heart medicine	_____
☐ Blood pressure medicine	_____
☐ Blood cholesterol medicine	_____
☐ Hormones	_____
☐ Birth control pills	_____
☐ Medicine for breathing or lungs	_____
☐ Insulin	_____
☐ Other medicines for diabetes	_____

(continued)

❑ Arthritis medicine _____

❑ Medicine for depression _____

❑ Medicine for anxiety _____

❑ Thyroid medicine _____

❑ Medicine for ulcers _____

❑ Painkiller medicine _____

❑ Allergy medicine _____

❑ HIV/AIDS medicine _____

❑ Hepatitis medicine _____

❑ Other (please specify) _____

Physical Fitness, Physical Activity, and Exercise

In general, compared with other persons your age, rate how physically fit you are:

1	2	3	4	5	6	7	8	9	10
❑	❑	❑	❑	❑	❑	❑	❑	❑	❑

Not at all fit   Somewhat fit   Extremely fit

Outside of your normal work or daily responsibilities, how often do you engage in exercise that at least moderately increases your breathing and heart rate, and makes you sweat, for at least 20 minutes (e.g., brisk walking, cycling, swimming, aerobic dance, stair climbing, rowing, basketball, racquetball, vigorous yardwork).

❑ 5 or more times per week   ❑ 3 or 4 times per week   ❑ 1 or 2 times per week

❑ Less than once a week   ❑ Seldom or never

How much hard physical work is required on your job?

❑ A great deal   ❑ A moderate amount   ❑ A little   ❑ None

How long have you exercised or played sports regularly?

❑ I do not exercise regularly   ❑ Less than 1 year   ❑ 1 to 2 years

❑ 2 to 5 years   ❑ 5 to 10 years   ❑ More than 10 years

Diet

On average, how many servings of fruits do you eat per day?

(One serving = 1 medium apple, banana, or orange; ½ cup chopped, cooked, or canned fruit; ¾ cup fruit juice)

❑ None   ❑ 1   ❑ 2   ❑ 3   ❑ 4 or more

On average, how many servings of vegetables do you eat per day?

(One serving = 1/2 cup cooked or chopped raw, 1 cup raw leafy, 3/4 cup vegetable juice.)

❑ None   ❑ 1   ❑ 2   ❑ 3   ❑ 4 or more

On average, how many servings of bread, cereal, rice, or pasta do you eat per day?

❑ None   ❑ 1–3   ❑ 4–6   ❑ 7–9   ❑ 10 or more

When you use grain and cereal products, which of the following do you emphasize?

❑ Whole grain, high-fiber   ❑ Mixture of whole grain and refined   ❑ Refined, low-fiber

On average, how many servings of fish, poultry, lean meat, cooked dry beans, peanut butter, or nuts do you eat per day?

(One serving = 2–3 ounces of meat, 1/2 cup cooked dry beans, 2 tablespoons peanut butter, or 1/3 cup of nuts.)

    ❑ None   ❑ 1   ❑ 2   ❑ 3   ❑ 4 or more

On average, how many servings of dairy products do you eat per day?

(One serving = 1 cup milk or yogurt, 1.5 ounces natural cheese, 2 ounces processed cheese.)

    ❑ None   ❑ 1   ❑ 2   ❑ 3   ❑ 4 or more

When you use dairy products, which do you emphasize?

    ❑ Regular   ❑ Low-fat   ❑ Nonfat

How would you characterize your intake of fats and oils (e.g., regular salad dressings, butter or margarine, mayonnaise, vegetable oils)?

    ❑ High   ❑ Moderate   ❑ Low

Body Data

How tall are you (without shoes)? _____ feet _____ inches

How much do you weigh (with minimal clothing and without shoes)? _____ pounds

What is the most you have ever weighed? _____ pounds

Which of the following are you *currently* trying to do?

    ❑ Lose weight   ❑ Gain weight   ❑ Stay about the same   ❑ None of these

Psychological Health

How have you been feeling in general during the past month?

    ❑ In excellent spirits   ❑ In very good spirits   ❑ In good spirits

    ❑ I've been up and down in spirits a lot   ❑ In low spirits   ❑ In very low spirits

During the past month, how much stress would you say that you have experienced?

    ❑ A lot   ❑ Moderate   ❑ Relatively little   ❑ Almost none

In the past year, how much effect has stress had on your health?

    ❑ A lot   ❑ Some   ❑ Hardly any or none

On average, how many hours of sleep do you get in a 24-hour period?

    ❑ Fewer than 5 hours   ❑ 5 or 6 hours   ❑ 7–9 hours   ❑ More than 9 hours

Substance Use

Have you smoked at least 100 cigarettes in your entire life?   ❑ Yes   ❑ No

How would you describe your cigarette smoking habit?

    ❑ Never smoked

    ❑ Used to smoke—how many years has it been since you smoked? _____ years

    ❑ Currently smoke—how many cigarettes do you smoke on average? _____ cigarettes/day

(continued)

Form 2.3   (continued)

How many alcoholic drinks do you consume?
(A "drink" is a glass of wine, a wine cooler, a bottle or can of beer, a shot glass of liquor, or a mixed drink.)

- ❏ None
- ❏ Less than 1 drink per week
- ❏ 1–6 drinks per week
- ❏ 1 drink per day
- ❏ 2–3 drinks per day
- ❏ More than 3 drinks per day

Occupational Health

Please explain your main job duties: _____

_____

_____

_____

After a day's work, do you often have pain or stiffness that lasts for more than 3 hours?

- ❏ All the time
- ❏ Most of the time
- ❏ Some of the time
- ❏ Rarely or never

How often does your work entail repetitive pushing and pulling movements, or lifting while bending or twisting, leading to back pain?

- ❏ All the time
- ❏ Most of the time
- ❏ Some of the time
- ❏ Rarely or never

I hereby state that, to the best of my knowledge, my answers to the preceding questions are complete and correct.

_____	_____	_____
Printed name of respondent	Signature of respondent	Date (mm/dd/yy)
_____	_____	_____
Printed name of parent of guardian	Signature of parent or guardian	Date (mm/dd/yy)
_____	_____	_____
Printed name of witness	Signature of witness	Date (mm/dd/yy)

From G. Haff and C. Dumke, 2012, *Laboratory manual for exercise physiology* (Champaign, IL: Human Kinetics). Adapted, by permission, from D.C. Nieman, 2003, *Exercise testing and prescription: A health-related approach*, 5th ed. (New York: McGraw-Hill), 774. © The McGraw-Hill Companies.

Form 2.4 Checklist for Signs and Symptoms of Disease

Instructions: Ask your subject or client if he or she has any of the following conditions and risk factors. If so, refer him or her to a physician to obtain a signed medical clearance prior to any exercise testing or participation.

Last name	First name	Middle initial

Cardiovascular

YES	NO	Condition	Comments
❑	❑	Hypertension	
❑	❑	Hypercholesterolemia	
❑	❑	Heart murmur	
❑	❑	Myocardial infarction	
❑	❑	Fainting/dizziness	
❑	❑	Claudication	
❑	❑	Chest pain	
❑	❑	Palpitation	
❑	❑	Ischemia	
❑	❑	Tachycardia	
❑	❑	Ankle edema	
❑	❑	Stroke	

Pulmonary

YES	NO	Condition	Comments
❑	❑	Asthma	
❑	❑	Bronchitis	
❑	❑	Emphysema	
❑	❑	Nocturnal dyspnea	
❑	❑	Coughing up blood	
❑	❑	Exercise-induced asthma	
❑	❑	Breathlessness during or after mild exertion	

Metabolic

YES	NO	Condition	Comments
❑	❑	Diabetes	
❑	❑	Obesity	
❑	❑	Glucose intolerance	
❑	❑	McArdle syndrome	
❑	❑	Hypoglycemia	
❑	❑	Thyroid disease	
❑	❑	Cirrhosis	

(continued)

Musculoskeletal

YES	NO	Condition	Comments
❑	❑	Osteoporosis	
❑	❑	Osteoarthritis	
❑	❑	Low back pain	
❑	❑	Prosthesis	
❑	❑	Muscular atrophy	
❑	❑	Swollen joints	
❑	❑	Orthopedic pain	
❑	❑	Artificial joint	

Risk Factors

A client who has two or more of these factors should consult a physician for clearance to participate in exercise testing.

❑	❑	Male older than 45 y _____
❑	❑	Female older than 55 y or who had hysterectomy or is postmenopausal _____ _____
❑	❑	Smoking or quit smoking within past 6 months _____
❑	❑	Blood pressure >140/90 mmHg _____
❑	❑	Doesn't know blood pressure _____
❑	❑	Blood cholesterol >200 mg/dl _____
❑	❑	Doesn't know blood cholesterol _____
❑	❑	Close relative had heart attack before age 55 (father or brother) or age 65 (mother or sister) _____
❑	❑	Physically inactive (does not engage in >30 min of physical activity more than 4 days/wk) _____
❑	❑	Overweight by more than 20 lb (9 kg) _____

From G. Haff and C. Dumke, 2012, *Laboratory manual for exercise physiology* (Champaign, IL: Human Kinetics). Adapted, by permission, from V. Heyward, 2010, *Advanced fitness assessment and exercise prescription,* 6th ed. (Champaign, IL: Human Kinetics), 320-321.

Physical Activity Readiness
Medical Examination
(revised 2002)

PARmed-X PHYSICAL ACTIVITY READINESS MEDICAL EXAMINATION

The PARmed-X is a physical activity-specific checklist to be used by a physician with patients who have had positive responses to the Physical Activity Readiness Questionnaire (PAR-Q). In addition, the Conveyance/Referral Form in the PARmed-X can be used to convey clearance for physical activity participation, or to make a referral to a medically-supervised exercise program.

Regular physical activity is fun and healthy, and increasingly more people are starting to become more active every day. Being more active is very safe for most people. The PAR-Q by itself provides adequate screening for the majority of people. However, some individuals may require a medical evaluation and specific advice (exercise prescription) due to one or more positive responses to the PAR-Q.

Following the participant's evaluation by a physician, a physical activity plan should be devised in consultation with a physical activity professional (CSEP-Certified Personal Trainer™ or CSEP-Certified Exercise Physiologist™). To assist in this, the following instructions are provided:

PAGE 1: • Sections A, B, C, and D should be completed by the participant BEFORE the examination by the physician. The bottom section is to be completed by the examining physician.

PAGES 2 & 3: • A checklist of medical conditions requiring special consideration and management.

PAGE 4: • Physical Activity & Lifestyle Advice for people who do not require specific instructions or prescribed exercise.

• Physical Activity Readiness Conveyance/Referral Form - an optional tear-off tab for the physician to convey clearance for physical activity participation, or to make a referral to a medically-supervised exercise program.

This section to be completed by the participant

A PERSONAL INFORMATION:

NAME _____

ADDRESS _____

_____

TELEPHONE _____

BIRTHDATE _____ GENDER _____

MEDICAL No. _____

B PAR-Q: Please indicate the PAR-Q questions to which you answered YES

❏ Q 1 Heart condition
❏ Q 2 Chest pain during activity
❏ Q 3 Chest pain at rest
❏ Q 4 Loss of balance, dizziness
❏ Q 5 Bone or joint problem
❏ Q 6 Blood pressure or heart drugs
❏ Q 7 Other reason:

_____

C RISK FACTORS FOR CARDIOVASCULAR DISEASE:
Check all that apply

❏ Less than 30 minutes of moderate physical activity most days of the week.
❏ Currently smoker (tobacco smoking 1 or more times per week).
❏ High blood pressure reported by physician after repeated measurements.
❏ High cholesterol level reported by physician.

❏ Excessive accumulation of fat around waist.
❏ Family history of heart disease.

Please note: *Many of these risk factors are modifiable. Please refer to page 4 and discuss with your physician.*

D PHYSICAL ACTIVITY INTENTIONS:

What physical activity do you intend to do?

_____

_____

_____

This section to be completed by the examining physician

Physical Exam:

Ht	Wt	BP i)	/
		BP ii)	/

Conditions limiting physical activity:

❏ Cardiovascular ❏ Respiratory ❏ Other
❏ Musculoskeletal ❏ Abdominal

Tests required:

❏ ECG ❏ Exercise Test ❏ X-Ray
❏ Blood ❏ Urinalysis ❏ Other

Physical Activity Readiness Conveyance/Referral:

Based upon a current review of health status, I recommend:

Further Information:
❏ Attached
❏ To be forwarded
❏ Available on request

❏ No physical activity

❏ Only a medically-supervised exercise program until further medical clearance

❏ Progressive physical activity:

❏ with avoidance of: _____

❏ with inclusion of: _____

❏ under the supervision of a CSEP-Certified Exercise Physiologist™

❏ Unrestricted physical activity—start slowly and build up gradually

CSEP | SCPE
THE GOLD STANDARD IN EXERCISE
SCIENCE AND PERSONAL TRAINING

© Canadian Society for Exercise Physiology www.csep.ca

1

(continued)

Physical Activity Readiness
Medical Examination
(revised 2002)

PARmed-X   PHYSICAL ACTIVITY READINESS MEDICAL EXAMINATION

Following is a checklist of medical conditions for which a degree of precaution and/or special advice should be considered for those who answered "YES" to one or more questions on the PAR-Q, and people over the age of 69. Conditions are grouped by system. Three categories of precautions are provided. Comments under Advice are general, since details and alternatives require clinical judgement in each individual instance.

	Absolute Contraindications	Relative Contraindications	Special Prescriptive Conditions	
	Permanent restriction or temporary restriction until condition is treated, stable, and/or past acute phase.	Highly variable. Value of exercise testing and/or program may exceed risk. Activity may be restricted. Desirable to maximize control of condition. Direct or indirect medical supervision of exercise program may be desirable.	Individualized prescriptive advice generally appropriate: • limitations imposed; and/or • special exercises prescribed. May require medical monitoring and/or initial supervision in exercise program.	**ADVICE**

Cardiovascular	❏ aortic aneurysm (dissecting)
	❏ aortic stenosis (severe)
	❏ congestive heart failure
	❏ crescendo angina
	❏ myocardial infarction (acute)
	❏ myocarditis (active or recent)
	❏ pulmonary or systemic embolism—acute
	❏ thrombophlebitis
	❏ ventricular tachycardia and other dangerous dysrhythmias (e.g., multi-focal ventricular activity)
		❏ aortic stenosis (moderate)
		❏ subaortic stenosis (severe)
		❏ marked cardiac enlargement
		❏ supraventricular dysrhythmias (uncontrolled or high rate)
		❏ ventricular ectopic activity (repetitive or frequent)
		❏ ventricular aneurysm
		❏ hypertension—untreated or uncontrolled severe (systemic or pulmonary)
		❏ hypertrophic cardiomyopathy
		❏ compensated congestive heart failure
			❏ aortic (or pulmonary) stenosis—mild angina pectoris and other manifestations of coronary insufficiency (e.g., post-acute infarct)
			❏ cyanotic heart disease
			❏ shunts (intermittent or fixed)
			❏ conduction disturbances
			  • complete AV block
			  • left BBB
			  • Wolff-Parkinson-White syndrome
			❏ dysrhythmias—controlled
			❏ fixed rate pacemakers
				• clinical exercise test may be warranted in selected cases, for specific determination of functional capacity and limitations and precautions (if any).
				
				• slow progression of exercise to levels based on test performance and individual tolerance.
				
				• consider individual need for initial conditioning program under medical supervision (indirect or direct).

			❏ intermittent claudication	progressive exercise to tolerance
			❏ hypertension: systolic 160-180; diastolic 105+	progressive exercise; care with medications (serum electrolytes; post-exercise syncope; etc.)
Infections	❏ acute infectious disease (regardless of etiology)	❏ subacute/chronic/recurrent infectious diseases (e.g., malaria, others)	❏ chronic infections ❏ HIV	variable as to condition
Metabolic		❏ uncontrolled metabolic disorders (diabetes mellitus, thyrotoxicosis, myxedema)	❏ renal, hepatic & other metabolic insufficiency	variable as to status
			❏ obesity ❏ single kidney	dietary moderation, and initial light exercises with slow progression (walking, swimming, cycling)
Pregnancy		❏ complicated pregnancy (e.g., toxemia, hemorrhage, incompetent cervix, etc.)	❏ advanced pregnancy (late 3rd trimester)	refer to the "PARmed-X for PREGNANCY"

References:

Arraix, G.A., Wigle, D.T., Mao, Y. (1992). Risk Assessment of Physical Activity and Physical Fitness in the Canada Health Survey Follow-Up Study. **J. Clin. Epidemiol.** 45:4 419-428.

Mottola, M., Wolfe, L.A. (1994). Active Living and Pregnancy, In: A. Quinney, L. Gauvin, T. Wall (eds.), **Toward Active Living: Proceedings of the International Conference on Physical Activity, Fitness and Health**. Champaign, IL: Human Kinetics.

PAR-Q Validation Report, British Columbia Ministry of Health, 1978.

Thomas, S., Reading, J., Shephard, R.J. (1992). Revision of the Physical Activity Readiness Questionnaire (PAR-Q). **Can. J. Spt. Sci.** 17: 4 338-345.

The PAR-Q and PARmed-X were developed by the British Columbia Ministry of Health. They have been revised by an Expert Advisory Committee of the Canadian Society for Exercise Physiology chaired by Dr. N. Gledhill (2002).

No changes permitted. You are encouraged to photocopy the PARmed-X, but only if you use the entire form.

Disponible en français sous le titre
«Évaluation médicale de l'aptitude à l'activité physique (X-AAP)»

Continued on page 3...

2

	Special Prescriptive Conditions	ADVICE
Lung	❑ chronic pulmonary disorders	special relaxation and breathing exercises
	❑ obstructive lung disease	breath control during endurance exercises to tolerance; avoid polluted air
	❑ asthma	
	❑ exercise-induced bronchospasm	avoid hyperventilation during exercise; avoid extremely cold conditions; warm up adequately; utilize appropriate medication.
Musculoskeletal	❑ low back conditions (pathological, functional)	avoid or minimize exercise that precipitates or exasperates e.g., forced extreme flexion, extension, and violent twisting; correct posture, proper back exercises
	❑ arthritis—acute (infective, rheumatoid; gout)	treatment, plus judicious blend of rest, splinting and gentle movement
	❑ arthritis—subacute	progressive increase of active exercise therapy
	❑ arthritis—chronic (osteoarthritis and above conditions)	maintenance of mobility and strength; non-weightbearing exercises to minimize joint trauma (e.g., cycling, aquatic activity, etc.)
	❑ orthopaedic	highly variable and individualized
	❑ hernia	minimize straining and isometrics; stregthen abdominal muscles
	❑ osteoporosis or low bone density	avoid exercise with high risk for fracture such as push-ups, curl-ups, vertical jump and trunk forward flexion; engage in low-impact weight-bearing activities and resistance training
CNS	❑ convulsive disorder not completely controlled by medication	minimize or avoid exercise in hazardous environments and/or exercising alone (e.g., swimming, mountain climbing, etc.)
	❑ recent concussion	thorough examination if history of two concussions; review for discontinuation of contact sport if three concussions, depending on duration of unconsciousness, retrograde amnesia, persistent headaches, and other objective evidence of cerebral damage
Blood	❑ anemia—severe (< 10 Gm/dl)	control preferred; exercise as tolerated
	❑ electrolyte disturbances	
Medications	❑ antianginal ❑ antiarrhythmic ❑ antihypertensive ❑ anticonvulsant ❑ beta-blockers ❑ digitalis preparations ❑ diuretics ❑ ganglionic blockers ❑ others	NOTE: consider underlying condition. Potential for: exertional syncope, electrolyte imbalance, bradycardia, dysrhythmias, impaired coordination and reaction time, heat intolerance. May alter resting and exercise ECG's and exercise test performance.
Other	❑ post-exercise syncope	moderate program
	❑ heat intolerance	prolong cool-down with light activities; avoid exercise in extreme heat
	❑ temporary minor illness	postpone until recovered
	❑ cancer	if potential metastases, test by cycle ergometry, consider non-weight bearing exercises; exercise at lower end of prescriptive range (40-65% of heart rate reserve), depending on condition and recent treatment (radiation, chemotherapy); monitor hemoglobin and lymphocyte counts; add dynamic lifting exercise to strengthen muscles, using machines rather than weights.

*Refer to special publications for elaboration as required

The following companion forms are available online: http://www.csep.ca/forms

The **Physical Activity Readiness Questionnaire (PAR-Q)** - a questionnaire for people aged 15-69 to complete before becoming much more physically active.

The **Physical Activity Readiness Medical Examination for Pregnancy (PARmed-X for PREGNANCY)** - to be used by physicians with pregnant patients who wish to become more physically active.

For more information, please contact the:

Canadian Society for Exercise Physiology
370-18 Louisa Ottawa, ON K1R 6Y6
Tel. 1-877-651-3755 • FAX (613) 234-3565 • Online: www.csep.ca

Note to physical activity professionals...

It is a prudent practice to retain the completed Physical Activity Readiness Conveyance/Referral Form in the participant's file.

CSEP | SCPE
THE GOLD STANDARD IN EXERCISE
SCIENCE AND PERSONAL TRAINING

© Canadian Society for Exercise Physiology   www.csep.ca/forms

Continued on page 4...

3

(continued)

Form 2.5   *(continued)*

Physical Activity Readiness
Medical Examination
(revised 2002)

PARmed-X PHYSICAL ACTIVITY READINESS MEDICAL EXAMINATION

✂ ┄┄

PARmed-X Physical Activity Readiness Conveyance/Referral Form

Based upon a current review of the health status of _____ , I recommend:

❑   No physical activity

❑   Only a medically-supervised exercise program until further medical clearance

❑   Progressive physical activity

    ❑   with avoidance of: _____

    ❑   with inclusion of: _____

    ❑   under the supervision of a CSEP-Certified Exercise Physiologist™

❑   Unrestricted physical activity — start slowly and build up gradually

Further Information:
❑ Attached
❑ To be forwarded
❑ Available on request

Physician/clinic stamp:

_____ M.D.

_____ 20_____
                    (date)

NOTE: This physical activity clearance is valid for a maximum of six months from the date it is completed and becomes invalid if your medical condition becomes worse.

4

Source: Physical Activity Readiness Medical Examination (PARmed-X) © 2002. Reprinted with permission from the Canadian Society for Exercise Physiology. http://www.csep.ca/forms.asp.

Form 2.6 Risk Factors for Coronary Heart Disease

Risk factors	Criteria	Yes	No	Points
Positive risk factors				
Age	Men ≥45 yr; Women ≥55 yr	❏	❏	
Family history	Myocardial infarction, coronary revascularization, or sudden death before age 55 for father or first-degree male relative (brother or son) or before age 65 for mother or first-degree female relative (sister or daughter)	❏	❏	
Cigarette smoking	Current cigarette smoking, or smoking cessation within previous 6 months, or exposure to environmental tobacco smoke	❏	❏	
Hypertension	Systolic blood pressure ≥140 mmHg or diastolic blood pressure ≥90 mmHg measured on two separate occasions, or use of antihypertensive medication	❏	❏	
Dyslipidemia	Total cholesterol ≥200 mg $\cdot$ dl^{-1} or high-density lipoprotein-cholesterol (HDL-C) <40 mg $\cdot$ dl^{-1} or low-density lipoprotein-cholesterol (LDL-C) ≥130 mg $\cdot$ dl^{-1} or on lipid lowering medication	❏	❏	
Impaired fasting glucose	Fasting glucose ≥100 mg $\cdot$ dl^{-1} but <126 mg $\cdot$ dl^{-1} or 2 hour OGTT value ≥140 mg $\cdot$ dl^{-1} but <200 mg $\cdot$ dl^{-1} as measured on two separate occasions	❏	❏	
Obesity	Body mass index ≥30 kg $\cdot$ m^{-2} or waist circumference >102 cm (40 in) for men or >88 cm (35 in) for women	❏	❏	
Physical inactivity	Not participating in a regular exercise program or not meeting minimum physical activity recommendations from the U.S. Surgeon General's report (accumulating 30 minutes or more of moderate (40-60% $\dot{V}O_2R$) physical activity on most days of the week for at least 3 months)	❏	❏	
Total positive risk factor points =				
Negative risk factors				
High HDL-C	Serum HDL-C ≥60 mg $\cdot$ dl^{-1}	❏	❏	
Total negative risk factor points =				
Total risk factor points =				

Add 1 if the individual has the positive risk factor. If HDL-C is elevated, subtract 1 risk factor point from the sum of the positive risk factors.

From G. Haff and C. Dumke, 2012, *Laboratory manual for exercise physiology* (Champaign, IL: Human Kinetics). Adapted from American College of Sports Medicine (1).

Form 2.7 Fantastic Lifestyle Checklist

Instructions: Unless otherwise specified, place an 'X' beside the box which best describes your behavior or situation in the past month. Explanations of questions and scoring are provided at the end of the form.

Family and friends	I have someone to talk to about things that are important to me	Almost never		Seldom		Some of the time		Fairly often		Almost always	
	I give and receive affection	Almost never		Seldom		Some of the time		Fairly often		Almost always	
Physical activity	I am vigorously active for at least 30 min per day (e.g., running, cycling, etc.)	Less than once/week		1-2 times/week		3 times/week		4 times/week		5 or more times/week	
	I am moderately active (gardening, climbing stairs, walking, housework)	Less than once/week		1-2 times/week		3 times/week		4 times/week		5 or more times/week	
Nutrition	I eat a balance diet (see explanation)	Almost never		Seldom		Some of the time		Fairly often		Almost always	
	I often eat excess sugar, salt, animal fats, or junk food	Four of these		Three of these		Two of these		One of these		None of these	
	I am within _____ kg of my healthy weight	Not within 8 kg		8 kg (20 lb)		6 kg (15 lb)		4 kg (10 lb)		2 kg (5 lb)	
Tobacco and toxics	I smoke tobacco	More than 10 times/week		1–10 times/week		None in the past 6 months		None in the past year		None in the past 5 years	
	I use drugs such as marijuana or cocaine	sometimes								Never	
	I overuse prescribed or 'over the counter' drugs	Almost daily		Fairly often		Only occasionally		Almost never		Never	
	I drink caffeine-containing coffee, tea, or cola	More than 10/day		7-10/day		3-6/day		1-2/day		Never	

(continued)

Alcohol	My average alcohol intake per week is _____ (see explanation)	More than 20 drinks	13-20 drinks	11-12 drinks	8-10 drinks	0-7 drinks	
	I drink more than four drinks on an occasion	Almost daily	Fairly often	Only occasionally	Almost never	Never	
	I drive after drinking	Sometimes				Never	
Sleep, seatbelt use, stress, and safe sex	I sleep well and feel rested	Almost never	Seldom	Some of the time	Fairly often	Almost always	
	I use seatbelts	Never	Seldom	Some of the time	Most of the time	Always	
	I am able to cope with the stresses in my life	Almost never	Seldom	Some of the time	Fairly often	Almost always	
	I relax and enjoy leisure time	Almost never	Seldom	Some of the time	Fairly often	Almost always	
	I practice safe sex (see explanation)	Almost never	Seldom	Some of the time	Fairly often	Always	
Type of behavior	I seem to be in a hurry	Almost always	Fairly often	Some of the time	Seldom	Almost never	
	I feel angry or hostile	Almost always	Fairly often	Some of the time	Seldom	Almost never	
Insight	I am a positive or optimistic thinker	Almost never	Seldom	Some of the time	Fairly often	Almost always	
	I feel tense or uptight	Almost always	Fairly often	Some of the time	Seldom	Almost never	
	I feel sad or depressed	Almost always	Fairly often	Some of the time	Seldom	Almost never	
Career	I am satisfied with my job or role	Almost never	Seldom	Some of the time	Fairly often	Almost always	
Test scoring							
Step 1	Total the X's in each column	___	___	___	___	___	
Step 2	Multiply the totals by the numbers indicated (write your answer on the box below)	× 0	× 1	× 2	× 3	× 4	
Step 3	Add your scores across the bottom for your grand total		___ +	___ +	___ +	___	
	Grand total					=	___

A Balanced Diet

According to Canada's Food Guide to Healthy Eating (for people four years and over): *Different People Need Different Amounts of Food.* The amount of food you need every day from the four food groups and other foods depends on your age, body size, activity level, whether you are male or female, and if you are pregnant or breast feeding. That's why the Food Guide gives a lower and higher number of servings for each food group. For example, young children can choose the lower number of servings, while male teenagers can select the higher number. Most other people can choose servings somewhere in between.

Grain products	Vegetables and fruit	Milk products	Meat and alternatives	Other Foods
Choose whole grain and enriched products more often.	Choose dark green and orange vegetables more often.	Choose lower fat milk products more often.	Choose leaner meats, poultry and fish, as well as dried peas, beans and lentils more often.	Taste and enjoyment can also come from other foods and beverages that are not part of the 4 food groups. Some of these are higher in fat or calories, so use these foods in moderation.
Recommended number of servings per day				
5-12	5-10	Children 4-9 years: 2-3 Youth 10-16 years: 3-4 Adults: 2-4 Pregnant and breast-feeding women: 3-4	2-3	

Alcohol Intake

1 drink equals:		*Canadian*	*Metric*	*U.S.*
1 bottle of beer	5% alcohol	12 oz.	340.8 ml	10 oz.
1 glass wine	12% alcohol	5 oz.	142 ml	4.5 oz.
1 shot spirits	40% alcohol	1.5 oz.	42.6 ml	1.25 oz.

Safe Sex

Refers to the use of methods of preventing infection or conception.

What Does The Score Mean?

85-100	70-84	55-69	35-54	0-34
Excellent	Very good	Good	Fair	Needs improvement

A low total score does not mean that you have failed. There is always the chance to change your lifestyle. Look at the areas where you scored a 0 or 1 and decide which areas you want to work on first.

(continued)

Form 2.7 *(continued)*

Tips

1. Don't try to change all the areas at once. This will be too overwhelming for you.

2. Writing down your proposed changes and your overall goal will help you to succeed.

3. Make changes in small steps towards the overall goal.

4. Enlist the help of a friend to make similar changes and/or to support you in your attempts.

5. Congratulate yourself for achieving each step. Give yourself appropriate rewards.

6. Ask your personal trainer, coach, family physician, nurse or health department for more information on any of these areas.

From G.Haff and C. Dumke, 2012, *Laboratory manual for exercise physiology* (Champaign, IL: Human Kinetics). Adapted from D. Wilson, 1998, *Fantastic lifestyle assessment.*

Flexibility Testing

Objectives

1. Define *flexibility* and the factors that affect it.
2. Differentiate between direct and indirect methods of assessing flexibility.
3. Describe and conduct the six most common variants of the sit-and-reach test.
4. Learn the shoulder elevation and back-scratch methods for evaluating shoulder flexibility.
5. Compare individual and group results from tests performed in the laboratory activity with those represented in the normative literature.

Definitions

flexibility—Range of motion in a joint or related series of joints (3, 34).

goniometer—Device, similar to a protractor and containing two arms, used to measure joint angles (16).

inclinometer—Gravity-based goniometer used to measure ranges of motion in a joint (16).

percentile rank—Method for comparing how a subject performs with how other subjects perform; typically represented as being between 1 percent and 99 percent (19).

Flexibility is generally defined as the range of motion in a joint or related series of joints (34, 36) or as the ability to move a muscle or group of muscles through a range of motion (15). An individual's overall flexibility can be affected by the structure of a joint (2, 5), the age and sex of the individual (26, 40), the elasticity and plasticity of the connective tissue (15, 41), and the individual's activity level (26).

Ball-and-socket joints (e.g., shoulder, hip) enable the greatest range of motion and offer the capacity to move in all planes. Second most flexible are the ellipsoidal joints (e.g., wrist), which move primarily in the sagittal and frontal planes. The smallest range of motion is found in the hinge joints (e.g., knee, elbow), which move primarily in the sagittal plane. The range of motion of any joint is affected by its articulating surfaces and by

the soft tissues surrounding it, such as tendons, ligaments, fascial sheaths, joint capsules, and skin (12, 16, 26). Flexibility involves both elasticity, which is the ability to return to a resting length, and plasticity, which is the ability to change the length of the soft tissues.

Generally, younger individuals are more flexible than their older counterparts (26, 27), and women tend to be more flexible than men (26). Sex-specific differences in flexibility are most likely related to anatomical and structural differences, as well as differences in the physical activities typically undertaken by members of each sex (26). Regardless of sex, flexibility declines by 20% to 30% between the ages of 30 and 70 (10). This reduction may be attributed to a drop in physical activity as one ages (24); as a whole, more active individuals, regardless of age or sex, tend to be more flexible (26). This relationship highlights the importance of appropriately designed training programs and the potential for an individual to increase flexibility and induce improvement in overall fitness.

Direct and Indirect Range of Motion Assessment

As one of the five major components of fitness (4, 5), flexibility is commonly tested as part of health-related test batteries. Flexibility is joint specific, and no single test can be used to evaluate total body flexibility (5); however, flexibility testing can be conducted on any joint of the body thanks to a variety of direct and indirect methods (16, 18).

The most common method for directly assessing joint range of motion is to use a **goniometer** (figure 3.1), a device very similar to a protractor that is used to measure joint angles (18). Goniometers have two arms coupled with a protractor for measuring the degrees of joint displacements (38). They are rather simple to use (38) and are considered highly reliable when standardized procedures are employed (8, 29). Goniometer size varies according to the joint being measured (29, 38).

Despite its usefulness, the goniometer may not be the best tool for directly assessing spinal movement or complex movements such as supination, pronation, inversion, and eversion. For these movements, an **inclinometer** (figure 3.2) is considered more accurate (38). An inclinometer uses a universal center of gravity as a consistent

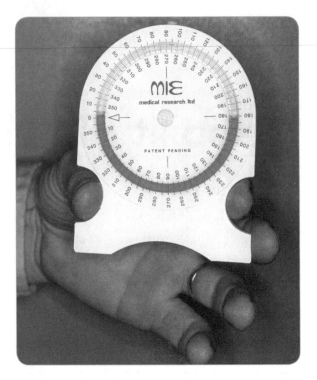

Figure 3.2 Inclinometer.

Reprinted from P. Maud and C. Foster, 2005, *Physiological assessment of human fitness*, 2nd ed. (Champaign, IL: Human Kinetics), 229. By permission of P. Maud.

starting point and a weighted needle and protractor to determine range of motion (29, 32). Positioning and securing the device to the subject can be difficult, but both electronic and mechanical inclinometers are reliable methods for assessing range of motion (29).

When direct methods for evaluating range of motion are unavailable, you can assess flexibility by means of indirect methods (30). These methods usually report results in terms of inches or centimeters (18), and they have been determined to be very reliable. The most common indirect methods for assessing flexibility in a health fitness testing battery are the sit-and-reach test (35), the shoulder elevation test (1), and the back-scratch test (16).

Body Area Considerations

The body areas typically assessed for flexibility in a health fitness testing battery are the back, hamstrings, and shoulders. The flexibility of these areas can be assessed by means of a variety of methods.

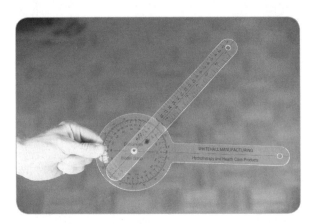

Figure 3.1 Goniometer.

Back and Hamstring Flexibility Tests

Poor flexibility of the low back extensors and the hamstrings has been associated with low back pain (28, 41) and poor performance in sporting activities requiring particular flexibility in these areas (24, 34). For these reasons, assessment of low back and hamstring flexibility is common in most exercise physiology laboratories and health fitness settings (7); this assessment can be achieved by means of direct or indirect methods.

Goniometer testing is considered to be a direct measure of hamstring flexibility when used in conjunction with a straight leg raise (13). This assessment procedure is considered to be the criterion, or "gold standard", for assessing hamstring flexibility. Even though goniometer-based testing has a subjective endpoint criterion (2), it is considered to be a highly reliable method for assessing hamstring flexibility (8).

Regardless of the joint being assessed, the basic methodology for using the goniometer is easy to implement (18, 38). The goniometer is placed with the center of the protractor fixed to the axis of rotation while one arm is lined up with the proximal articulating segment and the second arm corresponds to the distal segment (figure 3.3) (38). The movement of the second arm allows

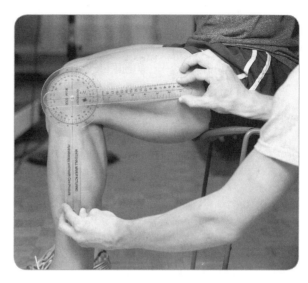

Figure 3.3 Goniometer placement for measuring range of motion.

quantification of the range of movement for the distal segment.

The most commonly used indirect test of hamstring and lower back flexibility is the sit-and-reach test (figure 3.4) (5, 19, 30). However, this test is generally considered to be a poor indicator of low back function and is more commonly accepted as an indicator of hamstring flexibility (20, 21, 23, 24, 30, 31, 33). Therefore, this assessment should be considered primarily as an indicator of

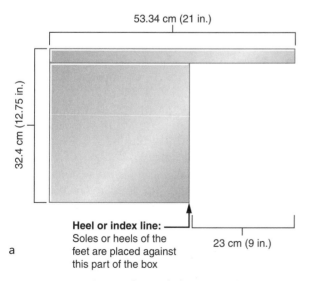

53.34 cm (21 in.)

32.4 cm (12.75 in.)

Heel or index line: Soles or heels of the feet are placed against this part of the box

23 cm (9 in.)

a

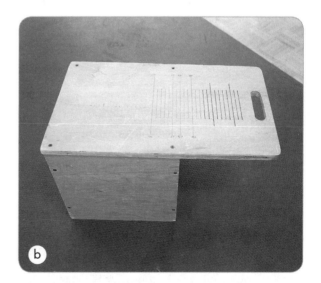

b

Figure 3.4 Sit-and-reach box: (a) diagram showing dimensions and (b) sample sit-and-reach box.

hamstring (i.e., semitendinosus, semimembranosus, and biceps femoris) flexibility and secondarily as a measurement of lower back (erector spinae), buttocks (gluteus maximus and gluteus medius), and calf (gastrocnemius) muscle flexibility (2).

The sit-and-reach test is considered by many to be a field-based assessment (30) but may involve some of the characteristics typically seen in laboratory-based testing (2). When performed with standardized methods, the sit-and-reach test has produced consistently reliable assessments ranging from 0.70 to 0.98 depending upon the population examined (20, 21, 30, 37). From a validity standpoint, the sit-and-reach test appears to produce a valid assessment only of hamstring flexibility ($r = 0.70-0.76$, $p < 0.05$) (30).

The sit-and-reach test can be used in several forms, including the traditional method first presented by the American Alliance of Health, Physical Education, Recreation and Dance (2, 3, 13). Modifications include the YMCA (2, 5, 34), Canadian (2, 22), wall (2, 17), V-sit (2, 11), and backsaver (20, 21) approaches.

Shoulder Flexibility Tests

Shoulder flexibility can affect a person's ability to perform activities of daily living, such as combing one's hair, dressing, and reaching for the seat belt in a car (16). In sport, deficiency in shoulder flexibility can limit performance and increase an athlete's risk of injury (39). As a result, assessment of shoulder flexibility is commonplace in both sport and fitness testing batteries. Two of the most common tests are the shoulder elevation test (1, 19) and the back-scratch test (16, 35).

References

1. Acevedo EO and Starks MA. *Exercise Testing and Prescription Lab Manual*. Champaign, IL: Human Kinetics, 2003.
2. Adams GM. *Exercise Physiology Laboratory Manual*. 3rd ed. Boston: McGraw-Hill, 1998.
3. American Alliance for Health, Physical Education, Recreation and Dance (AAHPERD). *Health Related Physical Fitness Test Manual*. Reston, VA.: AAHPERD, 1980.
4. American College of Sports Medicine. ACSM Position Stand. The Recommended Quantity and Quality of Exercise for Developing and Maintaining Cardiorespiratory and Muscular Fitness, and Flexibility in Healthy Adults. *Med Sci Sports Exerc* 30: 975–991, 1998.
5. American College of Sports Medicine. *ACSM's Guidelines for Exercise Testing and Prescription*. 8th ed. Philadelphia: Lippincott, Williams & Wilkins, 2010.
6. American College of Sports Medicine. *ACSM's Health-Related Physical Fitness Assessment Manual*, 3rd ed. Baltimore: Lippincott Williams & Wilkins, 2010, p. 180.
7. Baltaci G, Un N, Tunay V, et al. Comparison of Three Different Sit and Reach Tests for Measurement of Hamstring Flexibility in Female University Students. *Br J Sports Med* 37: 59–61, 2003.
8. Boone DC, Azen SP, Lin CM, et al. Reliability of Goniometric Measurements. *Phys Ther* 58: 1355–1390, 1978.
9. Cailliet R. *Low Back Pain Syndrome*. 5th ed. Philadelphia: Davis, 1995.
10. Chapman EA, deVries HA, Swezey R. Joint Stiffness: Effects of Exercise on Young and Old Men. *J Gerontol* 27: 218–221, 1972.
11. Cooper Institute for Aerobics Research (CIAR). *The Prudential FITNESSGRAM Test Administration Manual*. Dallas: CIAR, 1992.
12. de Vries HA, Housh TJ, Weir LL. *Physiology of Exercise for Physical Education, Athletics and Exercise Science*. 5th ed. Dubuque, IA: Brown, 1995.
13. Goeken LN and Hof AL. Instrumental Straight-Leg Raising: Results in Healthy Subjects. *Arch Phys Med Rehabil* 74: 194–203, 1993.
14. Golding LA, ed. *YMCA Fitness Testing and Assessment Manual*. 4th ed. Champaign, IL: Human Kinetics, 2000.
15. Haff GG, Cramer JT, Beck DT, et al. Roundtable Discussion: Flexibility Training. *Strength and Cond J* 28: 64–85, 2006.
16. Heyward VH. *Advanced Fitness Assessment and Exercise Prescription*. 6th ed. Champaign, IL: Human Kinetics, 2010.
17. Hoeger WW and Hopkins DR. A Comparison of the Sit and Reach and the Modified Sit and Reach in the Measurement of Flexibility in Women. *Res Q Exerc Sport* 63: 191–195, 1992.
18. Hoffman JR. *Norms for Fitness, Performance, and Health*. Champaign, IL: Human Kinetics, 2006.
19. Housh TJ, Cramer JT, Weir JP, Beck TW, and Johnson GO. *Physical Fitness Laboratories on a Budget*. Scottsdale, AZ: Holcomb-Hathaway, 2009, p. 234.
20. Hui SC, Yuen PY, and Morrow JR, Jr., et al. Comparison of the Criterion-Related Validity of Sit-and-Reach Tests With and Without Limb Length

Adjustment in Asian Adults. *Res Q Exerc Sport* 70: 401–406, 1999.

21. Hui SS and Yuen PY. Validity of the Modified Back-Saver Sit-and-Reach Test: A Comparison With Other Protocols. *Med Sci Sports Exerc* 32: 1655–1659, 2000.

22. Institute FaL. *Fitness and Lifestyle in Canada*. Ottawa, Ontario: Fitness and Amateur Sport, 1983.

23. Jackson AM and Langford NJ. The Criterion-Related Validity of the Sit and Reach Test: Replication and Extension of Previous Findings. *Res Q Exerc Sport* 60: 384–387, 1989.

24. Jackson AM and Baker AA. The Relationship of the Sit and Reach Test to Criterion Measures of Hamstring and Back Flexibility in Young Females. *Res Q Exerc Sport* 57: 183–186, 1986.

25. Jeffreys I. *Total Soccer Fitness*. Monterey: Coaches Choice, 2007, p. 233.

26. Jeffreys I. Warm-Up and Stretching. In: Baechle TR and Earle RW, eds., *Essentials of Strength Training and Conditioning*. 3rd ed. National Strength and Conditioning Association. Champaign, IL: Human Kinetics, 2008, pp. 296–276.

27. Kell RT, Bell G, and Quinney A. Musculoskeletal Fitness, Health Outcomes and Quality of Life. *Sports Med* 31: 863–873, 2001.

28. Kofotolis N and Kellis E. Effects of Two 4-Week Proprioceptive Neuromuscular Facilitation Programs on Muscle Endurance, Flexibility, and Functional Performance in Women With Chronic Low Back Pain. *Phys Ther* 86: 1001–1012, 2006.

29. Lea RD and Gerhardt JJ. Range-of-Motion Measurements. *J Bone Joint Surg Am* 77: 784–798, 1995.

30. Liemohn W, Sharpe GL, and Wasserman JF. Criterion Related Validity of the Sit-and-Reach Test. *J Strength Cond Res* 8: 91–94, 1994.

31. Macrae IF and Wright V. Measurement of Back Movement. *Ann Rheum Dis* 28: 584–589, 1969.

32. Maud PJ and Kerr KM. Static Techniques for the Evaluation of Joint Range of Motion and Muscle Length. In: PJ Maud and C Foster, eds., *Physiological Assessment of Human Fitness*. 2nd ed. Champaign, IL: Human Kinetics, 2006, pp. 227–252.

33. McAuley E, Hudash G, Shields K, et al. Injuries in Women's Gymnastics. The State of the Art. *Am J Sports Med* 16 Suppl 1: S124–131, 1988.

34. McNeal JR and Sands WA. Stretching for Performance Enhancement. *Curr Sports Med Rep* 5: 141–146, 2006.

35. Nieman DC. *Exercise Testing and Prescription: A Health-Related Approach*. 5th ed. New York: McGraw-Hill, 2003, p. 774.

36. Sands WA, McNeal JR, Stone MH, et al. Effect of Vibration on Forward Split Flexibility and Pain Perception in Young Male Gymnasts. *Int J Sports Physiol Perf* 3: 469–481, 2008.

37. Shephard RJ, Berridge M, and Montelpare W. On the Generality of the "Sit and Reach" Test: An Analysis of Flexibility Data for an Aging Population. *Res Q Exerc Sport* 61: 326–330, 1990.

38. Spring T, Franklin B, and deJong A. Muscular Fitness and Assessment. In: *ACSM's Resource Manual for Guidelines for Exercise Testing and Prescription*. Baltimore, MD: Lippincott, Williams & Wilkins, 2010, pp. 332–348.

39. Warner JJ, Micheli LJ, Arslanian LE, et al. Patterns of Flexibility, Laxity, and Strength in Normal Shoulders and Shoulders With Instability and Impingement. *Am J Sports Med* 18: 366–375, 1990.

40. Wilmore JH, Parr RB, Girandola RN, et al. Physiological Alterations Consequent to Circuit Weight Training. *Med Sci Sports* 10: 79–84, 1978.

41. Winters MV, Blake CG, Trost JS, et al. Passive Versus Active Stretching of Hip Flexor Muscles in Subjects With Limited Hip Extension: A Randomized Clinical Trial. *Phys Ther* 84: 800–807, 2004.

TRADITIONAL, WALL, AND V-SIT TEST COMPARISONS

EQUIPMENT

- Physician's scale or equivalent electronic scale
- Stadiometer
- Sit-and-reach box
- Yardstick or meterstick
- Individual and group data sheets
- Microsoft Excel or equivalent spreadsheet program

Find the group data sheets for this laboratory online at www.HumanKinetics.com/ LaboratoryManualForExercisePhysiology.

WARM-UP

Regardless of the sit-and-reach test method employed, all subjects should perform a structured warm-up beforehand to increase the test's reliability and validity. The warm-up session should include a 5 min general warm-up and a 5 min dynamic stretching warm-up (38).

The 5 min general warm-up can involve activities such as jogging, cycling, skipping, jumping rope, or doing calisthenics (26). The chosen activity should be undertaken with the specific goal of increasing heart rate, blood flow, muscle temperature, respiration rate, and perspiration while decreasing the viscosity of joint fluids (12, 26). Taken collectively, these responses prepare the subject for the more specific warm-up activities that are performed next.

After completing the general warm-up, the subject should perform dynamic warm-up activities focused on movements that work through the range of motions required for the testing activity (26). Dynamic warm-up activities appropriate for the sit-and-reach test include walking knee lifts, leg swings, trunk rotations, lunges, standing stiff-leg deadlifts, and the inchworm (25). See table 3.1 for a sample 10 min warm-up that includes 5 min of general warm-up and 5 min of dynamic stretching. After completing the warm-up period, the subject can begin the testing process by using the following protocol.

Table 3.1 Warm-Up for Sit-and-Reach Testing

Warm-up	Activity	Total time
General*	Cycling	5 minutes
Specific	Walking lunges	5 minutes
	Walking lunges with dynamic rotation	
	High-knee walk	
	Leg swings	
	Trunk rotations	
	Squats	

*The general warm-up activity can be selected based upon the equipment available.

TRADITIONAL SIT-AND-REACH TEST

The traditional sit-and-reach test requires the use of a sit-and-reach box for which the index or heel line is marked at 23 cm (9.1 in) (see figure 3.4).

Step 1: Place the sit-and-reach box against a wall or object that prevents it from slipping during the testing process.

Step 2: Have the subject perform a 10 min warm-up (see sample warm-up description in table 3.1).

Step 3: Instruct the subject to remove his or her shoes, then measure his or her height and weight. Record these numbers on the individual data sheet for laboratory activity 3.1. Also record the subject's name and sex, as well as the date, time, temperature, and barometric pressure.

Step 4: Have the subject sit on the floor with the heels and soles of his or her feet against the index or heel line (23 cm mark), with legs fully extended, and with the medial sides of the feet about 20 cm (8 in.) apart (see figure 3.5a).

Step 5: Place your hands across the subject's knees to ensure that he or she maintains full leg extension.

Step 6: Instruct the subject to extend his or her arms with one hand on top of the other and palms facing down (figure 3.5b).

Figure 3.5 Traditional sit-and-reach test: (a) start and (b) extended position.

Step 7: Have the subject bend forward and move his or her hands along the measuring scale on top of the sit-and-reach box. This position should be held for 1 to 2 s. If the subject's knees bend or the fingertips become uneven, the test does not count, and the process should be repeated. Record the resulting measure on the individual sit-and-reach data sheet.

Step 8: Perform steps 6 and 7 three more times.

Step 9: Record the results of the fourth trial in the appropriate location on the individual data sheet.

Step 10: Compare the fourth trial's results with the **percentile ranks** or normative data presented in table 3.2 and note the subject's ranking in the appropriate spot on the individual data sheet.

Table 3.2 Percentile Ranks for the Traditional Sit-and-Reach Test (cm)

Age (y)		20–29		30–39		40–49		50–59		60–69	
Sex		M	F	M	F	M	F	M	F	M	F
Percentile rank	90	39	40	37	39	34	37	35	37	32	34
	80	35	37	34	36	31	33	29	34	27	31
	70	33	35	31	34	27	32	26	32	23	28
	60	30	33	29	32	25	30	24	29	21	27
	50	28	31	26	30	22	28	22	27	19	25
	40	26	29	24	28	20	26	19	26	15	23
	30	23	26	21	25	17	23	15	23	13	21
	20	20	23	18	22	13	21	12	20	11	20
	10	15	19	14	18	9	16	9	16	8	15

May be used for the traditional, backsaver, and V-sit tests. If values fall between the percentile ranks presented, interpolate the data. For example, if a male subject is 21 years old and achieves a result of 37 centimeters, then his percentile rank is 85.

Reprinted, by permission, 1986, *The Canadian standardized test of fitness (CSTF) operations manual*, 3rd ed. (Public Health Agency of Canada).

WALL SIT-AND-REACH TEST

The wall sit-and-reach test is a modification in which the heel or index line is set according to the individual (17); specifically, the subject places his or her hips, back, and head against a wall when establishing what is considered the zero point. This step allows you to correct for the proportion of leg length to trunk length.

Step 1: Have the subject perform a 10 min warm-up (see sample in table 3.1).

Step 2: Instruct the subject to remove his or her shoes, then measure his or her height and weight. Record these numbers on the individual data sheet. Also record the subject's name and sex, as well as the date, time, temperature, and barometric pressure.

Step 3: Instruct the subject to sit on the floor with his or her hips, back, and head against a wall.

Step 4: Have the subject extend his or her legs with the feet roughly 20 to 30 cm (8 to 12 in.) apart.

Step 5: Position the sit-and-reach box against the subject's heels. In order to prevent the box from slipping, brace it with your feet or a suitable object.

Step 6: Instruct the subject to place one hand on top of the other with the palms facing down.

Step 7: Instruct the subject to reach forward as far as possible while keeping his or her hips, back, and head in contact with the wall. It is okay if the shoulder moves forward. Determine how far the person's fingertips reach and record this measurement to the nearest 1.25 cm (0.5 in.) (see figure 3.6a). This is called the *index line* or *zero position*.

Figure 3.6 Wall sit-and-reach test: *(a)* start and *(b)* extended.

Step 8: Having established the zero position, instruct the subject to reach forward three times during the same movement along the device while making sure to keep his or her palms against the measuring device (see figure 3.6b). The subject should hold the third movement for 2 s while you measure and record the distance.

Step 9: Subtract the measure recorded in step 7 from the value determined in step 8. Record this value on the data sheet.

Step 10: Repeat steps 7 through 9.

Step 11: Record the best of the two trials in the appropriate location on the data sheet.

Step 12: Interpret the results of the test by comparing them with the normative data and percentile ranks presented in table 3.3. Note the percentile rank in the appropriate box on the data sheet.

Laboratory Activity 3.1

Table 3.3 Percentile Ranks and Normative Data for the Wall Sit-and-Reach Test

Age (y)		<35				36–49				>50			
Sex		M		F		M		F		M		F	
Units		cm	in.	cm	in.	cm	in.	cm	in.	cm	in.	cm	in.
Percentile rank	90	45	17.7	45	17.7	41	16.1	44	17.3	38	15.0	38	15.0
	80	43	17.0	42	16.5	37	14.6	41	16.1	34	13.4	36	14.2
	70	40	15.7	41	16.1	35	13.8	39	15.4	31	12.2	35	13.8
	60	38	15.0	40	15.7	34	13.4	37	14.6	29	11.4	31	12.2
	50	37	14.6	38	15.0	32	12.6	34	13.4	26	10.2	28	11.0
	40	34	13.4	37	14.6	29	11.4	33	13.0	25	9.8	26	10.2
	30	33	13.0	35	13.8	27	10.6	31	12.2	24	9.4	23	9.1
	20	29	11.4	32	12.6	25	9.8	28	11.0	22	8.7	21	8.3
	10	23	9.1	26	10.2	21	8.3	25	9.8	20	7.9	19	7.5

If values fall between the percentile ranks presented, interpolate the data. The reference or heel line is set at 0.

Adapted from Adams (2).

V-SIT SIT-AND-REACH TEST

The V-sit method is another modification of the traditional sit-and-reach test (2, 11). The V-sit does not require a sit-and-reach box, but it does require the heels to be separated by 30 cm (about 12 in.) in order to create the V.

Step 1: Instruct the subject to perform a 10 min warm-up (see sample in table 3.1).

Step 2: Instruct the subject to remove his or her shoes, then measure his or her height and weight. Record these numbers on the individual data sheet. Also record the subject's name and sex, as well as the date, time, temperature, and barometric pressure.

Step 3: Have the subject sit on the floor and fully extend his or her legs with the feet separated by 30 cm (12 in.).

Step 4: Place a meterstick between the subject's legs so that the 23 cm (9.1 in.) mark aligns with the subject's heels. In order to prevent the meterstick from moving, tape it to the floor.

Step 5: Hold the subject's knees to ensure that his or her legs do not bend during the test.

Step 6: Have the subject place one hand on top of the other with the palms facing down and the fingertips aligned (see figure 3.7a).

Step 7: Instruct the subject to lean forward and move his or her hands along the meterstick until the subject is fully extended. This position should be held for 1 or 2 s. Record the distance achieved in the data sheet (figure 3.7b).

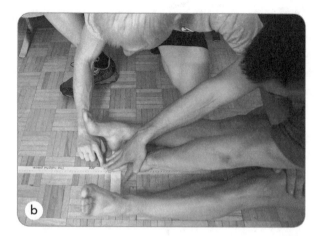

Figure 3.7   V-sit test: (a) start and (b) extended.

Step 8: Perform steps 5 through 7 three more times and consider the fourth trial as the maximal stretch. Record the results of the fourth trial in the appropriate part of the data sheet.

Step 9: Calculate the average of the four trials and record this value in the appropriate location on the individual data sheet.

Step 10: Compare the average achieved during the tests with the normative data and percentile ranks used for the traditional sit-and-reach test (see table 3.2), then note the percentile rank in the appropriate location on the data sheet

QUESTION SET 3.1

1. What major factors can affect flexibility?
2. What do sit-and-reach tests tell us about back and hamstring flexibility?
3. Compare the results of your sit-and-reach tests with those presented in the normative data. What do the results indicate?
4. Compare the class sit-and-reach test data as a whole with the normative data. What do these results indicate?
5. Create a bar graph that presents means for the class data collected for the traditional, wall, and V-sit tests.
6. Use a paired *t*-test to compare the results of the sit-and-reach tests for the five most flexible men to the results for the five most flexible women. How do these findings compare with what you would expect to find?
7. Fill in the table in the individual datasheet comparing the methods used in this lab. Indicate the best application for each test and describe the pros and cons of each test.

 Find the case studies for this laboratory online at www.HumanKinetics.com/ LaboratoryManualForExercisePhysiology.

Laboratory Activity 3.1

Laboratory Activity 3.1   Individual Data Sheet

Name or ID number: _____   Date: _____

Tester: _____   Time: _____

Sex: M / F (circle one)   Age: _____ y   Height: _____ in. _____ cm

Temperature: _____ °F _____ °C   Weight: _____ lb _____ kg

Barometric pressure: _____ mmHg   Relative humidity: _____ %

Sit-and-reach test results	Trial 1	Trial 2	Trial 3	Trial 4	Best	4th trial	Average	Index/heel line (cm)
Traditional								23
Wall								
V-sit								

Individual results ranking	Results (cm)	Ranking/category/percentile
Traditional		Percentile =
Wall sit		Percentile =
V-sit		Percentile =

YMCA, BACKSAVER, AND GONIOMETER TEST COMPARISONS

EQUIPMENT

- Physician's scale or equivalent electronic scale
- Stadiometer
- Goniometer
- Sit-and-reach box
- Yardstick or meterstick
- Tape
- Individual and group data sheets
- Microsoft Excel or equivalent spreadsheet program

Find the group data sheets for this laboratory online at www.HumanKinetics.com/ LaboratoryManualForExercisePhysiology.

WARM-UP

Regardless of the sit-and-reach test method employed, all subjects should perform a structured warm-up beforehand to increase the test's reliability and validity. The warm-up session should include a 5 min general warm-up and a 5 min dynamic stretching warm-up (38).

The 5 min general warm-up can involve activities such as jogging, cycling, skipping, jumping rope, or doing calisthenics (26). The chosen activity should be undertaken with the specific goal of increasing heart rate, blood flow, muscle temperature, respiration rate, and perspiration while decreasing viscosity of joint fluids (12, 26). Taken collectively, these responses prepare the subject for the more specific warm-up activities that are performed next.

After completing the general warm-up, the subject should perform dynamic warm-up activities focused on movements that work through the range of motions required for the testing activity (26). Dynamic warm-up activities appropriate for the sit-and-reach test might include walking knee lifts, leg swings, trunk rotations, lunges, standing stiff-leg deadlifts, and the inchworm (25). See table 3.1 for a sample 10 min warm-up that includes 5 min of general warm-up and 5 min of dynamic stretching. After completing the warm-up period, the subject can begin the testing process.

YMCA SIT-AND-REACH TEST

The YMCA sit-and-reach test is very similar to the traditional test except that no sit-and-reach box is used (5, 14). In this test, use a yardstick or meterstick placed on the floor so that the zero point is directed toward the subject. As a result of not using the sit-and-reach box, the YMCA test generally produces about a 2.5 cm (1 in.) difference in result when compared with the traditional sit-and-reach test (2).

Step 1: Place a yardstick or meterstick on the floor so that the zero end faces the subject. Place a piece of tape so that it intersects with the stick at 15 in. (38 cm).

Step 2: Instruct the subject to remove his or her shoes, then measure his or her height and weight. Record these numbers on the individual data sheet for laboratory activity 3.2. Also record the subject's name and sex, as well as the date, time, temperature, and barometric pressure.

Step 3: Have the subject perform a 10 min warm-up (see sample in table 3.1).

Step 4: Instruct the subject to remove his or her shoes and sit on the floor with the stick between his or her legs. The legs should be separated by 25 to 30 cm (10 to 12 in.) and extended at right angles to the taped line The heels should touch the taped line (see figure 3.8a).

Step 5: Place your hands across the subject's knees to ensure that he or she maintains full leg extension.

Step 6: Instruct the subject to extend his or her arms with one hand on top of the other and with the palms facing down (figure 3.8b).

Step 7: Have the subject bend forward and move his or her hands along the stick while keeping them parallel to one another. This position should be held for 1 or 2 s. Measure the distance reached and record it on the individual sit-and-reach data sheet. If the subject's knees bend or

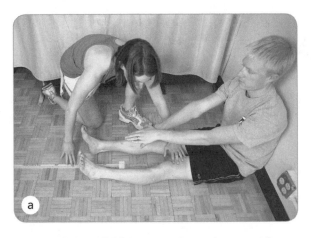 

Figure 3.8 YMCA sit-and-reach test: (a) start and (b) extended.

the fingertips become uneven, the test does not count, and it should be repeated.

Step 8: Perform steps 5 through 7 three more times.

Step 9: Record the best result from the four trials on the individual data sheet, then compare this value with the percentile ranks or normative data presented in table 3.4. Note the results in the appropriate spot on the individual data sheet.

Table 3.4 Percentile Ranks and Normative Data for the YMCA Sit-and-Reach Test (cm)

Age (y)			18–25		26–35		36–45		46–55		56–65		>65	
Sex			**M**	**F**	**M**	**F**	**M**	**F**	**M**	**F**	**M**	**F**	**M**	**F**
	Classification	**Percentile**												
Percentile Rank	Well above average	90	56	61	53	58	53	56	48	53	43	51	43	51
		80	51	56	48	53	48	53	43	51	38	48	38	46
	Above average	70	48	53	43	51	43	48	38	46	33	43	33	43
		60	46	51	43	51	41	46	36	43	33	41	30	43
	Average	50	43	48	38	48	38	43	33	41	28	38	25	38
		40	38	46	36	43	33	41	28	36	23	36	23	36
	Below average	30	36	43	33	41	33	38	25	36	23	33	20	33
		20	33	41	28	38	28	36	23	30	18	28	18	28
	Well below average	10	28	36	23	33	18	30	15	25	13	23	10	23

Note: The index or heel line for the YMCA sit-and-reach test is 38 cm (15 in.). If values fall between the percentile ranks presented, interpolate the data.

Reprinted from YMCA, 2000, *YMCA fitness and testing and assessment manual*, 4th ed., edited by L.A. Golding (Champaign, IL: Human Kinetics).

BACKSAVER SIT-AND-REACH TEST

The backsaver test is another modification of the traditional sit-and-reach test. It has been developed to reduce the posterior compression typically seen when bending forward with both legs extended (2, 9, 11). Instead, the backsaver test requires the subject to bend forward with one leg extended and the other bent (2, 11, 21). Even so, some subjects report discomfort, in this case in the hip joint of the bent leg (20, 21). Other than the altered leg positions and the fact that each leg is measured separately, the backsaver test is performed in a manner similar to that of the traditional sit-and-reach test.

Step 1: Place the sit-and-reach box against a wall or an object that will prevent it from slipping during the testing process.

Step 2: Have the subject perform a 10 min warm-up (see sample in table 3.1).

Step 3: Instruct the subject to remove his or her shoes, then measure his or her height and weight. Record these numbers on the individual data sheet. Also record the subject's name and sex, as well as the date, time, temperature, and barometric pressure.

Step 4: Instruct the subject to sit on the floor with the right leg extended so that the sole of the foot is placed against the sit-and-reach box at the index or heel line (at the 23 cm or 9.1 in mark). The subject's left leg should be bent at approximately 90° with the sole of the left foot placed flat on the floor (21) about 5 to 7.5 cm (2 to 3 in) from the sit-and-reach box (2).

Step 5: Have the subject extend his or her arms with one hand on top of the other and with the palms facing down.

Step 6: For the test, the subject should bend forward while moving his or her hands along the measuring scale on top of the sit-and-reach box. This position should be held for 1 or 2 s. The subject may have to move the bent leg to the side while extending forward to allow movement through the range of motion. If the extended knee bends or the fingertips become uneven, the test does not count, and it should be repeated. Record the resulting measurement on the individual data sheet.

Step 7: Perform steps 4 through 6 three more times with the right leg.

Step 8: Compare the fourth trial's results with the percentile ranks or normative data presented in table 3.2 and note the results in the appropriate spot on the individual data sheet.

Step 9: Have the subject change position so that the left leg is extended and the sole of the left foot is placed against the sit-and-reach box at the index or heel line (at the 23 cm or 9.1 in. mark). The right leg should be bent to approximately 90° with the sole of the right foot placed flat on the floor (21) about 5 to 7.5 cm (2 to 3 in.) from the sit-and-reach box (2). See figure 3.9a.

Step 10: Have the subject extend his or her arms with one hand on top of the other and with the palms facing down. See figure 3.9b.

Step 11: Instruct the subject to bend forward while moving his or her hands along the measuring scale on top of the sit-and-reach box. This position should be held for 1 or 2 s. The subject may have to move the bent leg

Figure 3.9 Backsaver test: (a) start and (b) extended.

Laboratory Activity 3.2

to the side while extending forward to allow movement through the range of motion. If the subject's extended knee bends or the fingertips become uneven, the test does not count, and it should be repeated. Record the resulting measurement on the individual sit-and-reach data sheet.

Step 12: Perform steps 9 through 11 three more times with the right leg.

Step 13: Compare the fourth trial's results with the percentile ranks or normative data presented in table 3.2 and note the results in the appropriate spot on the individual data sheet.

GONIOMETER TEST OF HAMSTRING FLEXIBILITY

The goniometer test of hamstring flexibility is often referred to as the criterion measure or "gold standard" for evaluating hamstring flexibility (21). It evaluates hamstring flexibility during a passive straight-leg raise and is performed separately on each leg (2, 21). The criterion score achieved for this test is registered as the maximal angle in degrees achieved when maximum hip flexion occurs (21). Although the test can generally be performed by one tester, you may prefer to use two testers—one to lift the subject's leg while the other uses the goniometer to measure hip flexion (2).

Step 1: Direct the subject through a 10 min warm-up (see sample in table 3.1).

Step 2: Instruct the subject to remove his or her shoes, then measure his or her height and weight. Record these numbers on the individual data sheet. Also record the subject's name and sex, as well as the date, time, temperature, and barometric pressure.

Step 3: Have the subject lie on his or her back on a table or on the floor.

Step 4: Align the goniometer with the axis of the subject's right hip.

Step 5: Align the stationary arm of the goniometer with the subject's trunk and position the moveable arm in line with the subject's right femur.

Step 6: Once the stationary and moveable arms are aligned, move the right leg toward hip flexion while making sure that the knee of the moving leg remains straight. The end point of flexion is determined when you feel tightness. Once this position is achieved, hold it for 1 or 2 s while measuring the angle achieved to the nearest degree.

Step 7: Return the right leg to a resting position and record the measurement on the data sheet.

Step 8: Align the goniometer with the axis of the subject's left hip (see figure 3.10a).

Step 9: Align the stationary arm of the goniometer with the subject's trunk and position the moveable arm in line with the subject's left femur.

Step 10: Once the stationary and moveable arms are aligned, move the left leg toward hip flexion while making sure that the knee of the moving leg remains straight. The end point of flexion is determined when you feel tightness. Once this position is achieved, hold it for 1 or 2 s while measuring the angle achieved to the nearest degree (see figure 3.10b).

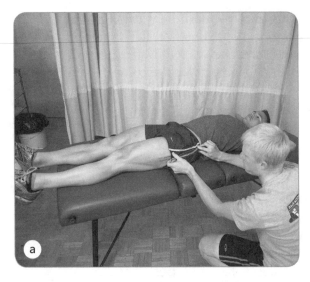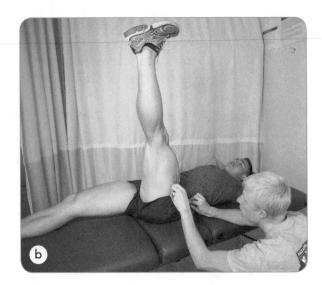

Figure 3.10 Goniometer hamstring test: *(a)* starting position and *(b)* with full hip flexion.

Step 11: Return the left leg to a resting position and record the measurement on the data sheet.

Step 12: Perform steps 3 through 11 three more times.

Step 13: Calculate the average of the four trials for both the right and left legs and record this information in the appropriate location on the data sheet. When interpreting the degree of flexion achieved, a smaller angle indicates better flexibility.

QUESTION SET 3.2

1. How does the YMCA test differ from the backsaver version of the sit-and-reach test?
2. Compare the results of your sit-and-reach test with the normative data. What do these results indicate?
3. Compare the class data for the sit-and-reach test as a whole with the normative data. What do these results indicate?
4. Create a bar graph that presents means for your data and for the class data.
5. Use a paired *t*-test to compare the results of the sit-and-reach tests for the five most flexible men with the results for the five most flexible women. How do these findings compare with what you would expect to find?
6. How do the percentile ranks compare between the three tests performed in this laboratory activity?
7. Fill in the table in the individual datasheet comparing the methods used in the present lab. Indicate the best application for each test and describe the pros and cons of each test.

Find the case studies for this laboratory online at www.HumanKinetics.com/ LaboratoryManualForExercisePhysiology.

Laboratory Activity 3.2

Laboratory Activity 3.2 Individual Data Sheet

Name or ID number: _____ Date: _____

Tester: _____ Time: _____

Sex: M / F (circle one)   Age: _____ y Height: _____ in. _____ cm

Temperature: _____ °F _____ °C Weight: _____ lb _____ kg

Barometric pressure: _____ mmHg Relative humidity: _____ %

Sit-and-reach test results		Trial 1	Trial 2	Trial 3	Trial 4	Best	4th trial	Index/heel line (cm)
YMCA								38
Backsaver	Right leg							23
	Left leg							23

Goniometer test results		Trial 1	Trial 2	Trial 3	Average		Degrees
Goniometer	Right leg						
	Left leg						

Test		Results (cm)	Ranking/category/percentile
YMCA			Category =
Backsaver	Right leg		Percentile =
	Left leg		Percentile =

Test		Degrees	Percentile
Goniometer	Right leg		Percentile =
	Left leg		Percentile =

CANADIAN SIT-AND-REACH TEST

EQUIPMENT

- Physician's scale or equivalent electronic scale
- Stadiometer
- Sit-and-reach box (as shown in figure 3.4)
- Yardstick or meterstick
- Individual and group data sheets
- Microsoft Excel or equivalent spreadsheet program

Find the group data sheets for this laboratory online at www.HumanKinetics.com/ LaboratoryManualForExercisePhysiology.

CANADIAN SIT-AND-REACH TEST

In comparison with the traditional sit-and-reach test, the Canadian variant differs in several distinct ways (2, 11). First, it involves a standardized warm-up protocol that includes static stretching. Second, the subject's feet are separated by 5 cm (2 in.)—significantly less than the 20 cm (7.9 in.) distance required by the traditional sit-and-reach test. Finally, the height of the sit-and-reach box is adjusted so that the meterstick or measuring device is positioned at the height of the toes, and the heel or index line is set at 26 cm (10.2 in.). Otherwise, the basic procedure is very similar to that used for the traditional test.

Step 1: Set up the testing area by securing the sit-and-reach box against a wall or an object that prevents it from slipping.

Step 2: Instruct the subject to remove his or her shoes, then measure his or her height and weight. Record these numbers on the individual data sheet for laboratory activity 3.3. Also record the subject's name and sex, as well as the date, time, temperature, and barometric pressure.

Step 3: Instruct the subject to perform a modified hurdle stretch with one leg extended and the opposite leg bent so that the sole of that foot is positioned against the extended leg (see figure 3.11). The stretch is held for 20 s and is repeated twice for each leg.

Step 4: Have the subject extend both legs so that the heels are separated by 5 cm (2 in.).

Step 5: Adjust the sit-and-reach box so that it is positioned at the height of the toes and so that the heels are positioned against the heel or index line,

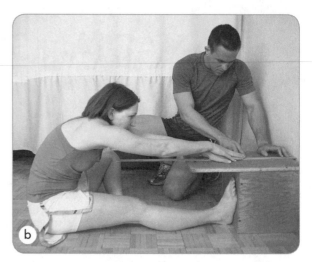

Figure 3.11 Modified hurdle stretch used for the Canadian sit-and-reach test: *(a)* start and *(b)* extended.

which is at 26 cm (10.2 in.). If you do not have access to an adjustable sit-and-reach box, place a meterstick at the subject's toe level with the heels crossing the meter stick at the 26 cm mark.

Step 6: Instruct the subject to place one hand on top of the other with the palms facing down. Have the subject slowly bend forward while running his or her hands along the meterstick. It is important that the subject keeps his or her head down and his or her hands placed evenly with one on top of the other. Once the subject reaches a point at which he or she can no longer extend, the position is held for 2 s. If the knees bend or the fingertips become uneven, the test does not count, and it should be repeated.

Step 7: Read the distance achieved to the nearest cm and record it on the individual data sheet.

Step 8: Perform steps 4 through 7 three more times and note the best of the four trials on the laboratory data sheet.

Step 9: Interpret the results of the test by comparing them with the normative data and percentile ranks presented in table 3.5. Note the subject's percentile ranking on the individual data sheet.

Table 3.5 Percentile Ranks and Norms for the Canadian Sit-and-Reach Test (cm)

Age (y)			15–19		20–29		30–39		40–49		50–59		60–69	
Sex			M	F	M	F	M	F	M	F	M	F	M	F
Percentile rank	Classification	Percentile												
	High	81–100	38	42	39	40	37	40	34	37	34	38	32	34
	Above average	61–80	34	38	34	37	33	36	29	34	28	33	25	31
	Average	41–60	29	34	30	33	28	32	24	30	24	30	20	27
	Below average	21–40	24	29	25	28	23	27	18	25	16	25	15	23
	Low	1–20	24	29	25	28	23	27	18	25	16	25	15	23

The index for the Canadian sit-and-reach test is 26 cm (10.2 in.). If using these norms with a traditional sit-and-reach test, subtract 3 cm (1.2 in.) from the norms presented here.

Source: The Canadian Physical Activity, Fitness & Lifestyle Aproach: CSEP-Health & Fitness Program's Health-Related Appraisal and Counselling Strategy, 3rd Edition © 2003. Reprinted with permission of the Canadian Society for Exercise Physiology.

Laboratory Activity 3.3

TRADITIONAL SIT-AND-REACH TEST

The traditional sit-and-reach test requires the use of a sit-and-reach box for which the index or heel line is marked at 23 cm (9.1 in.) (see figure 3.4).

Step 1: Place the sit-and-reach box against a wall or object that prevents it from slipping during the testing process.

Step 2: Have the subject perform a 10 min warm-up (see sample in table 3.1).

Step 3: Instruct the subject to remove his or her shoes, then measure his or her height and weight. Record these numbers on the individual data sheet for laboratory activity 3.3. Also record the subject's name and sex, as well as the date, time, temperature, and barometric pressure.

Step 4: Have the subject sit on the floor with the heels and soles of his or her feet against the index or heel line (23 cm or 9.1 in mark), with legs fully extended, and with feet about 20 cm (8 in.) apart.

Step 5: Place your hands across the subject's knees to ensure that he or she maintains full leg extension (see figure 3.5a).

Step 6: Instruct the subject to extend his or her arms with one hand on top of the other and palms facing down (figure 3.5b).

Step 7: Have the subject bend forward and move his or her hands along the measuring scale on top of the sit-and-reach box. This position should be held for 1 to 2 s. If the subject's knees bend or the fingertips become uneven, the test does not count, and the process should be repeated. Record the resulting measure on the individual sit-and-reach data sheet.

Step 8: Perform steps 6 and 7 three more times.

Step 9: Record the results of the fourth trial in the appropriate location on the individual data sheet.

Step 10: Compare the fourth trial's results with the percentile ranks or normative data presented in table 3.2 and note the subject's ranking in the appropriate spot on the individual data sheet.

BACKSAVER SIT-AND-REACH TEST

The backsaver sit-and-reach test is a modification of the traditional sit-and-reach test. It has been developed to reduce the posterior compression typically seen when bending forward with both legs extended (2, 6, 34). Instead, the backsaver test requires the subject to bend forward with one leg extended and the other bent (2, 11, 21). Even so, some subjects report discomfort, in this case in the hip joint of the bent leg (20, 21). Other than the altered leg positions and the fact that each leg is measured separately, the backsaver test is performed in a manner similar to that of the traditional sit-and-reach test.

Step 1: Place the sit-and-reach box against a wall or an object that will prevent it from slipping during the testing process.

Step 2: Have the subject perform a 10 min warm-up (see sample in table 3.1).

Step 3: Instruct the subject to remove his or her shoes, then measure his or her height and weight. Record these numbers on the individual data sheet. Also record the subject's name and sex, as well as the date, time, temperature, and barometric pressure.

Step 4: Instruct the subject to sit on the floor with the right leg extended so that the sole of the foot is placed against the sit-and-reach box at the index or heel line (23 cm or 9.1 in. mark). The subject's left leg should be bent at approximately 90° with the sole of the left foot placed flat on the floor (21) about 5 to 7.5 cm (2 to 3 in.) from the sit-and-reach box (2).

Step 5: Have the subject extend his or her arms with one hand on top of the other and with the palms facing down.

Step 6: For the test, the subject should bend forward while moving his or her hands along the measuring scale on top of the sit-and-reach box. This position should be held for 1 or 2 s. The subject may have to move the bent leg to the side while extending forward to allow movement through the range of motion. If the extended knee bends or the fingertips become uneven, the test does not count, and it should be repeated. Record the resulting measurement on the individual sit-and-reach data sheet.

Step 7: Perform steps 4 through 6 three more times with the right leg.

Step 8: Compare the fourth trial's results with the percentile ranks or normative data presented in table 3.2 and note the results in the appropriate spot on the individual data sheet.

Step 9: Have the subject change position so that the left leg is extended and the sole of the left foot is placed against the sit-and-reach box at the index or heel line (23 cm or 9.1 in. mark). The right leg should be bent to approximately 90° with the sole of the right foot placed flat on the floor (21) about 5 to 7.5 cm (2 to 3 in.) from the sit-and-reach box (2). See figure 3.9a.

Step 10: Have the subject extend his or her arms with one hand on top of the other and with the palms facing down. See figure 3.9b.

Step 11: Instruct the subject to bend forward while moving his or her hands along the measuring scale on top of the sit-and-reach box. This position should be held for 1 or 2 s. The subject may have to move the bent leg to the side while extending forward to allow movement through the range of motion. If the subject's extended knee bends or the fingertips become uneven, the test does not count, and it should be repeated. Record the resulting measurement on the individual sit-and-reach data sheet.

Step 12: Perform steps 9 through 11 three more times with the right leg.

Step 13: Compare the fourth trial's results with the percentile ranks or normative data presented in table 3.2 and note the results in the appropriate spot on the individual data sheet.

QUESTION SET 3.3

1. How does the Canadian sit-and-reach test differ from the traditional version? How does it differ from the various other tests that could be used when evaluating lower body flexibility?

2. How do the percentile ranks for the Canadian sit-and-reach test differ from those for the backsaver variant?

3. Compare the results of your sit-and-reach test with the normative data. What do these results indicate?

4. Compare the class data as a whole for the sit-and-reach test with the normative data. What do these results indicate?

5. Create a bar graph of the means and standard deviations for the five most flexible men and the five least flexible men.

6. Create a bar graph of the means and standard deviations for the five most flexible women and the five least flexible women.

7. Fill in the table in the individual datasheet comparing the methods used in this lab. Indicate the best application for each test and describe the pros and cons of each test.

Find the case studies for this laboratory online at www.HumanKinetics.com/ LaboratoryManualForExercisePhysiology.

Name or ID number: _____ Date: _____

Tester: _____ Time: _____

Sex: M / F (circle one)   Age: _____ y   Height: _____ in. _____ cm

Temperature: _____ °F _____ °C   Weight: _____ lb _____ kg

Barometric pressure: _____ mmHg   Relative humidity: _____ %

Sit-and-reach test results		Trial 1	Trial 2	Trial 3	Trial 4	Best	4th trial	Index/heel line (cm)
Canadian								26
Traditional								23
Backsaver	Right leg							23
	Left leg							23

Individual results ranking		Results (cm)	Ranking/category/percentile
Canadian			Percentile =
Traditional			Percentile =
Backsaver	Right leg		Percentile =
	Left leg		Percentile =

SHOULDER FLEXIBILITY TEST COMPARISONS

EQUIPMENT

- Mat
- 18 in. (46 cm) ruler
- Yardsticks or metersticks
- Individual and group data sheets

 Find the group data sheets for this laboratory online at www.HumanKinetics.com/ LaboratoryManualForExercisePhysiology.

WARM-UP

Have the subject perform a structured warm-up beforehand to increase the test's reliability and validity. The warm-up session should include a 5 min general warm-up and a 5 min dynamic stretching warm-up (37).

The 5 min general warm-up can involve activities such as jogging, cycling, skipping, jumping rope, or doing calisthenics (26). The chosen activity should be undertaken with the specific goal of increasing heart rate, blood flow, muscle temperature, respiration rate, and perspiration while decreasing viscosity of joint fluids (12, 26). Taken collectively, these responses prepare the subject for the more specific warm-up activities that are performed next.

After completing the general warm-up, the subject should perform dynamic warm-up activities focused on movements that work through the range of motions required for the testing activity (26). Dynamic warm-up activities appropriate for the shoulder flexibility test might include arm swings, arm circles, shoulder rotations, or other activities (25). After completing the warm-up period, the subject can begin the testing process.

SHOULDER ELEVATION TEST

This is a commonly used test of shoulder and chest flexibility (1).

Step 1: Set up the testing area. Place a mat on the floor in an area with enough room for testing to be performed.

Step 2: Instruct the subject to perform a 10 min warm-up that includes dynamic activities such as arm swings and shoulder rotations.

Step 3: Have the subject grasp a meterstick in front of his or her body with both hands using a pronated grip (knuckles facing forward). Ensure that the subject keeps his or her arms relaxed in this position.

Step 4: Determine arm length by measuring the distance from the acromion process to the top of the meterstick being held by the subject (figure 3.12a). Measure to the nearest cm. Record the result on the individual data sheet for laboratory activity 3.4.

Step 5: Have the subject assume a prone position (lying facedown) on the mat so that his or her chin touches the floor and his or her arms are extended overhead while holding the meter stick with the same grip used in step 2 (see figure 3.12b).

Step 6: Have the subject slowly raise the meterstick as high as possible while maintaining chin contact with the floor and keeping the elbows extended. Once the subject achieves the highest possible position, encourage him or her to hold this position for 1 or 2 s.

Step 7: Measure the distance between the floor and the bottom of the meterstick when the subject achieves the highest position possible (figure 3.12b). Record this measure in the individual data sheet.

Step 8: Perform steps 4 through twice more.

Step 9: After completing three testing trials, calculate the shoulder elevation score with the following formula (1, 16) and record it on the individual data sheet:

$$\text{Shoulder elevation score} = \frac{\text{Shoulder elevation (in.)} \times 100}{\text{arm length (in.)}}$$

Step 10: Compare the shoulder elevation score with the normative data and ratings presented in table 3.6 and record the result on the individual data sheet.

 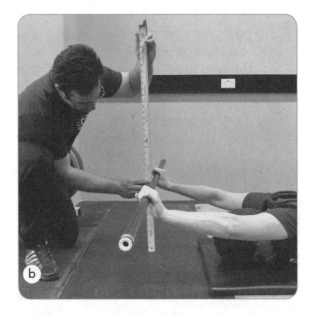

Figure 3.12 Shoulder extension test: *(a)* When measuring the subject's arm length, make sure that he or she is holding a meterstick with a pronated grip and that his or her arms are completely relaxed at his or her sides. *(b)* Have the subject lie facedown on the floor with his or her chin resting on the floor and arms extended overhead while holding a meterstick. Have the subject raise his or her arms as high as possible while keeping them extended. Make sure that the subject's chin remains in contact with the floor.

Table 3.6 Percentile Ranks and Norms for Shoulder Elevation

	Sex		Male		Female	
	Classification	Percentile	in.	cm	in.	cm
Percentile rank	Well above average	90	106–123	269–312	105–123	267–312
	Above average	70	88–105	224–267	86–104	218–264
	Average	50	70–87	178–221	68–85	173–216
	Below average	30	53–69	135–175	50–67	127–170
	Well below average	10	35–52	89–132	31–49	79–124

Adapted from Acevedo (1). Adapted from B.L. Johnson and J.K. Nelson, 1986, *Practical measurement for evaluation in physical education*, 4th ed. (Minneapolis, MN: Lea & Febiger).

BACK-SCRATCH TEST

The back-scratch test is a simple shoulder flexibility test that requires only minimal equipment.

Step 1: Have the subject perform a 10 min warm-up that includes dynamic activities such as arm swings and shoulder rotations.

Step 2: After the warm-up, have the subject raise his or her right arm, bent at the elbow, and reach across his or her back as far as possible. At the same time, have him or her place the left arm down and behind the back. The subject should attempt to cross the fingers of the left and right hands behind the back (figure 3.13a).

Step 3: Measure the distance of the finger overlap or gap to the nearest 0.5 in. (1.3 cm) (figure 3.13b). An overlap is considered to produce a positive score, whereas a gap between the hands is considered a negative score. If the fingers merely touch, the score is considered to be zero. Record the test result in the appropriate location on the individual data sheet.

Step 4: Perform steps 2 and 3 twice more.

Step 5: Have the subject raise his or her left arm, bent at the elbow, and reach across his or her back as far as possible. At the same time, have him or her place the right arm down and behind the back. The subject

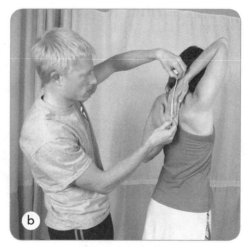

Figure 3.13 Back-scratch test: *(a)* arm position and *(b)* measurement.

Laboratory Activity 3.4

should attempt to cross the fingers of the left and right hands behind the back.

Step 6: Measure the distance of the finger overlap or gap to the nearest 0.5 in. (1.3 cm). An overlap is considered to produce a positive score, whereas a gap between the hands is considered a negative score. If the fingers merely touch, the score is considered to be zero. Record the test result in the appropriate location on the individual data sheet.

Step 7: Perform steps 5 and 6 twice more.

Step 8: Calculate the average for the raised right arm and raised left arm tests and record the results on the individual data sheet. Compare the average results for each arm with the normative data presented in table 3.7.

Table 3.7   Back-Scratch Test Normative Data for College-Age Students

Sex	Male		Female	
Classification	in.	cm	in.	cm
Excellent	≥5	≥12.7	≥5	≥12.7
Above average	2.0–4.8	5.1–12.2	2.0–4.75	5.1–12.1
Average	0.0–1.8	0.0–4.6	0.0–1.75	0.0–4.6
Below average	−1.0−−0.25	−2.5−−0.6	−1.0−−0.25	−2.5−−0.6
Poor	<−1	<−2.5	<−1	<−2.5

Adapted from Nieman (35).

QUESTION SET 3.4

1. What are the differences between the back-scratch and shoulder elevation tests?

2. Compare the results of your shoulder elevation and back-scratch tests with the normative data. What do these results indicate?

3. Compare the class data as a whole for the shoulder elevation and back-scratch tests with the normative data. What do these results indicate?

4. Create bar graphs of the means and standard deviations for the back-scratch and shoulder elevation tests—one for each sex and one with both sexes combined.

5. Use a paired *t*-test to compare the results of the shoulder elevation and back-scratch tests for the five most flexible men with the results for the five most flexible women. How do these findings compare with what you would expect to find?

Find the case studies for this laboratory online at www.HumanKinetics.com/ LaboratoryManualForExercisePhysiology.

Laboratory Activity 3.4 Individual Data Sheet

Name or ID number: _____ Date: _____

Tester: _____ Time: _____

Sex: M / F (circle one)   Age: _____ y   Height: _____ in. _____ cm

Temperature: _____ °F _____ °C   Weight: _____ lb _____ kg

Barometric pressure: _____ mmHg   Relative humidity: _____ %

Shoulder Extension Test

Arm length test: _____ in. _____ cm

Shoulder elevation		[Shoulder elevation (cm) × 100] / arm length (cm)]	Calculation method	Shoulder elevation score (cm)	Flexibility rating
Trials	1		(_____ × 100) /_____		
	2		(_____ × 100) /_____		
	3		(_____ × 100) /_____		
	Best		(_____ × 100) /_____		

Overall percentile rank: _____

Back-Scratch Test

	Trial 1 (cm)	Trial 2 (cm)	Trial 3 (cm)	Average (cm)
Right arm				
Left arm				
Average				

Ranking: _____

Laboratory Activity 3.4

Blood Pressure Measurements

Objectives

1. Understand the concept of systemic blood pressure fluctuations during the cardiac cycle.
2. Define the terms *cardiac cycle*, *systolic blood pressure*, and *diastolic blood pressure*.
3. Develop the skill of taking blood pressures during rest and exercise.
4. Examine blood pressure responses to changes in body position and aerobic and isometric exercise.
5. Understand risk stratifications of stages of hypertension and contraindications to exercise testing.

Definitions

blood pressure (BP)—Pressure in the peripheral circulation.

cardiac cycle—Mechanical and electrical events occurring during one heartbeat.

cardiac output (CO or $\dot{Q}$)—Amount of blood produced by the heart every minute ($L \cdot min^{-1}$).

diastole—Phase of the cardiac cycle during ventricular relaxation.

diastolic blood pressure—Pressure (mmHg) in the large arteries during diastole.

heart rate (HR)—Number of heartbeats per minute (beats $\cdot min^{-1}$).

laminar blood flow—Type of blood flow not resulting in audible sound; defined by streamline flow or flow in parallel layers without turbulence.

orthostatic hypotension—Reduction in blood pressure as the result of body position changes.

stroke volume (SV)—Amount of blood per heartbeat ($ml \cdot beat^{-1}$).

systole—Phase of cardiac cycle during ventricular contraction.

systolic blood pressure—Pressure (mmHg) in the large arteries during systole.

turbulent blood flow—Type of blood flow resulting in audible sound; defined by chaotic flow due to variations of pressure and velocity.

vascular resistance—Resistance to blood flow, determined by Poiseuille's law, according to which resistance equals the length of the vessel multiplied by the blood viscosity divided by vessel radius to the fourth power.

vasovagal syncope—Fainting due to low blood pressure perfusion of the brain.

Blood pressure (BP) is one of the basic measures in human physiology, and it provides a good measure of the work of the heart. It is considered so important that every trip to the doctor's office includes a BP measurement! BP can be defined as the pressure exerted against the walls of the various vessels of the circulatory system by the heart as it pumps blood to the body. BP depends on two basic factors: the volume of blood delivered to a vessel ($\dot{Q}$) and the amount of resistance exerted against the blood being delivered (vascular resistance or total peripheral resistance). **Cardiac output (CO or $\dot{Q}$)** is determined by **heart rate (HR)** and **stroke volume (SV)**: CO = HR × SV. **Vascular resistance** during exercise is determined largely by the diameter of the vessels; a small decrease in the radius of the vessels results in large increases in vascular resistance.

A change in either vascular resistance or cardiac output causes a change in BP.

Blood pressure is usually recorded as systolic pressure over diastolic pressure. Arterial **systolic blood pressure** is the maximum pressure attained during peak ventricular ejection or **systole** during the **cardiac cycle**. **Diastolic blood pressure** is the minimum arterial pressure within the cardiac cycle during ventricular relaxation or **diastole**. Average normal resting blood pressure is considered to be <120 mmHg over <80 mmHg. Tables 4.1 and 4.2 present resting blood pressure norms and their classifications.

BP depends on many factors, including body position, hydration, sex, muscle actions, stress, diet, and disease (e.g., atherosclerosis). This lab will demonstrate the response of BP to changes in body position, which are often called *orthostatic changes*.

Table 4.1 Percentile Norms for Blood Pressure in Active Men and Women

	Resting blood pressure (SBP/DBP; mmHg)					
	Women			Men		
Rank (%)	20-49 years	50-59 years	60+ years	20-49 years	50-59 years	60+ years
Very low (>80%)	<104 / <70	<110 / <70	<120 / <75	<111 / <75	<116 / <78	<120 / <76
Low (60-80%)	106-112 / 70-75	110-120 / 70-79	120-128 / 75-80	112-120 / 74-80	116-122 / 78-80	120-130 / 76-80
Average (40-60%)	110-120 / 72-80	120-130 / 79-82	128-136 / 80	120-126 / 80-84	122-130 / 80-86	130-140 / 80-84
High (20-40%)	118-130 / 78-82	130-140 / 82-90	136 -142 / 80-88	127-138 / 84-90	130-140 / 86-90	140-150 / 84-90
Very high (<20%)	<130 / >82	>140 / >90	>142 / >88	>138 / >90	>140 / >90	>150 / >90

Adapted from M.L. Pollock, J.H. Wilmore, and S.M. Fox, 1978, *Health and fitness through physical activity* (New York, NY: John Wiley and Sons).

Table 4.2 Classification of Blood Pressure for Adults 18 Years or Older*

Category	Systolic BP (mmHg)**	Diastolic BP (mmHg)
Normal	<120	<80
Prehypertension	120–139	80–89
Stage 1 hypertension	140–159	90–99
Stage 2 hypertension	≥160	≥100

*This classification applies to individuals not taking antihypertensive medication and not acutely ill. It is based on taking the average of two or more readings on two or more occasions.

**When systolic and diastolic pressures fall into different categories, use the higher category for classification.

Adapted from NHLBI, 2003, *The seventh report of the joint national committee on detection, evaluation, and treatment of high blood pressure.* NIH Publication No. 04-5230.

When a person is supine, the heart can work less hard to deliver blood to the periphery, especially the brain, since it does not have to work against gravitational forces; in addition, HR is typically lower due to the benefit of enhanced venous return and thus increased SV. When a person changes position from supine to standing, venous return is reduced, and thus blood pressure drops. This common occurrence is often referred to as **orthostatic hypotension**. Many people are familiar with the dizziness or light-headedness one can experience upon standing up; if severe, this effect can result in fainting, or **vasovagal syncope**, which is more common in people with low BP. Baroreceptors in the carotid arch sense this lowering of BP and stimulate the cardiovascular brain centers to adjust BP by increasing venous return and strength of heart contraction. As a result of these processes, BP is often lower (by 10–20 mmHg) immediately after standing than it is while supine, whereas steady-state standing BP is higher than supine BP.

Again, many factors can influence BP besides body position and exercise. Even the mere act of taking an individual's BP may cause it to rise due to a feeling of stress experienced by the subject or patient. This occurrence, often referred to as "white coat syndrome," can be exacerbated in a hospital or other medical environment. Thus stress can artificially elevate BP measurements. As a

result, providing a relaxing environment—along with other techniques, such as ensuring standardized body position, allowing a few minutes of rest prior to measurement, and having the subject avoid crossed legs or undue muscular actions during measurement—can be essential for accurate measurement of BP.

Blood Pressure Responses to Exercise

In the transition from rest to exercise, systolic blood pressure initially rises rapidly, then levels off once steady state is attained (1, 3). Typical systolic pressures during submaximal exercise range from 140 to 160 mmHg. Graded dynamic exercise normally produces a progressive increase in systolic pressure, and values can reach as high as 250 mmHg during maximal exercise.

Figure 4.1 shows systolic and diastolic blood pressure responses to exercise of increasing intensity. The rise in systolic pressure reflects the increased force of contraction of the ventricles in order to increase cardiac output from sympathetic stimulation. Diastolic pressure may show either no change or a very slight increase or decrease during graded exercise due to a redistribution of blood flow to the capillary beds in the large exercising muscle groups (1, 3).

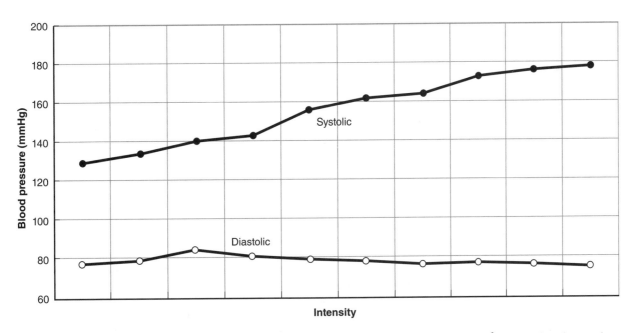

Figure 4.1 Systolic and diastolic pressure change in response to exercise of increasing intensity.

Following heavy exercise, systolic blood pressure drops rapidly. This rapid drop (hypotension) results from pooling of blood in vessels that were dilated during the heavy exercise. The pooling reduces venous return and thus decreases cardiac output. If the reduction in CO is severe enough, the subject or athlete may experience light-headedness due to insufficient blood flow to the brain. For this reason, a cool-down period is recommended following intense exercise in order to maintain HR and SV and allow the cardiovascular system to readjust gradually.

BP responds differently during isometric exercise (or slow concentric actions). Sustained muscle actions that occlude blood flow can increase resistance tremendously and can result in extremely high systolic and diastolic BP. Thus, even though muscle is contracting and in need of blood flow, continuous muscular contraction surrounding the blood vessels can significantly increase resistance and decrease flow. In an effort to overcome this occlusion, the heart increases pressure.

Accurate Blood Pressure Checks

Measuring blood pressure is a basic skill in exercise physiology since it provides a good indication of the work of the heart; as a result, this skill is critical for properly prescribing exercise for at-risk individuals. Indirect measurement of blood pressure is done with a stethoscope and a sphygmomanometer. The indirect measurement of blood pressure monitors the sounds of blood flow in the brachial artery (Korotkoff phases) that are audible through the stethoscope (for more information about the Korotkoff phases, see the accompanying highlight box). **Laminar blood flow** makes little or no sound in the arteries, whereas **turbulent blood flow** due to occlusion from the blood pressure cuff can make a variety of sounds. Although blood pressure can differ slightly in the left and right arms, the difference is minimal; for the purposes of skill acquisition in this lab, students should use the most convenient arm. Note, however, that in clinical assessment both arms should be measured, and the highest recording should be used as the BP measurement (2). When doing serial measurements, the same arm should be used. The arm should be as free of clothing as possible, since clothing can muffle sounds during auscultation; however, simply rolling up or bunching a sleeve can, if done too tightly, mimic a BP cuff and thus occlude blood flow. Different-size cuffs are available for use with different-size arms; specifically, cuffs are available for children, adults, and large adults. Index lines on the cuff should indicate whether the cuff is appropriate for a given subject or patient. The

Korotkoff Phases

When pressure in the cuff is greater than systolic blood pressure, blood flow is not allowed in the distal brachial artery. As pressure in the cuff is slowly released, blood is allowed to flow through the widening aperture. The amount of turbulent flow creates differing sounds, which are described by the Korotkoff phases:

- Phase I—The systolic pressure is indicated by the first tapping sound that becomes audible—read the dial or mercury column to the nearest mmHg.
- Phase II—A murmur or swishing sound is heard.
- Phase III—Sound increases in intensity.
- Phase IV—A distinct, abrupt muffling of sound is noted, though it may *not* be heard at rest. This is diastolic pressure during exercise.
- Phase V—At rest, the diastolic pressure is marked by the disappearance of the sound; record this pressure to the nearest mmHg. During exercise, you may hear sounds all the way down to zero mmHg; in this case, record Phase IV for diastolic pressure.

You can listen to the Korotkoff sounds at www.thinklabsmedical.com/stethoscope_community/sound_library/other-sounds/content/blood-pressure-korotkoff-sounds-2.

bladder should cover about 80% of the arm to enable it to occlude blood flow effectively when inflated.

BP measurement can be affected by the testing situation and by preparation. Ideally, subjects refrain from using stimulants such as caffeine prior to testing. They should wear loose clothing, be normally hydrated, and avoid strenuous exercise for several hours prior to the test. While being tested, they should not cross their legs, and they should avoid any form of isometric muscle action, such as pressing down on their legs, dangling their feet off the ground, or sitting erect with the back unsupported. In addition, the environment should be free of stimuli such as loud music or unnecessary activity (2).

References

1. Brooks GA, Fahey TD, and Baldwin KM. *Exercise Physiology: Human Bioenergetics and Its Applications.* 4th ed. New York: McGraw-Hill, 2005.

2. Kenney WL, Wilmore JH, and Costill DL. *Physiology of Sport and Exercise.* 5th ed. Champaign, IL: Human Kinetics, 2012.

3. Perloff D, Grim C, Flack J, Frohlich ED, Hill M, McDonald M, and Morgenstern BZ. Human Blood Pressure Determination by Sphygmomanometry. *Circulation* 88: 2460–2470, 1993.

EFFECTS OF BODY POSITION ON BP

EQUIPMENT

- Sphygmomanometer
- Stethoscope
- Stopwatch
- Individual and group data sheets

Find the group data sheets for this laboratory online at www.HumanKinetics.com/ LaboratoryManualForExercisePhysiology.

RESTING BLOOD PRESSURE

Step 1: Have the subject sit in a chair for 2 min to attain a steady state.

Step 2: Place the BP cuff on the subject's arm (see figure 4.2b).

Step 3: Palpate for radial pulse and measure HR.

Step 4: The subject should be comfortably seated with the arm slightly flexed, the palm facing up, and the forearm supported at the heart level by a flat surface.

Step 5: Place the cuff with the lower margin about 1 in. (2.5 cm) above the antecubital space. Ensure that the bladder covers the brachial artery. During the measurement, the arm should be supported at the level of the heart (see figure 4.2b).

Step 6: Palpate to find the brachial artery in the antecubital space (see figure 4.2a). Place the stethoscope bell lightly over the brachial artery. Too much pressure on the bell can affect blood flow in the artery and cause sounds independent of the cuff pressure. Too little pressure can reduce your ability to hear Korotkoff phases. Be sure the bell is flat against the skin.

Step 7: With the stethoscope held in place, close the air release screw (clockwise) and inflate the blood pressure cuff to 200 mmHg or more by pumping the air bulb. The pressure closes off the flow to the brachial artery, thus causing the pulse sound to stop; at rest, you may need only to inflate to 180 mmHg to be above systolic blood pressure (see figure 4.2c).

Step 8: Release pressure slowly and smoothly at a rate of 2 or 3 mmHg·s⁻¹ by releasing the air release screw (counterclockwise).

Step 9: As pressure is released, the Korotkoff sounds become audible. The appearance of a faint tapping sound marks the systolic blood pressure; at this point, note the mmHg reading. This sound will increase in intensity before fading.

Step 10: When the subject is at rest, such as sitting or standing, the diastolic blood pressure is marked the disappearance of sound in the brachial artery, which indicates laminar blood flow. During exercise, the diastolic pressure is marked by an abrupt muffling of the pulse sound. Note the mmHg at the appropriate Korotkoff phase.

Step 11: Once systolic and diastolic measurements are obtained, release all pressure out of the cuff.

Step 12: Record your measurements on the data sheet for laboratory activity 4.1 in the following format: systolic blood pressure over diastolic blood pressure (SBP/DBP).

Step 13: Repeat steps 1 through 12 with the same subject for trial 2 data collection. Each lab member should act as both the subject and the BP data collector. Each member should fill out the individual data sheet for his or her own BP measurements.

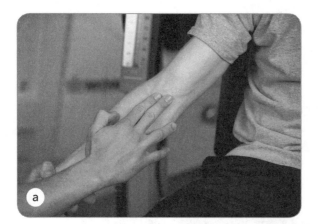
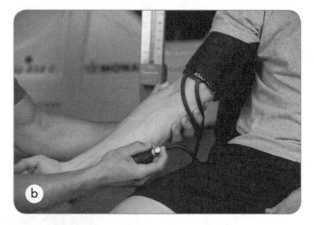

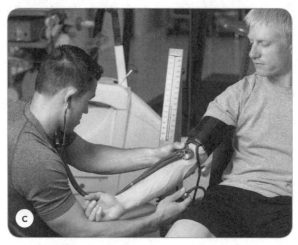

Figure 4.2 Measuring resting BP: (a) palpation of the brachial artery, (b) placement of the blood pressure cuff, and (c) measuring resting BP.

Laboratory Activity 4.1

SUPINE AND STANDING BLOOD PRESSURE

Step 1: Have the subject quietly maintain a supine position for 2 min with the BP cuff already appropriately attached to his or her arm (see figure 4.2b).

Step 2: After 2 min, measure and record the palpated HR and BP as described in the procedure for measuring resting blood pressure. While taking the BP, have the subject keep his or her arm still, since lifting it would affect the measurement. Have the subject let the cuffed arm rest comfortably next to him or her with the palm up in order to allow access to the antecubital space.

Step 3: Start pumping to 200 mmHg just prior to having the subject stand up.

Step 4: Have the subject stand up and immediately record HR and BP.

Step 5: Have the subject remain standing for at least 2 min and then repeat the HR and BP measurements.

Step 6: Repeat steps 1 through 5 to collect data for trial 2. Record the data in the appropriate location on the data sheet.

QUESTION SET 4.1

1. Prepare a data table reporting the subject's name, height, and weight, as well as his or her supine, sitting, and standing HR and BP collected during the laboratory session.

2. What changes did you observe in HR and BP with body position? What accounts for these changes?

3. What is happening physiologically when a person stands up quickly and gets light-headed?

 Find the case studies for this laboratory online at www.HumanKinetics.com/ LaboratoryManualForExercisePhysiology.

Laboratory Activity 4.1 Individual Data Sheet

Name or ID number: _____ Date: _____

Tester: _____ Time: _____

Sex: M / F (circle one) Age: _____ y Height: _____ in. _____ cm

Temperature: _____ °F _____ °C Weight: _____ lb _____ kg

Barometric pressure: _____ mmHg Relative humidity: _____ %

Trial	HR (beats·min⁻¹)	Blood pressure (mmHg)					
		Supine position		Initial standing		>2 min standing	
		Systolic	Diastolic	Systolic	Diastolic	Systolic	Diastolic
1							
2							
Mean							
SD							

EFFECTS OF DYNAMIC EXERCISE ON BP

EQUIPMENT

- Sphygmomanometer
- Cycle ergometer
- Stethoscope
- Stopwatch
- Individual data sheet

TECHNIQUES FOR HR AND BP MEASUREMENT DURING EXERCISE

A moving subject complicates the measurement of HR and BP because of the added background noise. Here are some tricks to help you perform HR palpation and BP auscultation during exercise on a cycle ergometer (see figures 4.3a and b):

- Avoid looking at the pedals while taking HR—the cadence can throw off your counting.
- While counting the HR for 15 s, you might close your eyes to concentrate.
- Control the noise in the room as much as you can—keep your labmates' talking to a minimum!
- While taking BP, be conscious of the stethoscope and sphygmomanometer tubing. Any bumping or rubbing against knees, arms, or the ergometer will add noise to your auscultation.
- Control and support the subject's arm to avoid motion and undue muscle actions.
- Instruct the subject to refrain from gripping the handlebars tightly, since engaging in this isometric action can increase BP.
- Initial cuff pressure needs to be increased as intensity increases in order to ensure that cuff pressure exceeds systolic blood pressure. At low levels of exercise, 200 mmHg is sufficient, but at higher intensities you may have to go to 250 mmHg or more before releasing pressure.
- Since the subject will have a higher HR during exercise, you can afford to release air pressure faster (at a rate of 5 or 6 mmHg $\cdot$ s^{-1}).
- Continue monitoring the subject's adherence to the prescribed cadence and be aware of the subject's subjective symptoms (e.g., pallor, dyspnea).

Because students are generally new to taking exercise HR and BP, do not worry about sticking to the 3 min stage time at this point. If you miss a HR or BP, just try again. It is more important to practice the skill.

SUBMAXIMAL EXERCISE BP

Step 1: Record the subject's characteristics (height, weight) for the lab 4.2 data sheet.

Step 2: Fit the subject on the cycle ergometer comfortably with a slight knee angle (5° to 15°) when the leg is fully extended (see figure 4.3a).

Step 3: Prior to the test, read the subject the rating of perceived exertion (RPE) instructions (see lab 9 for a full discussion of RPE):

> *During the test, we want you to pay close attention to how hard you feel the exercise work rate is. This feeling should reflect your total amount of exertion and fatigue, combining all sensations and feelings of physical stress, effort, and fatigue. Don't concern yourself with any one factor, such as leg pain, shortness of breath, or exercise intensity, but try to concentrate on your total, inner feeling of exertion. Try not to underestimate or overestimate your feelings of exertion; be as accurate as you can.*

Step 4: Allow the subject to ask questions prior to being connected to the BP equipment.

Step 5: Have the subject begin pedaling at 70 revolutions per minute (rev·min⁻¹). If necessary, use a metronome to maintain constant cadence, which can affect power on the bike and thus intensity (BP is a function of exercise intensity). Add 0.5 kg of resistance to the ergometer for stage 1.

Step 6: Allow 3 min of pedaling at 70 rev·min⁻¹ to attain steady state, then measure HR, RPE, and BP. (Use the steps in lab 4.1 for BP measurement.)

Figure 4.3 Measuring BP on cycle ergometer: *(a)* proper positioning of subject with slight knee angle (5-15%) with the leg fully extended and *(b)* technique for measuring BP.

Laboratory Activity 4.2

Step 7: Increase the ergometer resistance to 1 kg (2.2 lb) for stage 2. Continue to have the subject pedal at 70 rev · min⁻¹.

Step 8: Allow 3 min to attain steady state, then measure HR, RPE, and BP.

Step 9: Increase the ergometer resistance to 1.5 kg (3.3 lb) for stage 3. Continue to have the subject pedal at 70 rev · min⁻¹.

Step 10: Allow 3 min to attain steady state, then measure HR, RPE, and BP.

Step 11: Decrease the ergometer resistance to 0.5 kg (1.1 lb) in stage 4 for recovery. Continue to have the subject pedal at 70 rev · min⁻¹.

Step 12: Allow 3 min to attain steady state, then measure HR, RPE, and BP.

Step 13: Perform steps 1 through 12 on a second test subject and record the data on the individual data sheet.

QUESTION SET 4.2

1. Prepare a data table reporting the subject's name, height, and weight, as well as the HR and BP data collected during the laboratory session.

2. How do systolic blood pressure and diastolic blood pressure respond to exercise of increasing intensity?

3. What are the relative contraindications relating to BP response during an exercise test?

Find the case studies for this laboratory online at www.HumanKinetics.com/ LaboratoryManualForExercisePhysiology.

Laboratory 4.2 Individual Data Sheet

Date: _____ Time: _____

Tester: _____

Temperature: _____ °F _____ °C Barometric pressure: _____ mmHg

Relative humidity: _____ %

Subject A

Name or ID number: _____

Sex: M / F (circle one)   Age: _____ y Height: _____ in. _____ cm

Weight: _____ lb _____ kg

Age-predicted HRmax: _____ beats · min^{-1}

Resting BP: _____ / _____ mmHg

Resting HR: _____ beats · min^{-1}

85% age-predicted HRmax: _____ beats · min^{-1}

	Time (min)	Workload (kg)	HR (beats · min^{-1})	BP (mmHg)	RPE
Stage 1		0.5			
Stage 2	6	1.0			
Stage 3	9	1.5			
Stage 4	12	0.5			

Subject B

Name or ID number: _____

Sex: M / F (circle one)   Age: _____ y   Height: _____ in. _____ cm

Weight: _____ lb _____ kg

Age-predicted HRmax: _____ beats·min^{-1}

Resting BP: _____ / _____ mmHg

Resting HR: _____ beats·min^{-1}

85% age-predicted HRmax:_____ beats·min^{-1}

	Time (min)	Workload (kg)	HR (beats·min^{-1})	BP (mmHg)	RPE
Stage 1	3	0.5			
Stage 2	6	1.0			
Stage 3	9	1.5			
Stage 4	12	0.5			

EFFECTS OF ISOMETRIC MUSCLE ON BP

EQUIPMENT

- Leg extension machine (or, if unavailable, a wall)
- Hand dynamometer
- Sphygmomanometer
- Stethoscope
- Stopwatch
- Individual and group data sheets

UPPER BODY ISOMETRIC EXERCISE BP

Step 1: Record the subject's characteristics (height, weight) for the data table. Two subjects will complete this lab activity to fill out the individual data sheet.

Step 2: Place the BP cuff on the subject's arm.

Step 3: Have the subject sit quietly for 2 min.

Step 4: After 2 min, measure and record the resting HR (palpated) and BP.

Step 5: Instruct the subject to perform a maximum voluntary contraction (MVC) on the handgrip dynamometer with the dominant hand (see figure 4.4; also see lab 11 on page 256 for handgrip instructions).

Step 6: Calculate and record 30% of MVC in kg.

Step 7: After a minute or two of rest, have the subject hold the handgrip at 30% of MVC for at least 1 min.

Step 8: At the end of the minute of isometric action, measure HR, RPE, and BP while the subject is still holding 30% MVC. Measure BP in the nonexercising, nondominant arm. It is important that the nonexercising arm is relaxed during the BP measurement.

Step 9: Record the data in the appropriate locations on the data sheet.

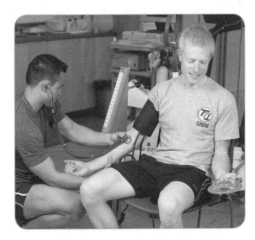

Figure 4.4 BP measurement during handgrip exercise.

LOWER BODY ISOMETRIC EXERCISE BP

Step 1: The second part of this lab involves lower body isometric exercise, which can be done with a leg extension machine or against a wall (see figures 4.5 and 4.6).

Step 2: If using the leg extension machine, have the subject select a weight on the machine that is approximately 50% of 1RM.

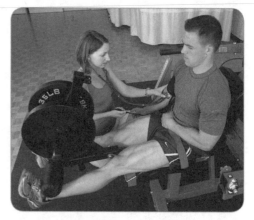

Figure 4.5 BP measurement during isometric leg extension.

Step 3: Have the subject hold the weight at full extension for as long as he or she can or for 3 min, whichever comes first. When the subject is close to muscle failure, have him or her indicate to you that it is time to take HR and BP.

Step 4: While the subject continues to hold the 50% of 1RM, measure HR, RPE, and BP. Be sure that the subject relaxes the arm while you take the BP measurement.

Step 5: Record the data in the appropriate locations on the data sheet.

If a leg extension machine is not available, the subject can perform a wall sit, which is sometimes referred to as "waiting for the bus" (see figure 4.6).

Step 1: Have the subject do a wall sit for as long as he or she can or for 3 min, whichever comes first. When the subject is close to muscle failure, have him or her indicate to you that it is time to take HR, RPE, and BP.

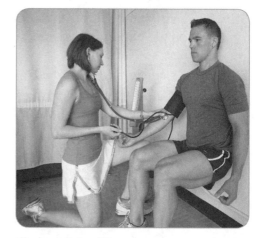

Figure 4.6 BP measurement during wall sit.

Step 2: While the subject continues to hold the wall sit, measure HR and BP. Be sure that the subject relaxes the arm while you take the BP measurement.

Step 3: Record the data in the appropriate locations on the data sheet.

QUESTION SET 4.3

1. Complete a data table reporting the subject's name, height, and weight, as well as the HR and BP data collected during the laboratory session.

2. How does isometric exercise affect systolic and diastolic blood pressure? Why?

3. How does the BP response during arm exercise compare with the response during leg exercise? What explains this?

4. What does this lab illustrate to you about exercise prescription in hypertensive clients?

5. What do peak blood pressures during isometric actions depend on?

Find the case studies for this laboratory online at www.HumanKinetics.com/ LaboratoryManualForExercisePhysiology.

Laboratory 4.3 Individual Data Sheet

Date: _____ Time: _____

Tester: _____

Temperature: _____°F _____°C Barometric pressure: _____mmHg

Relative humidity: _____ %

Subject A

Name or ID number: _____

Sex: M / F (circle one)   Age: _____ y Height: _____ in. _____ cm

Weight: _____ lb _____ kg

Handgrip MVC: _____ kg

30% handgrip MVC: _____ kg

Leg extension 50% 1RM:_____ kg

	Time (min)	Workload (kg)	HR (beats · min⁻¹)	BP (mmHg)	RPE
Resting		0			
Handgrip					
Leg extension					
Wall sit					

Subject B

Name or ID number: _____

Sex: M / F (circle one)   Age: _____ y   Height: _____ in. _____ cm

Weight: _____ lb _____ kg

Handgrip MVC: _____ kg

30% handgrip MVC: _____ kg

Leg extension 50% 1RM: _____ kg

	Time (min)	Workload (kg)	HR (beats · min⁻¹)	BP (mmHg)	RPE
Resting		0			
Handgrip					
Leg extension					
Wall sit					

LABORATORY 5

Resting Metabolic Rate Determinations

Objectives

1. Understand the indirect measurement of energy metabolism.
2. Understand the respiratory parameters indicating metabolic status and energy utilization ($\dot{V}O_2$, $\dot{V}CO_2$, RER).
3. Determine resting metabolic rate (RMR) by measuring oxygen uptake.
4. Compare the determination of RMR via oxygen uptake with the determination of RMR via estimation equations.

Definitions

adenosine triphosphate (ATP)—Molecule with energy harnessed in phosphate bonds used to perform mechanical or metabolic work.

aerobic metabolism—Production of ATP in the presence of oxygen utilizing oxidative phosphorylation.

anaerobic metabolism—Production of ATP in the absence of oxygen.

basal metabolic rate (BMR)—Minimal level of energy required to sustain the body's vital functions in the waking state; measured via indirect calorimetry in a supine position after 8 h of sleep and a 12 h fast with the subject having spent the night in the laboratory to avoid movement prior to measurement.

bioenergetics—The metabolic processes that result in stored energy being converted into a useable form of energy (ATP).

carbon dioxide production ($\dot{V}CO_2$)—Amount of CO_2 produced by the body in the conversion of carbon-containing food to usable energy.

indirect calorimetry—Measurement of energy expenditure from oxygen consumption and carbon dioxide production.

metabolic cart—Automated machine that measures expired CO_2 and O_2 in order to calculate $\dot{V}O_2$ and $\dot{V}CO_2$.

metabolic equivalent (MET)—Amount of energy expended at rest ($3.5 \ ml \cdot kg^{-1} \cdot min^{-1}$).

oxygen consumption ($\dot{V}O_2$)—Amount of oxygen taken up and utilized by an individual in a given time.

respiratory exchange ratio (RER or R)—Ratio of carbon dioxide production ($\dot{V}CO_2$) to oxygen consumption ($\dot{V}O_2$).

resting metabolic rate (RMR)—Energy expenditure at rest in a fasted, well-rested state in a supine position but, unlike BMR, not requiring the subject to stay overnight in the laboratory.

steady state—Plateau in metabolic parameters at a constant amount of power.

total energy expenditure (TEE)—The sum of the major energy consuming processes that contribute to an individual being in energy balance. The three main components of TEE are thermic effect of food, physical activity, and resting metabolic rate.

The conversion of food into the usable form of energy for our metabolism—**adenosine triphosphate (ATP)**—is called **bioenergetics**. ATP is a high-energy phosphate used for the energy required during muscular contractions.

$$\text{Food} + O_2 \rightarrow CO_2 + H_2O + ATP + \text{heat}$$

When this conversion is completed in the presence of oxygen (O_2), it is referred to as **aerobic metabolism**. Humans are also capable of **anaerobic metabolism**—energy production in the absence of oxygen, which is discussed further in laboratory 12. As exercise physiologists, we measure **oxygen consumption ($\dot{V}O_2$)** and **carbon dioxide production ($\dot{V}CO_2$)** by using **indirect calorimetry** to estimate energy expenditure. As a result, these measurements are often called *respiratory gas exchange*. This process can be conducted while subjects are at rest or during exercise. We must assume that these measurements include little or no anaerobic contribution to energy expenditure, which typically holds true during low-intensity and **steady state** exercise. *Steady state* indicates that parameters of metabolism are in equilibrium. To ensure steady state measurements during rest and exercise, physiologists typically require several minutes to be completed at a given intensity.

Measurement of $\dot{V}O_2$ and $\dot{V}CO_2$ requires special equipment such as a **metabolic cart** (see appendix D). It uses the knowledge that our atmosphere contains 20.93 percent oxygen, 0.03 percent carbon dioxide, and 79.04 percent nitrogen and that the concentrations of these gases change in an expired breath. The amount of oxygen decreases as we consume oxygen, whereas the amount of carbon dioxide increases and nitrogen remains unchanged in our expired breath; for a more detailed explanation of these calculations and of the Haldane transformation see Kenney (8) and appendix C. Thus the consumption of oxygen and the production of carbon dioxide become functions of the fractions of these gases in the air and the expired breath, as well as the volume of air per breath. Therefore, metabolic carts measure only three critical components—the percentages of oxygen and carbon dioxide in the expired breath and the volume of that breath. As we know from physics, the volume of a gas is affected by both temperature and pressure, which means that the local barometric pressure and ambient tempera-

ture are important for these measurements. Many metabolic carts can make the measurement of expired gases, ventilation, and temperature and pressure for every breath, whereas others use a mixing chamber to average the numbers over a discrete period of time (about 15 s). Part of this lab requires knowledge of how to use your institution's metabolic equipment (see appendix D), and your lab instructor will guide you through its use.

Aerobic Metabolism and Respiratory Exchange Ratio

One important measurement that can be accomplished with knowledge of $\dot{V}O_2$ and $\dot{V}CO_2$ is the type of fuel being used. On this point, the respiratory quotient (RQ) or respiratory exchange ratio (RER) provides information regarding the nutrient mixture being oxidized at any given time. RQ and RER differ only in the place in which they are measured—that is, at the cellular level and in the expired gases, respectively. For the purposes of this lab, we will refer to RER.

$$RQ = \frac{\dot{V}CO_2}{\dot{V}O_2} \text{ (at the cellular level)}$$

$$RER = \frac{\dot{V}CO_2}{\dot{V}O_2} \text{ (measured in expired gases)}$$

These ratios measure the quantity of CO_2 produced in relation to the quantity of O_2 consumed. Because of inherent differences in the chemical composition of carbohydrate, fat, and protein, each requires a different amount of oxygen to completely oxidize the molecules to the end products of metabolism—ATP, CO_2, and H_2O. Thus, because the caloric equivalent for oxygen differs somewhat depending on the nutrient oxidized, one must know the RER and the amount of oxygen consumed in order to precisely estimate the body's energy expenditure in $kcal \cdot min^{-1}$ (see table 5.1). The number of kcal produced for each L of oxygen consumed depends on the substrate being utilized or RER.

The RQ is 1.0 for carbohydrate, 0.70 for fat, and 0.82 for protein. The protein contribution to energy production has been measured, and unless the subject is in a starved state it is very low (generally less than 5%). To get a true understanding of the contribution of protein to energy production, urine must be collected, since urea is the metabolic end point for the amino groups of amino acids that have been oxidized. However, because of the complexity of analysis and the inconvenience of collecting urine, as well as the small contribution of protein to energy expenditure, RER is generally considered to be "nonprotein" RER. At any given time, a mixture of carbohydrate and fat is being oxidized by the mitochondria for energy needs, and the kcal produced per L of O_2 consumed vary accordingly. Table 5.1 reflects the percentage of fat and carbohydrate contributing to energy production at a given respiratory exchange ratio.

Normally, individuals consuming a diet of mixed carbohydrate and fat have an RER value of about 0.85 at rest, which means that for every L of oxygen consumed, approximately 4.86 kcal are produced. In measuring **resting metabolic rate (RMR)**, the researcher must have a measure of oxygen consumption and an RER value for each minute of the measurement period in order to determine the kcal produced per minute (see the accompanying highlight box for a sample calculation).

Many exercise physiology students never realize *why* carbohydrate has an RER of 1.0 and fat has an RER of 0.7. Here are the equations for the overall oxidation of carbohydrate and fat, respectively:

Predicting Energy Expenditure Using $\dot{V}O_2$ and RER

For this example, our test subject Greg, who weighs 200 lb (90.9 kg), has walked up a 3% grade at 3 miles per hour. Using indirect calorimetry, you have measured his $\dot{V}O_2$ at 15.2 ml·kg^{-1}·min^{-1} and his RER at 0.92.

$$200 \text{ lb} \times \frac{1 \text{ kg}}{2.2 \text{ lb}} = 90.9 \text{ kg}$$

$$15.2 \frac{\text{ml}}{\text{kg}\cdot\text{min}} \times 90.9 \text{ kg} \times \frac{1 \text{ L}}{1000 \text{ ml}} = 1.38 \frac{\text{L}}{\text{min}}$$

$$1.38 \frac{\text{L}}{\text{min}} \times 4.95 \frac{\text{kcal}}{\text{L}} = 6.83 \frac{\text{kcal}}{\text{min}}$$

If Greg walked at this intensity for 20 min, he would burn 137 kilocalories (kcal). Often, RER is not known, in which case these values are rounded to 5.0 kcal·L^{-1} of O_2 consumed. This is often the case when using equipment found in health clubs or clinics to predict energy expenditure for clients.

Carbohydrate

$$6 O_2 + C_6H_{12}O_6 \rightarrow 6 CO_2 + 6 H_2O + 38 \text{ ATP}$$

$$RER = \frac{\dot{V}CO_2}{\dot{V}O_2} = \frac{6 CO_2}{6 O_2} = 1.0$$

Fat

$$C_{16}H_{32}O_2 + 23 O_2 \rightarrow 16 CO_2 + 16 H_2O + 129 \text{ ATP}$$

$$RER = \frac{\dot{V}CO_2}{\dot{V}O_2} = \frac{16 CO_2}{23 O_2} = 0.7$$

Table 5.1 Caloric Equivalence of the Respiratory Exchange Ratio (RER) and % kcal From Carbohydrate and Fat

RER	Energy kcal·L^{-1}	% kcal Carbohydrate	Fat
0.71	4.69	0	100
0.75	4.74	16	84
0.80	4.80	33	67
0.85	4.86	51	49
0.90	4.92	68	32
0.95	4.99	84	16
1.00	5.05	100	0

Reprinted, by permission, from W.L. Kenney, J.H. Wilmore, and D.L. Costill, 2012, *Physiology of sport and exercise*, 5th ed. (Champaign, IL: Human Kinetics), 118.

The 6 carbons contained in glucose require 6 molecules of oxygen for complete aerobic combustion in oxidative phosphorylation. The production of 6 carbon dioxides yields a ratio of 1.0. Conversely, 23 oxygen molecules are required in the example of a 16 carbon fatty acid, which results in a ratio of 0.7. Knowing that an RER value of 0.7 implies 100 percent of fat burned, we can calculate percentages of carbohydrate and fat burned from RER.

$$\% \text{ fat burned} = \frac{1 - \text{RER}}{1 - 0.7}$$

Percent carbohydrate is then calculated simply by subtracting percent fat burned from 100, since we are assuming nonprotein RER.

Although RQ and RER are measured by the same parameters ($\dot{V}O_2$ and $\dot{V}CO_2$), these ratios often yield different results based upon the location of the measurement. Under resting or submaximal exercise conditions, the measurement of RER by expired gases reflects the actual gas exchange at the cellular level. An RQ of 1.0 reflects 100 percent utilization of carbohydrate and is the highest value possible. However, at maximal exhaustive exercise, when anaerobic energy production increases, these ratios may vary considerably. Because more CO_2 is expired due to high lactic acid levels associated with maximal exercise, the RER will be greater than the RQ under these conditions. RER is often recorded above 1.15 during a $\dot{V}O_2$max test. In fact, an RER of above 1.1 is a criterion for a valid $\dot{V}O_2$max test (these concepts are elaborated further in labs 9 and 10).

Total Energy Expenditure and Resting Metabolic Rate

The minimal level of energy required to sustain the body's vital functions in the waking state is termed the **basal metabolic rate (BMR)**, which is usually expressed in $kcal \cdot d^{-1}$. True BMR is difficult to obtain, since the subject must sleep for at least 8 h and fast for at least 12 h. This typically requires staying overnight in a metabolic laboratory. As a result, most researchers measure the resting metabolic rate (RMR), which is often also called resting energy expenditure (REE).

RMR still requires a rested, fasted individual, but it does not necessitate staying overnight in the lab. The RMR should be measured under the following conditions. The individual should

- abstain from strenuous exercise for at least 48 h prior to measurement,
- abstain from eating for 12 to 14 h prior to measurement,
- lie or recline quietly in a dark, quiet environment (but awake!) for 30 min prior to the measurement and during the O_2 sampling, and
- have a normal body temperature (absent a fever).

The test itself is conducted by measuring the oxygen consumed during a 10 to 15 min sampling period to ensure steady state at rest. Values for oxygen consumption during the RMR test usually range between 200 and 300 $ml \cdot min^{-1}$ (0.2–0.3 $L \cdot min^{-1}$ or 0.8 to 1.43 $kcal \cdot min^{-1}$), depending on the individual's size and age and a few other factors. Expressed as relative oxygen consumption, values should approximate 3.5 $ml \cdot kg^{-1} \cdot min^{-1}$, or 1 **metabolic equivalent (MET)**.

The assessment of RMR can be important because of its contribution to the total energy expended every day. RMR is one of three components that contribute to an individual's **total energy expenditure (TEE)**. Approximately 67 percent of TEE in a 24 h period is a result of the RMR. The other two components are thermic effect of food (TEF), sometimes called dietary-induced thermogenesis (DIT), and energy expenditure from physical activity (PA). TEF and DIT are both terms used for the energy needed to digest our energy intake through diet (TEI or total energy intake). This amount typically ranges only from about 100 to 200 $kcal \cdot d^{-1}$ (7.5% of TEI), so it is often ignored. PA, however, can be a very large component of an individual's TEE. It is the energy expended above rest during any purposeful activity—whether simply walking across campus or playing a vigorous game of basketball. Thus PA can vary tremendously between individuals and can range from fewer than 300 kcal to more than 3,000 kcal.

$$\text{TEE } (kcal \cdot d^{-1}) = \text{RMR} + \text{TEF} + \text{PA}$$

Interest in RMR typically pertains to weight loss or weight gain. Practical use of RMR includes establishing a baseline of energy expenditure for constructing a sound program of weight control or weight management. Since RMR is only one of two main components of TEE, TEE often needs to be estimated from predicted levels of physical activity. The Institute of Medicine (the health arm

Table 5.2 Physical Activity Level Categories

Category	Physical activity level (PAL)	Physical activity coefficient (PAC) for males/females
Sedentary	>1.0–<1.4	1.00/1.00
Low active	>1.4–<1.6	1.11/1.12
Active	>1.6–<1.9	1.25/1.27
Very active	>1.9–<2.5	1.48/1.45

Adapted from the Food and Nutrition Board, 2005, *Dietary reference intakes for energy, carbohydrate, fiber, fat, fatty acids, cholesterol, protein, and amino acids (macronutrients).*

of the National Academy of Sciences) developed four different physical activity levels (PALs) as a means of predicting the PA component of TEE (Brooks et al. 2004). PAL is the ratio of TEE divided by RMR. See table 5.2.

If the physical activity level (PAL) is 1.0, then the subject's TEE is essentially the same as his or her RMR. Conversely, for a very active female with a PAL of 1.95, TEE is 1.95 times RMR. Thus if her RMR is estimated to be 1,650 kcal, then her TEE would be 1.95 × 1,650 = 3,217.5 kcal · d^{-1}.

Another option for predicting the PA component of TEE is to use the category descriptor (male or female) to determine the physical activity coefficient (PAC) in the following equations for estimating TEE, with the results expressed in kcal · d^{-1}.

Males 19+ years old

$$TEE = 662 - 9.53 \times age\ (y) + [PAC \times (15.91 \times BW\ (kg) + 539.6 \times HT\ (m))]$$

Females 19+ years old

$$TEE = 354 - 6.91 \times age\ (y) + [PAC \times (9.361 \times BW\ (kg) + 726 \times HT\ (m))]$$

Sample Calculations for TEE

Our test subject is a 21-year-old male who weighs 80 kg (176 lb), stands 1.85 m (6 ft 1 in.) tall, and considers himself active. The following calculations illustrate that he would have a TEE of 3,301 kcal · d^{-1}.

$$TEE = 662 - 9.53 \times 21 + [1.25 \times (15.91 \times 80 + 539.6 \times 1.85)]$$

$$= 662 - 200.13 + [1.25 \times (1,272.8 + 998.26)]$$

$$= 662 - 200.13 + 2,838.8$$

$$= 3,301\ kcal \cdot d^{-1}$$

Estimations of TEE are more accurate if the caloric expenditure of physical activity is determined more precisely; however, using the PAL or PAC may provide a starting point for programs involving weight loss or gain.

RMR varies between individuals, which is why the previously mentioned pretest instructions are essential for obtaining a precise measurement. The individual must be fasted and rested, since digesting a meal or recovering from a recent bout of exercise can affect RMR. Even if a subject is compliant with the pretest instructions, large individual differences can occur.

RMR can be affected by factors including the following:

1. RMR is likely to be increased by an increase in lean body mass (LBM) and decreased by a loss of LBM. But these changes can be small, meaning that very large increases in LBM need to occur in order to see even small increases in RMR. Remember that at rest, muscle consumes very little energy. However, energy conservation, as experienced during a low-calorie diet, is capable of decreasing RMR.

2. Endurance training may decrease RMR due to decreases in HR, BP, and respiration associated with a training regimen—the body may become more efficient in daily energy needs. Others have found that being fit may increase RMR. Thus research in this area is equivocal.

3. RMR increases with body weight.

4. A person with a large body surface area will have a higher RMR; thus, a tall person will have a higher RMR than a short person.

5. Males typically have a higher RMR than females.

6. In general, a younger person has a higher RMR. Children and adolescents require large amounts of energy for growth.

7. As body temperature increases, so does RMR. A high body temperature can raise RMR by 30 percent to 40 percent.

8. From age 20 to age 70, RMR can decrease by 10% to 20%. The decline has been estimated at about 2 percent per decade. Some of this decline may be due to loss in LBM (sarcopenia).

9. Fasting (negative energy balance) decreases RMR by 10% to 25%, depending on the extent of the fast or diet.

10. An increase in ambient temperature causes RMR to decrease, whereas cold weather adaptation increases RMR (often referred to as *cold-induced thermogenesis*).

11. Stress increases RMR by mobilizing the sympathetic nervous system. Similarly, circulating catecholamines (epinephrine and norepinephrine) from the adrenal glands can increase RMR.

12. The amount of thyroxine produced by the thyroid gland is an important hormonal control of RMR.

For further reading on factors that influence RMR, see the following items from this chapter's reference list: Carpenter et al. 1995 (2), Compher et al. 2006 (3), Haugen et al. 2003 (5), Heshka et al. 1990 (6), Jorgensen et al. 1998 (7), Lemmer et al. 2001 (9), Schmidt et al. 1994 (12), Sparti et al. 1997 (13), and Speakman and Selman 2003 (14). Because of the current understanding of RMR from the direct measurement utilizing a metabolic cart, RMR can also be estimated with reasonable validity. The following lab activities involve both estimating RMR and measuring it directly.

References

1. Brooks GA, Butte NF, Rand WM, Flatt JP, and Caballero B. Chronicle of the Institute of Medicine Physical Activity Recommendation: How a Physical Activity Recommendation Came to Be Among Dietary Recommendations. *Am J Clin Nutr* 79: 921S–930S, 2004.

2. Carpenter WH, Poehlman ET, O'Connell M, and Goran MI. Influence of Body Composition and Resting Metabolic Rate on Variation in Total Energy Expenditure: A Meta-Analysis. *Am J Clin Nutr* 61: 4–10, 1995.

3. Compher C, Frankenfield D, Keim N, and Roth-Yousey L. Best Practice Methods to Apply to Measurement of Resting Metabolic Rate in Adults: A Systematic Review. *J Am Diet Assoc* 106: 881–903, 2006.

4. Harris JA and Benedict FG. *A Biometric Study of Basal Metabolism in Man*. Washington, DC: Carnegie Institution of Washington, 1919

5. Haugen HA, Melanson EL, Tran ZV, Kearney JT, and Hill JO. Variability of Measured Resting Metabolic Rate. *Am J Clin Nutr* 78: 1141–1145, 2003.

6. Heshka S, Yang MU, Wang J, Burt P, and Pi-Sunyer FX. Weight Loss and Change in Resting Metabolic Rate. *Am J Clin Nutr* 52: 981–986, 1990.

7. Jorgensen JO, Vahl N, Dall R, and Christiansen JS. Resting Metabolic Rate in Healthy Adults: Relation to Growth Hormone Status and Leptin Levels. *Metabolism* 47: 1134–1139, 1998.

8. Kenney WL, Wilmore JH, and Costill DL. *Physiology of Sport and Exercise*. 5th ed. Champaign, IL: Human Kinetics, 2012.

9. Lemmer JT, Ivey FM, Ryan AS, Martel GF, Hurlbut DE, Metter JE, Fozard JL, Fleg JL, and Hurley BF. Effect of Strength Training on Resting Metabolic Rate and Physical Activity: Age and Gender Comparisons. *Med Sci Sports Exerc* 33: 532–541, 2001.

10. Mifflin MD, St Jeor ST, Hill LA, Scott BJ, Daugherty SA, and Koh YO. A New Predictive Equation for Resting Energy Expenditure in Healthy Individuals. *Am J Clin Nutr* 51: 241–247, 1990.

11. Owen OE, Kavle E, Owen RS, Polansky M, Caprio S, Mozzoli MA, Kendrick ZV, Bushman MC, and Boden G. A Reappraisal of Caloric Requirements in Healthy Women. *Am J Clin Nutr* 44:1-19, 1986.

12. Schmidt WD, Hyner GC, Lyle RM, Corrigan D, Bottoms G, and Melby CL. The Effects of Aerobic and Anaerobic Exercise Conditioning on Resting Metabolic Rate and the Thermic Effect of a Meal. *Int J Sport Nutr* 4: 335–346, 1994.

13. Sparti A, DeLany JP, de la Bretonne JA, Sander GE, and Bray GA. Relationship Between Resting Metabolic Rate and the Composition of the Fat-Free Mass. *Metabolism* 46: 1225–1230, 1997.

14. Speakman JR and Selman C. Physical Activity and Resting Metabolic Rate. *Proc Nutr Soc* 62: 621–634, 2003.

15. WHO. 1985. *Energy and protein requirements: Report of a joint FAO/WHO/UNU expert consultation*. WHO Technical Report Series No. 724. Geneva.

PREDICTING RMR

EQUIPMENT

- RMR prediction equations
- Calculator
- Individual and group data sheets

Find the group data sheets for this laboratory online at www.HumanKinetics.com/ LaboratoryManualForExercisePhysiology.

Because the direct measurement of oxygen uptake is impractical in many situations, various formulas and nomograms have been developed to estimate RMR. Of course, certain information must be known: height (HT), weight (WT), age, and sex. In general, equations that use more factors known to affect RMR are considered to be more accurate. Some sample equations are given in table 5.3.

Table 5.3 Equations for Estimating RMR

Number	Male	Female	Source
1	66.473 + 13.7516(BW) + 5.0033(HT) – 6.755(age)	655.0955 + 9.5634(BW) + 1.8496(HT) – 4.6756(age)	(4)
2	879 + 10.2(BW)	795 + 7.18(BW)	(10, female only)
3	BSA × 38 kcal · hr⁻¹ × 24 h	BSA × 35 kcal · hr⁻¹ × 24 h	
4	BW × 24.2 kcal · kg⁻¹	BW × 22.0 kcal · kg⁻¹	
5	BW × 1.0 × 24 h	BW × 0.9 × 24 h	
6	15.3 × BW + 679	14.7 × BW + 496	(15)
7	9.99(BW) + 6.25(HT) – 4.92(age) + 5.0	9.99(BW) + 6.25(HT) – 4.92(age) – 161	(9)

All answers are given in kilocalories per day. HT = height in centimeters. BW = body weight in kilograms. Age = age in years. BSA in m^2 is the square root of (BW × HT / 3600). These equations vary in accuracy but typically predict RMR within a 10 to 20 percent error.

RMR CALCULATIONS

Step 1: Determine your own RMR by using the prediction equations (show all calculations). Express RMR in kcal per minute, per hour, and per day. Complete your own data sheet for each equation.

Step 2: Record these values in the group data table.

Step 3: Compute the mean value for males and for females in the class, and then compare your values to the class average.

QUESTION SET 5.1

1. Which of the equations do you believe best predicts your RMR? Why?

2. What difference did you find between males and females in the class? What explains this difference?

3. Calculate your predicted TEE using the equations above (from the PAL and PAC in table 5.2). What assumptions are made with these equations?

Find the case studies for this laboratory online at www.HumanKinetics.com/ LaboratoryManualForExercisePhysiology.

Laboratory Activity 5.1

Name or ID number: _____   Date: _____

Tester: _____   Time: _____

Sex: M / F (circle one)   Age: _____ y   Height: _____ in. _____ cm

Weight: _____ lb _____ kg

Body Surface Area (BSA): _____ m²

		kcal · min⁻¹	kcal · hr⁻¹	kcal · d⁻¹
RMR prediction equations	1			
	2			
	3			
	4			
	5			
	6			
	7			

MEASURING RMR

EQUIPMENT

- Quiet room with place for subject to lie down
- Metabolic cart including gas flow meter and O_2 and CO_2 analyzers
- Hans Rudolph valve, mouthpiece, and nose clips
- RMR prediction equations provided in lab activity 5.1
- Calculator
- Individual data sheets

METABOLIC CART RMR TEST

Step 1: After two students volunteer to be measured for RMR, other students split and assist in measuring one of the two subjects. Report data for the subject with whom you assisted.

Step 2: If possible in your lab, the test should be done in the early morning, and the subjects should have fasted.

Step 3: If at all possible, the subjects should abstain from eating for at least 3 h prior to testing and avoid strenuous exercise for 24 h beforehand.

Step 4: Determine your subject's weight, height, and age for data entry into the metabolic cart and for use in the prediction equations.

Step 5: Have the individual lie quietly for 15 min.

Step 6: During this time, calculate the subject's predicted RMR and fill out the appropriate data sheet from these prediction equations.

Step 7: Outfit the subject with the breathing apparatus from the calibrated metabolic cart for a 10 min measurement period. Do not disturb the subject while collecting data.

Step 8: Record the $\dot{V}O_2$ in $L \cdot min^{-1}$ and RER at the end of each minute of the 10 min measurement period.

Step 9: Complete the calculations for the remainder of the table. Include $kcal \cdot min^{-1}$ and $kcal \cdot d^{-1}$, as well as percent carbohydrate and percent fat, using the $kcal \cdot L^{-1}$ of O_2 provided in table 5.1.

QUESTION SET 5.2

1. Compare the values for the subject measured from the metabolic cart to the values predicted from the prediction equations. Which equation do you think gives the best prediction? Why?

2. From your knowledge of the subject's adherence to the typical RMR pretest instructions, which of the pretest instructions do you feel had an effect on the measurement?

3. How much of the RMR is from carbohydrate? From fat? What does this information suggest about the prevailing fuel of choice for our bodies at rest?

 Find the case studies for this laboratory online at www.HumanKinetics.com/ LaboratoryManualForExercisePhysiology.

Laboratory Activity 5.2 Individual Data Sheet

Date: _____ Time: _____

Tester: _____

Temperature: _____°F _____°C Barometric pressure: _____mmHg

Relative humidity: _____ %

Subject A

Name or ID number: _____

Sex: M / F (circle one)   Age: _____ y Height: _____ in. _____ cm

Weight: _____ lb _____ kg

BSA: _____ m²

		kcal · min⁻¹	kcal · hr⁻¹	kcal · d⁻¹
RMR prediction equations	1			
	2			
	3			
	4			
	5			
	6			
	7			

Subject B

Name or ID number: _____

Sex: M / F (circle one)   Age: _____ y   Height: _____ in. _____ cm

Weight: _____ lb _____ kg

BSA: _____ m²

		kcal · min⁻¹	kcal · hr⁻¹	kcals · d⁻¹
RMR prediction equations	1			
	2			
	3			
	4			
	5			
	6			
	7			

Measuring RMR Using Indirect Calorimetry: Subject A

Minutes	$\dot{V}O_2 (L \cdot min^{-1})$	RER	% carbohydrate	% fat	$kcal \cdot L^{-1}$	$kcal \cdot hr^{-1}$	$kcal \cdot d^{-1}$
1							
2							
3							
4							
5							
6							
7							
8							
9							
10							
Mean							

Laboratory Activity 5.2

Measuring RMR Using Indirect Calorimetry: Subject B

Minutes	$\dot{V}O_2 (L \cdot min^{-1})$	RER	% carbohydrate	% fat	kcal · L^{-1}	kcal · hr^{-1}	kcal · d^{-1}
1							
2							
3							
4							
5							
6							
7							
8							
9							
10							
Mean							

Oxygen Deficit and EPOC Evaluations

Objectives

1. Demonstrate how data are collected for the measurement of O_2 deficit and EPOC.
2. Draw and label an O_2 deficit and EPOC graph.
3. Calculate the O_2 deficit and EPOC from laboratory-collected data.
4. Discuss biochemical concepts related to O_2 deficit and EPOC.

Definitions

anaerobic glycolysis—Metabolic pathway that can provide energy quickly; does not require oxygen but produces lactate.

excess post-oxygen consumption (EPOC)—Amount of oxygen consumption above rest after cessation of exercise (synonymous with *oxygen debt*).

oxidative phosphorylation—Production of adenosine triphosphate (ATP) using oxygen as the final electron acceptor in the electron transport chain.

oxygen debt—Amount of oxygen consumption above rest after cessation of exercise (synonymous with *excess post-oxygen consumption*).

oxygen deficit—Lag in oxygen consumption in the transition from rest to submaximal exercise.

phosphocreatine (PCr)—High-energy phosphate that can donate a phosphate group to adenosine diphosphate (ADP) to resynthesize ATP very quickly; does not require oxygen but does not have the storage capacity to support ATP provision for longer than about 10 seconds.

steady state—Plateau in metabolic parameters at a constant amount of power.

Now that we have a general understanding of oxygen consumption and energy expenditure from laboratory 5, we can apply these concepts to the transition from rest to exercise. Oxygen consumption implies that energy (ATP) is being provided through the oxidative phosphorylation pathway. Oxidative phosphorylation has no trouble meeting the energy need for resting energy metabolism. When you transition to exercise, however, energy requirements increase, and this new demand for ATP is not met immediately by oxidative phosphorylation.

Transition From Rest to Exercise

When a person makes the transition from rest to a submaximal work task, ATP demand immediately increases to the level required in order for the work to be done; however, there is a delay in the measured oxygen uptake responses. During this period of time, early in exercise, some of the ATP is provided by the immediate and short-term sources (i.e., **phosphocreatine [PCr]** and **anaerobic glycolysis**). By about 3 min into the activity, oxygen consumption plateaus to reach "steady state," which indicates that the ATP requirement is now being met aerobically by oxidative phosphorylation. The volume of oxygen "missing" in those first minutes of work is called the **oxygen deficit** as shown in figure 6.1.

During exercise of submaximal intensity (<75%–80% of maximal), the task's energy requirement can be met by means of oxygen consumption ($\dot{V}O_2$) or aerobic metabolism through **oxidative phosphorylation**. Steady state $\dot{V}O_2$ then represents the energy expenditure of the submaximal intensity. When exercise is stopped, oxygen uptake does not immediately fall to the resting value; instead, it decreases over a period of several minutes. The volume of oxygen consumed during recovery from exercise, above the resting value, has been called the **oxygen debt** or **excess post-oxygen consumption (EPOC)**, as shown in figure 6.3.

Say, for example, that a subject straddles the treadmill in preparation for exercise. The energy demand for standing on the treadmill is relatively low—for most people, it approximates 3.5 $ml \cdot kg^{-1} \cdot min^{-1}$, or one metabolic equivalent (MET). When the subject steps on the treadmill and beings moving at a moderate pace, the energy requirement or ATP demand increases immediately in a square wave fashion. Oxidative phosphorylation, however, does not immediately meet this new energy requirement, and, as a result, the body relies on anaerobic ATP pathways to meet the demand until the aerobic system can be stimulated to the new level of ATP production. The magnitude or amount of the deficit depends on a number of factors, including intensity of exercise, genetics, and fitness level. Thus a trained person incurs less oxygen deficit than an untrained person in reaching the same oxygen consumption (see figure 6.2). The trained person relies less on anaerobic pathways and thus causes less **lactate accumulation** and **PCr depletion**, both of which can contribute to fatigue in certain situations and also contribute to the magnitude of the deficit and EPOC.

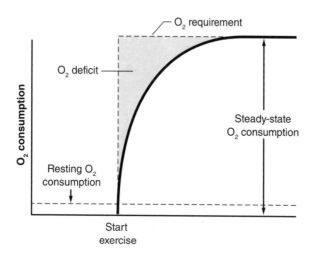

Figure 6.1 Oxygen consumption during rest-to-exercise transition, steady state, and recovery.

Reprinted, by permission, from J.H. Wilmore, D.L. Costill, and W.L. Kenney, 2008, *Physiology of sport and exercise*, 4th ed. (Champaign, IL: Human Kinetics), 109.

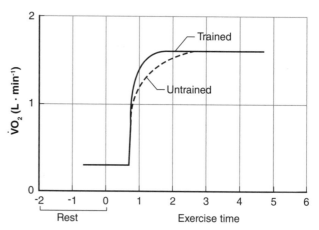

Figure 6.2 The relationship between exercise time and oxygen uptake, illustrating the time to reach a set level of $\dot{V}O_2$max in a trained and an untrained man.

Oxygen Uptake During Exercise and Recovery

The steep initial increase in oxygen consumption and the subsequent slower increase to steady state are often referred to as the *fast* and *slow* portions of oxygen uptake kinetics. For further reading in this active area of research, see the following items in this chapter's reference list: 3, 5, 6, 7, 8, 9, and 10.

During steady state oxygen consumption (figure 6.3*a*), the energy needed to support submaximal exercise is provided by the aerobic system or by oxidative phosphorylation. Therefore, the $\dot{V}O_2$ represents the total energy expenditure of that exercise. Once exercise ceases, oxygen uptake (oxygen debt or EPOC) falls quickly in the first 2 to 4 min, then more slowly, toward the pre-exercise value. The fast component is believed to be due to rapid reestablishment of the PCr and oxygen stores. The slow component was at one time believed to be related to the oxygen needed to clear lactic acid from the blood by the liver, where it was converted back into glucose (gluconeogenesis through the Cori cycle). Now, however, it is thought that only about 20% of recovery oxygen uptake is used for this process and that the vast majority relates to other processes. It is also clear

that oxygen uptake during recovery can remain elevated for a wide variety of reasons unrelated to restoration of the high-energy phosphate stores—for example, elevation of heart rate, breathing rate, body temperature, muscle inflammation, hormone levels, sodium potassium pump activity, and the reestablishment of ion gradients (2, 4).

The magnitude of the oxygen debt (or EPOC) is therefore a function of how elevated these factors become, which is determined mainly by exercise intensity. A higher intensity of exercise (figure 6.3*b*) causes greater changes in heart rate, ventilation, body temperature, and hormone levels. It is important to understand that very high intensities also result in much greater deficit and EPOC. Indeed, when anaerobic pathways cover the majority of ATP requirement *during* the exercise bout, then nearly the entire bout is considered oxygen deficit. Take, for example, a 100 m dash. Figure 6.4 is a prediction of the anaerobic and aerobic contributions during and after this intense bout of exercise.

Note that oxygen consumption contributes little to energy production *during* the 100 m dash; however, following the bout of exercise $\dot{V}O_2$ continues to rise, and it may take a while to return to resting baseline. Again, a large part of the EPOC is due to the replenishment of PCr

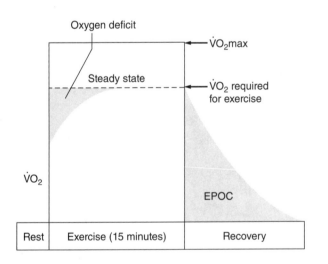

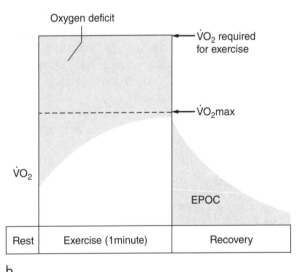

a   b

Figure 6.3 Oxygen consumption and requirements during (*a*) low-intensity, steady-state exercise and (*b*) high-intensity, non-steady-state exercise.

Adapted, by permission, from NSCA, 2008, *Essentials of strength training and conditioning*, 3rd ed. (Champaign, IL: Human Kinetics), 35-36.

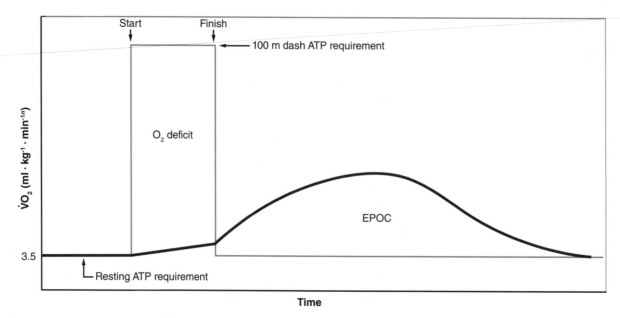

Figure 6.4 Oxygen deficit and EPOC for a 100 m dash.

and oxygen stores. Since this bout of exercise, which lasts about 10 to 13 s, relies mostly on the ATP-PCr pathway, the reestablishment of PCR would contribute to the majority of the EPOC. But remember that almost every exercise situation imaginable involves a mix of the three energy pathways: ATP-PCr, fast or anaerobic glycolysis, and oxidative phosphorylation. This would hold true for other seemingly "anaerobic sports" such as weightlifting and for intermittent sports such as football, soccer, and rugby. In fact, the muscle repair that follows muscle-damaging exercise can contribute to elevated oxygen consumption at rest for up to 48 hours or more (1).

References

1. Dolezal BA, Potteiger JA, Jacobsen DJ, and Benedict SH. Muscle Damage and Resting Metabolic Rate After Acute Resistance Exercise With an Eccentric Overload. *Med Sci Sports Exer* 32: 1202–1207, 2000.

2. Gaesser GA and Brooks GA. Metabolic Bases of Excess Post-Exercise Oxygen Consumption: A Review. *Med Sci Sports Exer* 16: 29–43, 1984.

3. Grassi B. Regulation of Oxygen Consumption at Exercise Onset: Is It Really Controversial? *Exer Sport Sci Rev* 29: 134–138, 2001.

4. Kenney WL, Wilmore JH, and Costill DL. *Physiology of Sport and Exercise*. 5th ed. Champaign, IL: Human Kinetics, 2012.

5. Korzeniewski B and Zoladz JA. Biochemical Background of the $\dot{V}O_2$ On-Kinetics in Skeletal Muscles. *J Physiol Sci* 56: 1–12, 2006.

6. Maizes JS, Murtuza M, and Kvetan V. Oxygen Transport and Utilization. *Respir Care Clin N Am* 6: 473–500, 2000.

7. Poole DC, Barstow TJ, Gaesser GA, Willis WT, and Whipp BJ. VO2 Slow Component: Physiological and Functional Significance. *Med Sci Sports Exer* 26: 1354–1358, 1994.

8. Whipp BJ. The Slow Component of O_2 Uptake Kinetics During Heavy Exercise. *Med Sci Sports Exer* 26: 1319–1326, 1994.

9. Xu F, Rhodes EC. Oxygen Uptake Kinetics During Exercise. *Sports Med* 27: 313–327, 1999.

10. Zoladz JA, Korzeniewski B. Physiological Background of the Change Point in VO2 and the Slow Component of Oxygen Uptake Kinetics. *J Physiol Pharmacol* 52: 167–184, 2001.

CALCULATION OF OXYGEN DEFICIT AND EPOC

EQUIPMENT

- Metabolic cart, including gas flow meter and O_2 and CO_2 analyzers
- Hans Rudolph valve, mouthpiece, and noseclips
- Treadmill
- Calculator
- Individual data sheet

STEADY STATE OXYGEN CONSUMPTION

Step 1: One or two students will volunteer as subjects; ideally, the subjects will have differing fitness levels, but you can also compare different intensities in similarly fit subjects. The remaining students should split into two groups and assist in the measurement of one of the subjects. Report data for the subject with whom you assisted.

Step 2: Determine the subject's weight, height, and age for data entry into the metabolic cart and for the data sheet.

Step 3: Have the subject sit quietly for 5 min.

Step 4: Outfit the subject with the breathing apparatus from the calibrated metabolic cart for a 5 min measurement period. Obtain sitting values (i.e., with the subject seated) for the data sheet every minute.

Step 5: Have the subject jog at a comfortable pace for 10 min. (Choose a submaximal intensity so that the subject can reach steady state. If you are comparing two subjects with different fitness levels, they should exercise at the same intensity. If you choose instead to compare intensity, then one subject should go at an easy to moderate intensity, whereas the other should be more intense.)

Step 6: Immediately from the start of exercise, record the metabolic data every 30 s for the first 4 min; then, record it every minute according to the individual data sheet.

Step 7: Have the subject immediately sit for 8 min.

Step 8: Immediately upon the cessation of exercise, record the metabolic data every 30 s for the first 3 min of recovery; then record it every minute according to the individual data sheet.

CALCULATING OXYGEN DEFICIT AND EPOC

Typically, you would use calculus to calculate the area under a curve in order to know the magnitude of deficit and EPOC. However, since not all exercise physiology students have this mathematical background, we will use a simpler method to estimate this calculation:

Step 1: Determine the average steady state exercise $\dot{V}O_2$ by taking the mean of the $\dot{V}O_2$ measurements of minutes 11 to 15. In the example given in table 6.1, the average steady state $\dot{V}O_2$ is assumed to be 25 $ml \cdot kg^{-1} \cdot min^{-1}$.

Step 2: Subtract each of the measurements from min 5.5 to 9.0 during the rest to exercise transition from the average steady state $\dot{V}O_2$ and record these differences.

Step 3: Sum these differences together to obtain an estimate of the oxygen deficit before reaching steady state (in table 6.1, this is 52.5 $ml \cdot kg^{-1} \cdot min^{-1}$). The reverse of this example—or the sum of the oxygen uptake during recovery (minutes 15.5 to 19)—minus the average pre-exercise seated oxygen uptake (which is approximately 3.5 $ml \cdot kg^{-1} \cdot min^{-1}$) is the EPOC.

Table 6.1 Sample Estimation of Oxygen Deficit

Exercise min	Measured $\dot{V}O_2$ $(ml \cdot kg^{-1} \cdot min^{-1})$	Average steady state $\dot{V}O_2$* – measured $\dot{V}O_2$
5.5	8.5	16.5
6.0	11.7	13.3
6.5	15.7	9.3
7.0	19.5	5.5
7.5	21.3	3.7
8.0	22.8	2.2
8.5	23.5	1.5
9.0	24.5	0.5
Estimated total oxygen deficit =		52.5 $ml \cdot kg^{-1} \cdot min^{-1}$

* In this example, the average steady state $\dot{V}O_2$ is assumed to be 25 $ml \cdot kg^{-1} \cdot min^{-1}$.

QUESTION SET 6.1

1. What was the steady state O_2 uptake for your subject?
2. Calculate the O_2 deficit (energy not accounted for by $\dot{V}O_2$).
3. Which ATP pathways contribute to the energy needed for the O_2 deficit? What would you measure to determine the source of ATP?
4. Calculate the EPOC (total $\dot{V}O_2$ above rest postexercise).
5. Is the debt greater than the deficit? Why might they be equal in some exercise tests, whereas in others the EPOC would be greater than the deficit?
6. EPOC has two components—a fast component (large drop in O_2 uptake values in the first few minutes) and a slow component (gradual return to rest). What explanations are given for the existence of the two components?

7. Using graph paper or a spreadsheet, graph the values collected in class, identify the areas representing the O_2 deficit and EPOC, and shade in those areas. Your graph should look similar to the figure 6.1 in this lab. Include rest, exercise and recovery intervals, oxygen debt and deficit, and the fast and slow components of oxygen deficit.

8. How does the steady state oxygen consumption compare to the predicted oxygen consumption for the chosen speed and grade on the treadmill? Use the ACSM metabolic prediction equations found in appendix B.

Find the case studies for this laboratory online at www.HumanKinetics.com/ LaboratoryManualForExercisePhysiology.

Name or ID number: _____ Date: _____

Tester: _____ Time: _____

Sex: M / F (circle one)   Age: _____ y   Height: _____ in. _____ cm

Temperature: _____ °F _____ °C   Weight: _____ lb _____ kg

Barometric pressure: _____ mmHg   Relative humidity: _____ %

	Time (min)	$\dot{V}O_2$ (ml · kg⁻¹· min⁻¹)	$\dot{V}CO_2$ (ml · kg⁻¹· min⁻¹)	RER	HR (beats · min⁻¹)
Sitting	1				
	2				
	3				
	4				
	5				
Mean					
Exercise	5.5				
	6				
	6.5				
	7				
	7.5				
	8				
	8.5				
	9				
	10				
	11				
	12				
	13				
	14				
	15				
Steady state mean					

Laboratory Activity 6.1

	Time (min)	$\dot{V}O_2$ (ml · kg⁻¹· min⁻¹)	$\dot{V}CO_2$ (ml · kg⁻¹· min⁻¹)	RER	HR (beats · min⁻¹)
Sitting	13.5				
	14				
	14.5				
	15				
	15.5				
	16				
	17				
	18				
	19				
	20				
	21				

Submaximal Exercise Testing

Objectives

1. Reinforce the terminology and techniques associated with measuring blood pressure at rest and during exercise.

2. Learn the skills needed for administering three types of submaximal exercise test.

3. Reinforce the terminology and techniques associated with measuring heart rate at rest and during exercise.

4. Estimate $\dot{V}O_2$max by the measure of heart rate response to exercise.

Definitions

absolute work rate—Intensity of exercise measured on an ergometer by a power unit (i.e., watts or $kg \cdot m \cdot min^{-1}$); not dependent on fitness.

maximum heart rate—Maximum rate at which an individual's heart can beat per minute; often estimated by subtracting age in years from 220.

relative work rate—Intensity of exercise measured relative to an individual's maximal abilities, such as percent $\dot{V}O_2$max or percent of work rate (power) maximum; dependent on fitness.

submaximal—Exercise intensity below 85% of maximal effort.

Knowing a subject's cardiorespiratory capacity, or $\dot{V}O_2$max, can be useful for exercise prescription or for diagnostic purposes in at-risk populations. Tests for estimating $\dot{V}O_2$max can be maximal (involving incremental workloads to the point of failure) or **submaximal** (ending before extreme exertion). Submaximal estimates of cardiorespiratory capacity can be much more practical, since they do not require expensive automated metabolic laboratory equipment. They do, however, require some knowledge and skill in order to carry them out properly. For this reason, such estimates often form part of the practical portion of certification examinations (e.g., for the American College of Sports Medicine's Health Fitness Specialist certification); for more information about certifications, see appendix F.

Many types of submaximal tests exist, and they differ for different populations, by number of several stages, and by modality. Since quantifying work performed is necessary in order to determine submaximal effort, common modalities include several approaches that facilitate quantification, such as bench stepping, treadmill work, and cycle ergometer tests. In this lab, you will perform an example of each of these testing modalities and include a one-stage and a two-stage test. Submaximal tests require monitoring of heart rate (HR), blood pressure (BP), patient symptoms, time, ergometer work rates, and ratings of perceived exertion (RPEs). These lab activities will not involve direct measurement of respiratory gases.

Because of the linear nature of the response of HR to work rate (see figure 7.1), submaximal work rates can be extrapolated to the subject's

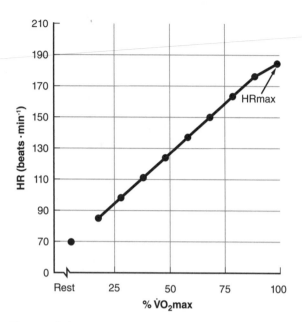

Figure 7.1 Heart rate response to exercise intensity.

Reprinted, by permission, from W. L. Kenney, J.H. Wilmore, and D.L. Costill, 2012, *Physiology of sport and exercise*, 5th ed. (Champaign, IL: Human Kinetics), 183.

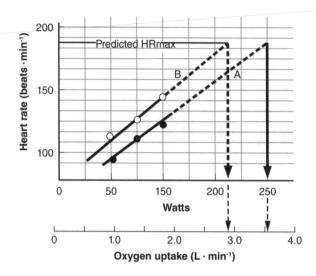

Figure 7.2 Effect of training on the heart rate response.

Reprinted, by permission, from P.O. Åstrand et al., 2003, *Textbook of work physiology*, 4th ed. (Champaign, IL: Human Kinetics), 285.

age-predicted maximum heart rate. Doing so has become one of the most popular assessment techniques for estimating a person's aerobic capacity, cardiorespiratory fitness, or $\dot{V}O_2$max.

Determining HR response to exercise is one of the most critical skills that an exercise physiology student can develop. This response's usefulness in predicting cardiorespiratory fitness involves the relationship between HR, age, and fitness level. As a person becomes fitter, his or her HR decreases at a given absolute submaximal work rate (see figure 7.2).

This response is one of the hallmarks of adaptation to training. Crucially, the reduction in HR at submaximal intensities following training results not from a decrease in $\dot{V}O_2$ or cardiac output (CO) but from an increase in stroke volume (SV), which allows a lower HR to accomplish the same CO and $\dot{V}O_2$. Another cardiovascular adaptation to training is the recovery of HR following an exercise bout. After exercise, HR returns to normal resting HR more quickly in a trained individual than in an untrained one. Thus, accurately timed postexercise HR can be used as a predictor of fitness, as in the step test included in laboratory activity 7.1.

It is important to consider the assumptions of submaximal testing in many situations. First, it is assumed that a subject attains steady state exercise at each workload so that the outcome measures of HR and BP are at steady state. It is also assumed that efficiency does not differ significantly between subjects; in other words, it is assumed that all subjects expend the same amount of energy ($\dot{V}O_2$) at a given **absolute work rate** independent of fitness. However, these subjects can be at vastly different percentages of their individual maximal capacity, or **relative work rates**. The most important assumption may be that **maximum heart rate** is the same across a given age. Error can be significant in the estimation of maximum HR according to the formula in which HR = 220 − age; one standard deviation is ±12 beats per minute (beats·min⁻¹). A more recent and accurate equation has been developed in which maximal HR = 208 − (0.7 × age) (8). Although error can arise from these assumptions in prediction of maximum capacity, error can also be attributed to inaccurate pedaling cadence (work rate), inaccurate load on the ergometer, and imprecise measures of heart rate and blood pressure—all of which are controlled by the testers.

Measurement of Heart Rate

Heart rate during exercise can be measured by palpation, auscultation, ECG, or use of an HR monitor. Measuring heart rate by palpation, as you will do in this lab, is an important skill. Heart rate can be palpated at several sites, including the radial, carotid, temporal, and femoral arteries. This is relatively easy to do when the subject is at rest but can be challenging during exercise due to extraneous noise. With a stopwatch, count the number of beats over 6, 10, or 15 seconds and multiply the number of beats by 10, 6, or 4, respectively, to determine the number of beats per minute. Heart rate can also be measured by timing the number of seconds for 30 beats, then calculating beats per minute by using the following equation:

$$HR \ (beats \cdot min^{-1}) = 1,800 \ / \ [time \ (s) \ for \ 30 \ beats]$$

This method is thought to be more accurate since it removes error due to partial beats. With either method, start counting heartbeats with zero upon the start of the timing.

Rating of Perceived Exertion

When a subject is doing an exercise test, it is useful to know how he or she is feeling, and testers often address this need by using scales for the rating of perceived exertion (RPE). This type of measure can be used to monitor progress toward maximal exertion. Two subjective scales have been developed (6), and both are appropriate to use during maximal exercise testing. Both scales were developed by Gunnar Borg, the difference being that one ranges from 0 to 10 while the other ranges from 6 to 20 (1).

The RPE scales include instructions to be read to the subject prior to the start of the test (5):

During the test, we want you to pay close attention to how hard you feel the exercise work rate is. This feeling should reflect your total amount of exertion and fatigue, combining all sensations and feelings of physical stress, effort, and fatigue. Don't concern yourself with any one factor, such as leg pain, shortness of breath or exercise intensity, but try to concentrate on your total, inner feeling of exertion. Try not to underestimate or overestimate your feelings of exertion; be as accurate as you can.

The following laboratory activities involve submaximal testing using the bench step, treadmill, and cycle ergometer. These protocols were

Rating of Perceived Exertion Scales

The 15-point Borg scale starts with 6, which is usually associated with sedentary behavior. Descriptors of exertion are assigned to every odd number on the scale, with 7 corresponding to "extremely light," 9 to "very light," 11 to "light," 13 to "somewhat hard," 15 to "hard," 17 to "very hard," and 19 to "extremely hard." Anecdotally, the numbers 15 or 16 are often associated with the anaerobic or ventilatory threshold. The scale continues up to 20, which represents the exertion at maximal intensity. The 15-point Borg scale ranges from 6 to 20 in the belief that adding a 0 to the RPE number approximates the HR; however, the previously mentioned differences in maximal and resting HR between individuals call the accuracy of this prediction into question. The Borg category-ratio scale, which can also be used to measure exertion, has values ranging from 0 ("nothing at all") to 10 ("extremely strong") and above, but has no fixed number indicating the maximum possible exertion (1, 6).

selected because of their use in occupational, clinical, and apparently healthy populations. In addition, this approach exposes you to three measures of work output by humans.

Submaximal testing is generally safe when appropriate procedures are followed, but some circumstances call for terminating a test before the predetermined endpoint has been reached. Here are some general indications for stopping exercise tests (both submaximal and maximal):

- Onset of angina or angina-like symptoms
- Drop in systolic BP of >10 mmHg from baseline BP despite an increase in workload
- Excessive rise in BP: systolic pressure >250 mmHg or diastolic pressure >115 mmHg
- Shortness of breath, wheezing, leg cramps, or claudication
- Signs of poor perfusion: light-headedness, confusion, ataxia, pallor, cyanosis, nausea, or cold and clammy skin
- Failure of heart rate to increase with increased exercise intensity
- Noticeable change in heart rhythm
- Subject requests to stop

- Physical or verbal manifestations of severe fatigue
- Failure of the testing equipment

From ACSM, 2010, *ACSM's guidelines for exercise testing and prescription*. 8th ed. Philadelphia: Lippincott Williams & Wilkins.

References

1. Ebbeling CB, Ward A, Puleo EM, Widrick J, and Rippe JM. Development of a Single-Stage Sub-maximal Treadmill Walking Test. *Med Sci Sports Exer* 23: 966–973, 1991.

2. Golding LA. *The Y's Way to Physical Fitness*. 4th ed. Champaign, IL: Human Kinetics, 2000.

3. McArdle WD, Katch FI, Pechar GS, Jacobson L, Ruck S. Reliability and interrelationships between maximal oxygen intake, physical work capacity and step-test scores in college women. *Med Sci Sports*. 1972;4:182-6.

4. McArdle WD, Pechar GS, Katch FI, Magel JR. Percentile norms for a valid step test in college women. *Res Q*. 1973;44:498-500.

5. Noble BJ, Borg GA, Jacobs I, Ceci R, and Kaiser P. A Category-Ratio Perceived Exertion Scale: Relationship to Blood and Muscle Lactates and Heart Rate. *Med Sci Sports Exer* 15: 523–528, 1983.

6. Robertson RJ and Noble BJ. Perception of Physical Exertion: Methods, Mediators, and Applications. *Exer Sport Sci Rev* 25: 407–452, 1997.

7. Sharkey BJ and Gaskill S. *Fitness and Health*. 6th ed. Champaign, IL: Human Kinetics, 2007.

8. Tanaka H, Monahan KD, and Seals DR. Age-Predicted Maximal Heart Rate Revisited. *J Am Coll Cardiol 37*: 153–156, 2001.

SUBMAXIMAL BENCH STEP TEST

EQUIPMENT

- Step—41.3 cm (16.25 in) high for men and women
- Metronome
- Stopwatch
- Individual data sheets

STEP BOX

Stair stepping is an easy and inexpensive alternative to using an exercise ergometer, since it enables you to easily calculate work and power. You can make a step box from two-by-fours and plywood. A longer bench allows several subjects to be tested at the same time. Some protocols call for different heights for men and women. In this case, the box or bench can be made in a way that allows it to be simply rotated from the male height to the female height. You can also use a supplemental step to shorten the step for women. Take care to measure accurately, since any error in step height modifies the amount of work accomplished and thus the metabolic response. Work and power during step testing can easily be calculated knowing body weight and the step rate and height.

$$\text{Work (kg} \cdot \text{m)} = \text{body weight} \times \text{distance} \cdot \text{step}^{-1} \times \text{steps} \cdot \text{min}^{-1} \times \text{min}$$

$$\text{Power (kg} \cdot \text{m} \cdot \text{min}^{-1}) = \text{body weight} \times \text{distance} \cdot \text{step}^{-1} \times \text{steps} \cdot \text{min}^{-1}$$

QUEENS COLLEGE STEP TEST

Step testing is convenient for both indoor and outdoor settings and for use with either one person or multiple people. Step tests come in many types, and perhaps one of the most popular is the Queens College Step Test (3, 4). Like most step tests, this test uses the measurement of recovery heart rate to estimate the subject's level of fitness (recall that heart rate returns to resting values more quickly following submaximal exercise in fitter people than it does in those who are less fit). Many of the available step tests were developed to estimate the fitness necessary for firefighting and other physically demanding occupations, but they are no longer used for occupational screening because participants sometimes used drugs (e.g., beta-blockers) to lower their heart rate and thus inflate their apparent fitness (you would not have been able to get one of these jobs unless your estimated $\dot{V}O_2$max was greater than 45 ml $\cdot$ kg^{-1} $\cdot$ min^{-1} [8]). The test remains useful, however, especially for groups of individuals participating in an exercise program.

Step 1: Since the accuracy of the test relies on the heart rate response, try to eliminate factors that might alter this outcome measure. Ideally, subjects will have avoided exercise for the previous 24 h, fasted for at least 2 h, and avoided the use of foods and drugs that alter heart rate (e.g., coffee, soda, energy drinks, diet pills, beta-blockers).

Step 2: Pair up with another student and find an appropriate space in which to conduct the test. Either you or your partner will start as the tester, and the other person will serve as the subject. You will then reverse these roles.

Step 3: Have the subject sit on the bench step and rest for 3 min, after which the tester should palpate the radial pulse for 15 s and record the resting HR.

Step 4: Set the metronome at 88 beats·min⁻¹ to allow the subject to make contact with a foot on each beep in an up-up-down-down manner. This cadence results in the necessary 22 steps·min⁻¹ necessary for the test on women. For men, set the metronome at 96 beats·min⁻¹ and thus 24 steps·min⁻¹.

Step 5: When the subject is ready, begin the 3 min test and start the stopwatch (see figure 7.3a).

Step 6: To avoid muscle fatigue, the subject should switch the leading leg at least once during the test.

Step 7: After exactly 3 min of stepping, the subject should stop. The tester should palpate for the radial pulse (see figure 7.3b). Begin counting at exactly 3:05 and count for 15 s (i.e., to 3:20).

Step 8: Calculate the predicted $\dot{V}O_2$max by using the recovery HR in the equations below, where HR is beats·min⁻¹.

Men: $\dot{V}O_2$max (ml·kg⁻¹·min⁻¹) = 111.33 - (0.42 × HR)

Women: $\dot{V}O_2$max (ml·kg⁻¹·min⁻¹) = 65.81 - (0.1847 × HR)

Step 9: Record your own data on the individual data sheet and on the group data sheet. Include your percentile rank from table 7.1.

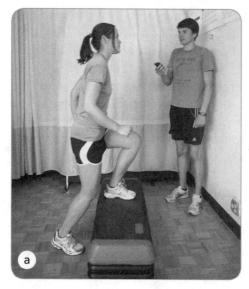 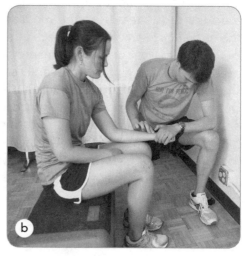

Figure 7.3 Step test: (a) starting position and (b) taking the pulse at the conclusion of the test.

Table 7.1 Percentile Values for Maximal Aerobic Power ($\dot{V}O_2$max, ml·kg^{-1}·min^{-1}) by Age and Sex

%	20-29 y	30-39 y	40-49 y	50-59 y	60-69 y	70-79 y	Rank
Men							
99	61.2	58.3	57.0	54.3	51.1	49.7	Superior
95	56.2	54.3	52.9	49.7	46.1	42.4	
90	54.0	52.5	51.1	46.8	43.2	39.5	Excellent
85	52.5	50.7	48.5	44.6	41.0	38.1	
80	51.1	47.5	46.8	43.3	39.5	36.0	
75	49.2	47.5	45.4	41.8	38.1	34.4	Good
70	48.2	46.8	44.2	41.0	36.7	33.0	
65	46.8	45.3	43.9	39.5	35.9	32.3	
60	45.7	44.4	42.4	38.3	35.0	30.9	
55	45.3	43.9	41.0	38.1	33.9	30.2	Fair
50	43.9	42.4	40.4	36.7	33.1	29.4	
45	43.1	41.4	39.5	36.6	32.3	28.5	
40	42.2	41.0	38.4	35.2	31.4	28.0	
35	41.0	39.5	37.6	33.9	30.6	27.1	Poor
30	40.3	38.5	36.7	33.2	29.4	26.0	
25	39.5	37.6	35.7	32.3	28.7	25.1	
20	38.1	36.7	34.6	31.1	27.4	23.7	
15	36.7	35.2	33.4	29.8	25.9	22.2	Very Poor
10	35.2	33.8	31.8	28.4	24.1	20.8	
5	32.3	31.1	29.4	25.8	22.1	19.3	
1	26.6	26.6	25.1	21.3	18.6	17.9	
Women							
99	55.0	52.5	51.1	45.3	42.4	42.4	Superior
95	50.2	46.9	45.2	39.9	36.9	36.7	
90	47.5	44.7	42.4	38.1	34.6	33.5	Excellent
85	45.3	42.5	40.0	36.7	33.0	32.0	
80	44.0	41.0	38.9	35.2	32.3	30.2	
75	43.4	40.3	38.1	34.1	31.0	29.4	Good
70	41.1	38.8	36.7	32.9	30.2	28.4	
65	40.6	38.1	35.6	32.3	29.4	27.6	
60	39.5	36.7	35.1	31.4	29.1	26.6	
55	38.1	36.7	33.8	30.9	28.3	26.0	Fair
50	37.4	35.2	33.3	30.2	27.5	25.1	
45	36.7	34.5	32.3	29.4	26.9	24.6	
40	35.5	33.8	31.6	28.7	26.6	23.8	
35	34.6	32.4	30.9	28.0	25.4	22.9	Poor
30	33.8	32.3	29.7	27.3	24.9	22.2	
25	32.4	30.9	29.4	26.6	24.2	21.9	
20	31.6	29.9	28.0	25.5	23.7	21.2	
15	30.5	28.9	26.7	24.6	22.8	20.8	Very Poor
10	29.4	27.4	25.6	23.7	21.7	19.3	
5	26.4	25.5	24.1	21.9	20.1	17.9	
1	22.6	22.7	20.8	19.3	18.1	16.4	

Adapted, by permission, from Cooper Institute, *Physical fitness assessments and norms for adults and law enforcement* (Dallas, TX: The Cooper Institute), 24. For more information: www.cooperinstitute.org.

Laboratory Activity 7.1

QUESTION SET 7.1

1. Complete the individual and group data sheets with the data collected in today's lab.

2. If the step test were being used for occupational screening and required a fitness score of 45 ml·kg⁻¹·min⁻¹, how many of your labmates would be able to get the job?

3. Do you think that using a fitness score as a screening tool for physically taxing occupations is fair to all participants? Why or why not? If yes, which occupations? Why?

4. In what other settings do you see a step test being useful? Why?

 Find the case studies for this laboratory online at www.HumanKinetics.com/ LaboratoryManualForExercisePhysiology.

Laboratory Activity 7.1 Individual Data Sheet

Name or ID number: _____ Date: _____

Tester: _____ Time: _____

Sex: M / F (circle one)   Age: _____ y   Height: _____ in. _____ cm

Temperature: _____°F _____°C   Weight: _____ lb _____ kg

Barometric pressure: _____ mmHg   Relative humidity: _____%

Raw Data

Age-predicted HRmax: _____ beats · min^{-1}

Resting 15 s pulse: _____   Resting HR: _____ beats · min^{-1}

3:05 to 3:20 pulse count: _____   Recovery HR: _____ beats · min^{-1}

$\dot{V}O_2$max Determination

Men

$$111.33 - (0.42 \times \underbrace{}_{\text{Recovery HR (beats·min}^{-1})}) = \underbrace{}_{\dot{V}O_2\text{max (ml·kg}^{-1}\cdot\text{min}^{-1})}$$

Women

$$65.81 - (0.1847 \times \underbrace{}_{\text{Recovery HR (beats·min}^{-1})}) = \underbrace{}_{\dot{V}O_2\text{max (ml·kg}^{-1}\cdot\text{min}^{-1})}$$

Percentile rank: _____

$\dot{V}O_2$max classification: _____

SUBMAXIMAL TREADMILL TEST

EQUIPMENT

- Treadmill
- Stethoscope and sphygmomanometer
- Stopwatch
- Individual and group data sheets

Find the group data sheets for this laboratory online at www.HumanKinetics.com/ LaboratoryManualForExercisePhysiology.

EBBELING SUBMAXIMAL TREADMILL TEST

The Ebbeling submax test is a single-stage treadmill test used to estimate cardiorespiratory fitness in low-risk men and women between 20 and 59 years of age (1). This test has shown high predictability ($R^2 = 0.86$) through multiple regression for men and women over a wide age range. It is also flexible in its chosen walking speed, thus allowing for a wide variety of fitness levels, and because it does not take much time it allows multiple tests to be conducted at a given time.

Step 1: Since the accuracy of the test relies on the heart rate response, try to eliminate factors that might alter this outcome measure. Ideally, subjects will have avoided exercise for the previous 24 h, fasted for at least 2 h, and avoided the use of foods and drugs that alter heart rate (e.g., coffee, soda, energy drinks, diet pills, beta-blockers).

Step 2: Select a walking speed between 2 and 4.5 mph based on your subject's age, sex, and anticipated fitness level.

Step 3: Warm up your subject at the selected speed and 0% grade for 4 min. At minutes 2 and 4, take HR, BP, and RPE. The warm-up should elicit a HR of 50% to 70% of age-predicted maximum. If the HR response is too low or too high, adjust the treadmill speed and repeat the warm-up.

Step 4: The 4 min single stage test begins by increasing the grade to 5% at the same speed established in the warm-up. At minutes 2 and 4 of this stage, collect HR, BP, and RPE.

Which HR to use?
HR last 30 sec 5:05 3:30-4:00
(avg HR last 30 sec 5:05 3:30-4:00 = 7:30 8)

Step 5: Cool down the subject by lowering the grade to 0% (as in the warm-up). At minutes 2 and 4 of the cool-down, collect HR, BP, and RPE data.

Step 6: Calculate $\dot{V}O_2$max by using the following equation:

$$\dot{V}O_2\text{max} = 15.1 + (21.8 \times \text{speed in mph}) - (0.327 \times \text{HR}) - 0.263 \times (\text{speed} \times \text{age in yr}) + 0.00504 \times (\text{HR} \times \text{age}) + 5.98 \times (\text{sex where } 0 = \text{female and } 1 = \text{male})$$

QUESTION SET 7.2

Consider for Conclusions of Submax tests

1. Complete the individual and group data sheets showing the data collected in today's lab.

2. How do the predictions of $\dot{V}O_2$max compare between the step test and the treadmill test? What could explain the differences? Which test do you feel is the most accurate? Why?

3. How do your values from the tests compare with the norms? $\dot{V}O_2$max norms are provided in table 7.1.

Find the case studies for this laboratory online at www.HumanKinetics.com/ LaboratoryManualForExercisePhysiology.

Laboratory Activity 7.2

Laboratory 7.2 Individual Data Sheet

Name or ID number: _____ Date: _____

Tester: _____ Time: _____

Sex: M / F (circle one)   Age: _____ y   Height: _____ in. _____ cm

Temperature: _____ °F _____ °C   Weight: _____ lb _____ kg

Barometric pressure: _____ mmHg   Relative humidity: _____ %

Raw Data

Age-predicted HRmax: _____ beats · min^{-1}

Resting BP: _____ / _____ mmHg

Resting HR: _____ beats · min^{-1}

50%–70% age-predicted HRmax: _____ – _____ beats · min^{-1}

	Time (min)	Speed (mph)	Grade (%)	HR (beats · min^{-1})	BP (mmHg)	RPE
Warm-up	2		0			
Warm-up	4		0			
Stage 1	6		5			
Stage 1	8		5			
Cool-down	10		0			
Cool-down	12		0			

$\dot{V}O_2$max Determination

$$15.1 + \left(21.8 \times \underline{\hspace{3cm}}_{\text{Speed in mph}}\right) - \left(0.327 \times \underline{\hspace{2cm}}_{\text{HR}}\right)$$

$$- 0.263 \times \left(\underline{\hspace{3cm}}_{\text{Speed in mph}} \times \underline{\hspace{2cm}}_{\text{Age (y)}}\right)$$

$$+ 0.0050 \times \left(\underline{\hspace{2cm}}_{\text{HR}} \times \underline{\hspace{2cm}}_{\text{Age (y)}}\right)$$

$$+ \left(5.98 \times \underline{\hspace{4cm}}_{\text{Sex where } 0 = \text{female and } 1 = \text{male}}\right)$$

$$= \underline{\hspace{4cm}}_{\dot{V}O_2 \max (\text{ml} \cdot \text{kg}^{-1} \cdot \text{min}^{-1})}$$

Percentile rank:_____

$\dot{V}O_2$max classification:_____

SUBMAXIMAL CYCLE ERGOMETER TEST

EQUIPMENT

- Monark cycle ergometer
- Stethoscope and sphygmomanometer
- Metronome
- Stopwatch
- Individual and group data sheets

Find the group data sheets for this laboratory online at www.HumanKinetics.com/ LaboratoryManualForExercisePhysiology.

YMCA BICYCLE ERGOMETER TEST

The YMCA protocol is a multistage submaximal cycle ergometer test used to estimate cardiorespiratory fitness in men and women (2). This test uses the near-linear relationship between heart rate and work output to extrapolate an age-predicted maximum HR. The YMCA protocol involves two to four 3 min stages of continuous exercise designed to raise the steady-state HR of the subject to between 110 beats·min^{-1} and 70% of heart rate reserve (HRR) or 85% of the age-predicted maximal HR (HRmax; 170 beats·min^{-1} for a 20-year-old) for at least two consecutive stages. Each work rate is performed for 3 min, and HR is recorded during the final 15 to 30 s of each minute. An HR that varies by more than 6 beats·min^{-1} within a stage indicates nonsteady state; in this case, the work rate should be maintained for an additional minute. The test administrator should recognize the error associated with age-predicted maximal HR and monitor the subject throughout the test to ensure that the test remains submaximal. The HR measured during the last minute of each steady state stage is plotted against work rate, which allows you to extrapolate maximum power output at the age-predicted HRmax. This allows you to predict maximum $\dot{V}O_2$.

Test administrators should use the protocol outlined in the following steps. Cadence must remain at 50 rev·min^{-1} for the entire test. Subjects should continue without interruption until the desired heart rate response is elicited (two heart rates from two different stages between 110 and about 170 beats·min^{-1}).

Step 1: One person from each group acts as the subject and performs the submaximal test while the remaining group members administer the test.

Step 2: Set the bicycle seat height so that the knee is bent at 5 to 15° when the leg is extended (figure 7.4*a*).

Step 3: Have the subject rest for 3 min, then, with the subject sitting on the bike, record his or her resting HR and BP before beginning the warm-up (figure 7.4*a*). The warm-up should include pedaling at 50 rev · min^{-1} at a resistance of 0 to 0.5 kg.

Step 4: Subject starts pedaling at 50 rev · min^{-1} at stage 1. If the cycle ergometers are not equipped with cadence indicators, then use a metronome set at either 50 or 100 beats · min^{-1} for the subject to follow with either one leg or both legs, respectively.

Step 5: Set the resistance on the ergometer to 0.5 kg. Start the stopwatch to monitor stage time.

Step 6: Follow Table 7.2 for the procedures within each stage. Take HR by palpation at each minute of the 3 min stages using a 15 s count. If available, use a HR monitor and compare its results with your palpation HR.

Step 7: Take BP and RPE beginning at 2 min into the stage (figure 7.4*b*).

Step 8: Throughout the stage, be aware of your subject's adherence to the 50 rev · min^{-1} cadence.

Step 9: Also be aware of the subject's subjective symptoms (e.g., pallor, dyspnea).

Step 10: If this is your first attempt at conducting the test, it may be difficult to complete all the necessary tasks within the 3 min stage. Thus, in this early phase of skill development, it is acceptable to extend the stage time in order to complete the necessary measurements. As you

Figure 7.4 YMCA bicycle ergometer test: (*a*) taking resting HR before the start of the test and (*b*) taking RPE after first stage.

become more practiced, strive to complete the data collection within the intended stage time.

Step 11: Following stage 1, adjust the resistance in accordance with the test progression provided in Figure 7.5 depending on the HR response in stage 1. Each test will start at the same 150 kg · m · min⁻¹. Following this stage, determine resistance from the HR in the last minute of stage 1. Once the second stage is determined, the test continues in this established track. A common mistake made by students is to jump between tracks in the third stage. If at any time the subject exceeds 85% of age-predicted maximum HR, you should stop the test. If an HR falls on the verge of two workloads, choose the higher workload.

Step 12: Complete data collection of HR, BP, and RPE for stages 2 and 3 in the same manner used for stage 1.

Step 13: Throughout the stages, be aware of your subject's adherence to the 50 rev · min⁻¹ cadence, as well as any subjective symptoms (e.g., pallor, dyspnea).

Step 14: Following stage 3, confirm that two steady state HRs are obtained between 110 beats · min⁻¹ and 85% of age-predicted HRmax (e.g., 170 beats · min⁻¹ for a 20-year-old). If not, then continue with stage 4.

Step 15: Cool the subject down at 0.5 kg resistance at 50 rev · min⁻¹. Record HR, BP, and RPE every 2 min. Continue the cool-down for 4 min or until the subject's HR is <120 beats · min⁻¹.

Step 16: Attempt to conduct 2 or 3 tests during the course of class. Rotate testing duties to practice all the necessary skills.

Step 17: Follow table 7.2 and figure 7.5 and record all data on the individual data sheet.

Table 7.2 Suggested Procedures for YMCA Submaximal Cycle Ergometer Test

Stage time (min:s)	Procedure
0:00–0:45	Monitor client's cadence, ergometer resistance, and subjective symptoms (e.g., pallor, dyspnea). Idle chatter can relax the client, but stay on task!
0:45–1:00	Take 15 s HR by palpation.
1:00–1:45	Monitor client's cadence, ergometer resistance, and subjective symptoms (e.g., pallor, dyspnea). Idle chatter can relax the client, but stay on task!
1:45–2:00	Take 15 s HR by palpation.
2:00–2:30	Take client's BP for this stage. Be sure the client relaxes his or her grip on the handlebars.
2:30–2:45	Take client's RPE for this stage.
2:45–3:00	Take 15 s HR by palpation.

Based on Golding (2).

HR
in 3rd min

• Stage 1: 150 kg·m·min⁻¹ (0.5 Kp)

>100

• Stage 2: 300 kg·m·min⁻¹ (1.0 Kp)
• Stage 3: 450 kg·m·min⁻¹ (1.5 Kp)
• Stage 4: 600 kg·m·min⁻¹ (2.0 Kp)

90-100

• Stage 2: 450 kg·m·min⁻¹ (1.5 Kp)
• Stage 3: 600 kg·m·min⁻¹ (2.0 Kp)
• Stage 4: 750 kg·m·min⁻¹ (2.5 Kp)

80-89

• Stage 2: 600 kg·m·min⁻¹ (2.0 Kp)
• Stage 3: 750 kg·m·min⁻¹ (2.5 Kp)
• Stage 4: 900 kg·m·min⁻¹ (3.0 Kp)

<80

• Stage 2: 750 kg·m·min⁻¹ (2.5 Kp)
• Stage 3: 900 kg·m·min⁻¹ (3.0 Kp)
• Stage 4: 1050 kg·m·min⁻¹ (3.5 Kp)

Figure 7.5 Guide to setting workloads on a bicycle ergometer.

Based on Golding (2).

FORMULAS

Calculating Power on the Bike

$$\text{Power } (\text{kg}\cdot\text{m}\cdot\text{min}^{-1}) = \text{rev}\cdot\text{min}^{-1} \times \text{resistance (kg)} \times 6 \text{ m}\cdot\text{rev}^{-1}$$

Prediction of Oxygen Consumption on Cycle Ergometer

$$\dot{V}O_2 \text{ (ml}\cdot\text{kg}^{-1}\cdot\text{min}^{-1}) = (1.8 \times \text{work rate [kg}\cdot\text{m}\cdot\text{min}^{-1}] \text{ / BW [kg]}) + 7 \text{ ml}\cdot\text{kg}^{-1}\cdot\text{min}^{-1}$$

Watts Conversion

$$\text{Watts} = (\text{kg}\cdot\text{m}\cdot\text{min}^{-1}) / 6.12$$

Laboratory Activity 7.3

Estimating $\dot{V}O_2$max

Body mass is less of a factor in work on a cycle ergometer. As a result, oxygen consumption is often expressed in $L \cdot min^{-1}$. However, maximal oxygen consumption is best expressed as $ml \cdot kg^{-1} \cdot min^{-1}$ when comparing data between subjects. We will try two methods for estimating $\dot{V}O_2$ max from the data you collected.

Method A

First, using graph paper or computer spreadsheet, plot a line using heart rate (dependent variable) and workload (independent variable) to establish estimated maximum work rate by linearly extrapolating to age-predicted maximum HR. Use the following equation to estimate maximum oxygen uptake ($\dot{V}O_2$ max) from this estimated maximum work rate.

$$\dot{V}O_2 \text{ max } (ml \cdot kg^{-1} \cdot min^{-1}) = (1.8 \times \text{max work rate } [kg \cdot m \cdot min^{-1}] \, / \, BW \, [kg]) + 7 \, ml \cdot kg^{-1} \cdot min^{-1}$$

The result is the estimated $\dot{V}O_2$ max.

Method B

This method estimates $\dot{V}O_2$ max numerically from YMCA test data without the need for graph paper.

First, determine submaximal $\dot{V}O_2$ (SM) for each steady state work output using the ACSM metabolic equation for leg cycling (for cycle workloads between 300 and 1,200 $kg \cdot m \cdot min^{-1}$):

$$\dot{V}O_2 \, (ml \cdot kg^{-1} \cdot min^{-1}) = (1.8 \times \text{work rate } [kg \cdot m \cdot min^{-1}] \, / \, BW \, [kg]) + 7 \, ml \cdot kg^{-1} \cdot min^{-1}$$

Second, determine the slope (m) of the *linear* HR and $\dot{V}O_2$ relationship:

$$\text{slope (m)} = (SMA - SMB) \, / \, (HRA - HRB)$$

where: SMA = submax predicted $\dot{V}O_2$ from stage A ($ml \cdot kg^{-1} \cdot min^{-1}$)

SMB = submax predicted $\dot{V}O_2$ from stage B ($ml \cdot kg^{-1} \cdot min^{-1}$)

HRA = steady state HR from stage A ($beats \cdot min^{-1}$)

HRB = steady state HR from stage B ($beats \cdot min^{-1}$)

Note: Stage A and stage B refer not to the first and second stages but to the two stages where target HR (between 110 $beats \cdot min^{-1}$ and 85% of age-predicted HRmax) was attained.

Finally, solve the following equation for $\dot{V}O_2$max ($ml \cdot kg^{-1} \cdot min^{-1}$):

$$\dot{V}O_2\text{max} = SMB + m \times (HRmax - HRB)$$

These equations are valid for estimating oxygen consumption at submaximal steady state workloads from 300 to 1,200 $kg \cdot m \cdot min^{-1}$ (50–200 W); therefore, caution must be used if extrapolating to workloads outside of this range. A large part of the error involved in estimating O_2 from submaximal HR responses occurs because the formula (220 − age) can provide only an estimate of maximal HR (+12 $beats \cdot min^{-1}$).

QUESTION SET 7.3

1. Calculate the estimated $\dot{V}O_2$ max (both absolute and relative) for the YMCA cycle test using both methods A and B.

2. Construct a graph illustrating the relationship between workload (kg·m·min^{-1}) and $\dot{V}O_2$ (ml·kg^{-1}·min^{-1}) on the cycle ergometer. Include SMA, SMB, and predicted $\dot{V}O_2$ max.

3. Prepare a graph illustrating the relationship between systolic BP, diastolic BP, and workload during the exercise test.

4. Give normative $\dot{V}O_2$ max values for both males and females of the following groups: elite marathon runners, sprinters, Olympic weightlifters, college-age males, and college-age females. You will have to use a reference source other than the lab manual (e.g., your text or the web).

5. What was *your* estimated $\dot{V}O_2$ value from the YMCA cycle test? Do you think this is an accurate measure of your fitness? Why or why not?

6. How do the values differ between techniques for calculating $\dot{V}O_2$ max (i.e., plotting vs. numerical, methods A and B)? For each method, name at least one source of error that might lead to an inaccurate estimation, based on the methods involved and the information required for each. (Hint: Think about the assumptions inherent in the test.)

 Find the case studies for this laboratory online at www.HumanKinetics.com/ LaboratoryManualForExercisePhysiology.

Laboratory 7.3   Individual Data Sheet

Name or ID number: _____   Date: _____

Tester: _____   Time: _____

Sex: M / F (circle one)   Age: _____ y   Height: _____ in. _____ cm

Temperature: _____°F _____°C   Weight: _____ lb _____ kg

Barometric pressure: _____ mmHg   Relative humidity: _____ %

Raw Data

Age-predicted HRmax: _____ beats·min^{-1}

Resting BP: _____ / _____ mmHg

Resting HR: _____ beats·min^{-1}

85% age-predicted HRmax: _____ beats·min^{-1}

	Time (min)	Workload (kg)	HR (beats·min^{-1})	HR monitor	BP (mmHg)	RPE
1st stage	1	0.5			____	____
	2	0.5				____
	3	0.5			____	
2nd stage	4				____	____
	5					____
	6				____	
3rd stage	7				____	____
	8					____
	9				____	
4th stage (if necessary)	10				____	____
	11					____
	12				____	

Recovery

Time (min)	Workload (kg)	HR (beats · min⁻¹)	HR monitor	BP (mmHg)	RPE
1–2	0.5				
3–4	0.5				

Work rate (kg·m·min⁻¹) = workload (kg) × rev·min⁻¹ × 6 m·rev⁻¹

Stage 1: 0.5 kg × 50 rev·min⁻¹ × 6 m·rev⁻¹ = 150 kg·m·min⁻¹

Stage 2: _____ kg × 50 rev·min⁻¹ × 6m·rev⁻¹ = _____ kg·m·min⁻¹

Stage 3: _____ kg × 50 rev·min⁻¹ × 6m·rev⁻¹ = _____ kg·m·min⁻¹

$\dot{V}O_2$max Determination

Method A

$$1.8 \times \frac{\text{Estimated max work rate (kg·m·min}^{-1})}{} \bigg/ \frac{\text{Body weight (kg)}}{} + 7\,\text{ml·kg}^{-1}\cdot\text{min}^{-1}$$

$$= \frac{}{\dot{V}O_2\text{max (ml·kg}^{-1}\text{·min}^{-1})}$$

Percentile rank:_____

$\dot{V}O_2$max classification:_____

Method B

$$1.8 \times \frac{\text{Stage A work rate (kg·m·min}^{-1})}{} \bigg/ \frac{\text{Body weight (kg)}}{} + 7\,\text{ml·kg}^{-1}\cdot\text{min}^{-1}$$

$$= \frac{}{\text{SMA}}$$

$$1.8 \times \frac{\text{Stage B work rate (kg·m·min}^{-1})}{} \bigg/ \frac{\text{Body weight (kg)}}{} + 7\,\text{ml·kg}^{-1}\cdot\text{min}^{-1}$$

$$= \frac{}{\text{SMB}}$$

$$\left(\frac{}{\text{SMA}} - \frac{}{\text{SMB}} \right) \bigg/ \left(\frac{}{\text{HRA}} - \frac{}{\text{HRB}} \right)$$

$$= \frac{}{\text{Slope (m)}}$$

$$\frac{}{\text{SMB}} + \frac{}{\text{Slope (m)}} \times \left(\frac{}{\text{HRmax}} - \frac{}{\text{HRB}} \right)$$

$$= \frac{}{\dot{V}O_2\text{max (ml·kg}^{-1}\text{·min}^{-1})}$$

Percentile rank:_____

$\dot{V}O_2$max classification:_____

Aerobic Power Field Assessments

Objectives

1. Explain how to perform several field assessments for aerobic power.
2. Conduct the Cooper 1.5-mile run/walk test, the Cooper 12-minute run/walk test, and the Rockport fitness walking test.
3. Estimate maximal aerobic power by means of several different field tests.
4. Evaluate the results of a field test for aerobic power.

Definitions

aerobic power—Maximal capacity for resynthesizing ATP with aerobic means; best determined with a graded exercise test to exhaustion ($\dot{V}O_2$max).

field test of aerobic power—Test, not requiring the individual to be in the laboratory, in which the subject either performs a timed completion of a set distance or completes a maximal distance in a set time.

graded exercise test—Test in which intensity is increased progressively in order to test cardiovascular health, wellness, and function (where intensity is increased by increasing speed, resistance, incline, or some combination thereof).

intraclass correlation—Indicator of the reliability of a measure.

maximal testing—Test requiring the subject to go to volitional failure in order to evaluate maximal heart rate, maximal oxygen consumption, or aerobic power.

palpation—Examination by touch; typically used to estimate heart rate when no heart rate monitor is available.

R—Multiple regression correlation coefficient that represents how well a dependent variable can be predicted from a combination of independent variables; ranges between -1 and 1.

regression equation—Statistical method used to relate two or more variables.

standard error of the estimate—Measure of the accuracy of a predictive equation (with a smaller value indicating a better estimate).

submaximal aerobic power test—Test used to estimate $\dot{V}O_2$max; typically performed in a laboratory setting and involving walking, jogging, running, or cycling.

$\dot{V}O_2$max—Maximal oxygen consumption or aerobic power.

Many consider the assessment of aerobic power, or fitness, to be one of the more important methods for determining an individual's overall cardiovascular fitness. You can choose from multiple methods for determining an individual's **aerobic power**—directly measure maximal oxygen consumption ($\dot{V}O_2\textbf{max}$) or estimate aerobic power from submaximal testing in the laboratory setting (4).

The best method for assessing overall cardio-respiratory capacity is generally considered to be direct measurement of oxygen consumption during a **graded exercise test** (5). However, though this method is viewed as the gold standard, it requires specialized equipment and trained personnel who can administer the test under controlled conditions (16). In addition, **maximal testing** may be contraindicated for some individuals (2).

One alternative method for determining aerobic power is the use of submaximal testing. These procedures typically require less medical supervision, personnel, and equipment than maximal tests call for (2). Submaximal tests also tend to require less time, and, because they use fixed workloads, they are ideal for assessing heart rate and blood pressure at predetermined time increments. At the same time, a **submaximal aerobic power test** does have its own limitations:

- $\dot{V}O_2$max is predicted rather than directly measured.
- Many of the equations used in conjunction with submaximal tests involve a 10% to 20% prediction error.
- This type of testing offers limited diagnostic capabilities for certain types of tests.
- Since these tests do not measure a maximal heart rate, they are of limited usefulness in formulating exercise prescriptions.

The theoretical basis for submaximal testing centers on the relationship between heart rate, oxygen consumption ($\dot{V}O_2$), and workload (5). Specifically, the linear relationship between heart rate, $\dot{V}O_2$, and workload (figure 8.1) makes it is possible to predict $\dot{V}O_2$ from the workload and heart rates achieved during a submaximal test. Numerous submaximal tests can be employed based on these relationships, but the need for specialized equipment makes them impractical in some situations.

An alternative method for estimating aerobic power is to use either a maximal or submaximal **field test of aerobic power**. These tests are easy alternatives to laboratory based assessments because they require minimal equipment and can be used with large groups of individuals at one

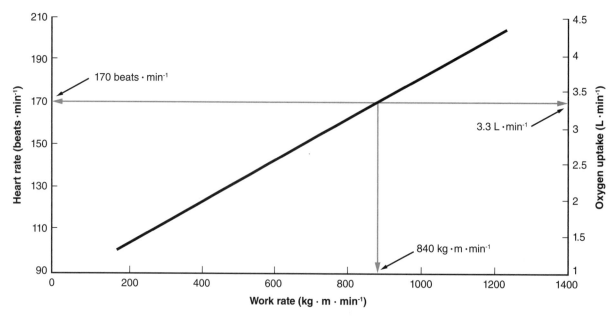

Figure 8.1 Relationship between heart rate, oxygen consumption, and workload.

time (1,2). This type of test comes in two major categories—those that measure the time needed to complete a set distance and those that measure the distance covered in a set amount of time (2). The results of a field test of aerobic power can be used to estimate the $\dot{V}O_2$max by means of specialized prediction equations. Major disadvantages of field-based assessment include the following:

- In some individuals, this type of test could be a maximal test.
- The individual's motivation and pacing strategy during the test can have a profound effect on the final outcome.
- This type of testing does not allow comprehensive monitoring of both heart rate and blood pressure during the test.

Field tests are generally not recommended for sedentary individuals who have been identified in pretesting screenings to be at moderate or high risk of cardiorespiratory or musculoskeletal complications (6).

The most common field tests of aerobic power are the Cooper 1.5-mile (2.4 km) run/walk test, the Cooper 12-minute run/walk test, and the Rockport fitness walking test. These tests can be used by a variety of individuals with minimal equipment (11). The Cooper 1.5-mile and Cooper 12-minute tests are considered to be maximal tests, whereas the Rockport test is considered to be a submaximal field test (13). They all allow you to make reasonable estimates of aerobic power.

Cooper 1.5-Mile Run/Walk Test

This test is one of the more popular field test approaches to estimating an individual's aerobic power (1, 2, 13). Because it is classified as a maximal field test, pre-exercise screening is necessary in order to determine whether the individual has known heart disease or risk factors indicative of heart disease. The American College of Sports Medicine suggests that this test is not appropriate for unconditioned individuals or those with heart disease or known heart disease risk factors (2).

The 1.5-mile (2.4 km) test estimates an individual's oxygen consumption by measuring the time it takes to run or walk 1.5 miles. Even though walking is allowed during the test, the objective is

to cover the 1.5-mile distance as fast as possible (13)—the faster the subject covers the distance, the higher his or her estimated $\dot{V}O_2$max. You can estimate the individual's $\dot{V}O_2$max by means of a **regression equation**:

Men

$$\dot{V}O_2\text{max (ml} \cdot \text{kg}^{-1} \cdot \text{min}^{-1}) =$$
$$91.736 - (0.1656 \times \text{body mass [kg]}) -$$
$$(2.767 \times 1.5\text{-mile run time in minutes})$$

Women

$$\dot{V}O_2\text{max (ml} \cdot \text{kg}^{-1} \cdot \text{min}^{-1}) =$$
$$88.020 - (0.1656 \times \text{body mass [kg]}) -$$
$$(2.767 \times 1.5\text{-mile run time in minutes})$$

These equations have been shown to be highly reliable, and the **standard error of the estimate** represents about 6.0% of the $\dot{V}O_2$max (9, 12). Both equations are also valid because of the high correlation ($\boldsymbol{R} = 0.90$) to the actual maximal testing results and the relatively low standard error of the estimate (13).

Another method for estimating $\dot{V}O_2$max from a 1.5-mile (2.4 km) run or walk time uses the following equation:

$$\dot{V}O_2\text{max (ml} \cdot \text{kg}^{-1} \cdot \text{min}^{-1}) =$$
$$65.404 + (7.707 \times \text{sex}) -$$
$$0.159 \times \text{body mass (kg)} - 0.843 \times \text{time (min)}$$

where sex = 1 if male, 0 if female.

This equation also produces valid results as indicated by the correlation coefficient ($R = 0.86$) and the standard error of the estimate (3.37 ml $\cdot$ kg^{-1} $\cdot$ min^{-1}). It is considered reliable because of the high **intraclass correlation** (ICC = 0.93).

Cooper 12-Minute Run/Walk Test

The 12-minute run/walk test was originally used in the development of a regression equation by Cooper (6) in order to estimate $\dot{V}O_2$max. The test was originally based on the linear relationship determined between a 12-minute "best effort" run and running velocity (150–300 m $\cdot$ min^{-1}) and steady state $\dot{V}O_2$ (3). The original investigation reported that the regression equation was highly correlated ($R = 0.897$) to the actual $\dot{V}O_2$max determined with the Balke protocol (4). While the original work clearly demonstrated that the test was valid, more recent work has reported greater variability in the validity coefficient, ranging from $r = 0.28$ to $r = 0.94$ for the measured $\dot{V}O_2$max and the estimated $\dot{V}O_2$max determined with a 12-min run/walk test (16). This greater variability may in part be explained by the type of criterion $\dot{V}O_2$max test used to determine the actual maximum capacity (15). Despite such discrepancy regarding validity, the 12-minute test remains widely accepted as accurately predicting $\dot{V}O_2$max (2, 6, 11, 15, 16).

Since the goal of this test is to cover as much distance as possible in the allotted time, it can be considered a maximal field test (11) and thus, like the 1.5-mile test, is generally not recommended by the American College of Sports Medicine for unconditioned individuals or those with heart disease or known heart disease risk factors (2). The total distance traveled during the 12-minute time period is used to estimate $\dot{V}O_2$max by means of the following equation:

$$\dot{V}O_2\text{max (ml} \cdot \text{kg}^{-1} \cdot \text{min}^{-1}) = 0.0268 \times \text{(distance in meters)} - 11.3$$

Rockport Fitness Walking Test

The Rockport test was originally developed by the Rockport Institute as a method for assessing cardiorespiratory fitness in men and women from age 20 to age 69. The test was further developed by Kline et al. (14) as a submaximal field test for estimating $\dot{V}O_2$max with the use of a 1-mile (1.6 km) walking protocol. This test was found to exhibit high validity coefficients ($r = 0.93$) and a low standard error of the estimate (SEE = 0.325 L·min^{-1}) (13), which suggests that this test yields a valid estimate of $\dot{V}O_2$max (11). Cross-validation of the original equations by Fenstermaker et al. (8) revealed that the Rockport test was valid and appropriate for 65- to 79-year-old women. The following equations can be used for men and women between 30 and 79:

Men (30–79 Years Old)

$$\dot{V}O_2\text{max (ml} \cdot \text{kg}^{-1} \cdot \text{min}^{-1}) = 139.168 - (0.3877 \times \text{age}) - (0.1692 \times \text{body mass in lbs}) - (3.2649 \times 1.0 \text{ mile time in min}) - (0.1565 \times \text{HR})$$

Note: This equation has a validation coefficient of 0.83 to 0.88 and an SEE of 4.5 to 5.3 ml·kg^{-1}·min^{-1}.

Women (30–79 Years Old)

$$\dot{V}O_2\text{max (ml} \cdot \text{kg}^{-1} \cdot \text{min}^{-1}) = 132.853 - (0.3877 \times \text{age}) - (0.1692 \times \text{body mass in lbs}) - (3.2649 \times 1.0 \text{ mile time in min}) - (0.1565 \times \text{HR})$$

Note: This equation has a validation coefficient of 0.59 to 0.88 and an SEE of 2.7 to 5.3 ml·kg^{-1}·min^{-1}.

While the Rockport fitness walking test generally produces valid and reliable data, several studies suggest that the equations used result in an overestimation of $\dot{V}O_2$max in college-age men and women (7, 9). Based on these findings, Dolgener et al. (5) modified the original equation to better estimate the $\dot{V}O_2$max achieved for 18- to 29-year-old men and women:

Men (18–29 Years Old)

$$\dot{V}O_2\text{max (ml} \cdot \text{kg}^{-1} \cdot \text{min}^{-1}) = 97.660 - (0.0957 \times \text{body mass in lb}) - (1.4537 \times 1.0 \text{ mile walk time in min}) - (0.1194 \times \text{HR})$$

Note: This equation has a validation coefficient of 0.50 to 0.85 and an SEE of 3.5 to 5.8 ml·kg^{-1}·min^{-1}.

Women (18–29 Years Old)

$$\dot{V}O_2\text{max (ml} \cdot \text{kg}^{-1} \cdot \text{min}^{-1}) = 88.768 - (0.0957 \times \text{body mass in lb}) - (1.4537 \times 1.0 \text{ mile walk time in min}) - (0.1194 \times \text{HR})$$

Note: This equation has a validation coefficient of 0.38 to 0.85 and an SEE of 3.0 to 4.8 ml·kg^{-1}·min^{-1}.

An additional method for calculating the $\dot{V}O_2$max of adults would be to use the following equation:

$$\dot{V}O_2\text{max (ml} \cdot \text{kg}^{-1} \cdot \text{min}^{-1}) = 132.853 + (\text{sex} \times 3.315) - (0.3877 \times \text{age}) - (0.1692 \times \text{body mass in kg}) - (3.2649 \times 1.0 \text{ mile walk time in min}) - (0.1565 \times \text{HR})$$

For college-age subjects, the following equation can be used:

$$\dot{V}O_2max\ (ml \cdot kg^{-1} \cdot min^{-1}) = 88.768 + (sex \times 8.892) - (0.2109 \times body\ mass\ in\ kg) - (1.4537 \times 1.0\ mile\ walk\ time\ in\ min) - (0.1194 \times HR)$$

In these equations, sex = 1 for males and 0 for females.

The Rockport fitness walking test is relatively easy to perform, but it does require the subject to palpate his or her heart rate for 15 s immediately after completing the test. Therefore, it is necessary for an individual who undertakes this test to able to perform this procedure (see the accompanying highlight box).

Basic Procedures for Palpating Heart Rate

Heart rate can be checked via **palpation** at several arterial sites: radial, carotid, temporal, and femoral (see table 8.1) (11). When palpating, use the tips of the middle and index fingers; never use the thumb, which has a pulse of its own and can produce erroneous heart rate counts. The most common palpation sites during exercise and at rest are at the carotid or radial arteries. Be careful when palpating the carotid artery because applying too much pressure to this site can result in a baroreceptor response that lowers heart rate dramatically and can cause the subject to pass out. If you start the stopwatch simultaneously with the initiation of the heartbeat count, use 0 as the first beat counted and then count beats for the preset duration. If, on the other hand, the stopwatch is already running, then initiate the count with 1 (1, 11). When using the Rockport fitness walking test, palpate for heart rate for 15 s immediately after cessation of the test. Once a total number of beats is counted for this 15 s time period, multiply that number by 4 and use the result in the Rockport test equations.

Table 8.1 Sites at Which Heart Rate Can Be Palpated

Site	Location
Brachial artery	About 2–3 cm above the antecubital fossa on the anteromedial aspect of the arm below the belly of the biceps brachii
Carotid artery	Located laterally to the larynx in the neck
Radial artery	In line with the base of the thumb on the anterolateral aspect of the thumb
Temporal artery	At the temple in line with the hairline of the head

Adapted, by permission, from Cooper Institute, *Physical fitness assessments and norms for adults and law enforcement* (Dallas, TX: The Cooper Institute). For more information: www.cooperinstitute.org.

References

1. American College of Sports Medicine. *ACSM's Guidelines for Exercise Testing and Prescription.* 8th ed. Philadelphia: Lippincott Williams & Wilkins, 2010.

2. American College of Sports Medicine. *ACSM's Health-Related Physical Fitness Assessment Manual.* 3rd ed. Baltimore: Lippincott Williams & Wilkins, 2010.

3. Balke B. *A Simple Field Test for the Assessment of Physical Fitness.* Civil Aeromedical Research Institute Report, 63-18. Oklahoma City: Federal Aviation Agency, 1963.

4. Beam WC and Adams GM. *Exercise Physiology Laboratory Manual.* 6th ed. Boston: McGraw-Hill, 2011.

5. Billat V and Lopes P. Indirect Methods for Estimation of Aerobic Power. In: Maud PJ and Foster C, eds., *Physiological Assessment of Human Fitness.* 2nd ed. Champaign, IL: Human Kinetics, 2006, pp. 19–38.

6. Cooper KH. A Means of Assessing Maximal Oxygen Intake: Correlation Between Field and Treadmill Testing. *JAMA* 203: 201–204, 1968.

7. Dolgener FA, Hensley LD, Marsh JJ, and Fjelstul JK. Validation of the Rockport Fitness Walking Test in College Males and Females. *Res Q Exerc Sport* 65: 152–158, 1994.

8. Fenstermaker K, Plowman SA, and Looney M. Validation of the Rockport Walking Test in Females

65 Years and Older. *Res Q Exer Sport* 63: 322–327, 1992.

9. George JD, Fellingham GW, and Fisher AG. A Modified Version of the Rockport Fitness Walking Test for College Men and Women. *Res Q Exer Sport* 69: 205–209, 1998.

10. George JD, Vehrs PR, Allsen PE, Fellingham GW, and Fisher AG. $\dot{V}O_2$max Estimation From a Submaximal 1-Mile Track Jog for Fit College-Age Individuals. *Med Sci Sports Exerc* 25: 401–406, 1993.

11. Heyward VH. *Advanced Fitness Assessment and Exercise Prescription*. 6th ed. Champaign, IL: Human Kinetics, 2010.

12. Hoffman JR. *Norms for Fitness, Performance, and Health*. Champaign, IL: Human Kinetics, 2006.

13. Housh TJ, Cramer JT, Weir JP, Beck TW, and Johnson GO. *Physical Fitness Laboratories on a Budget*. Scottsdale, AZ: Holcomb-Hathaway, 2009, p. 234.

14. Kline GM, Porcari JP, Hintermeister R, Freedson PS, Ward A, McCarron RF, Ross J, and Rippe JM. Estimation of $\dot{V}O_2$max From a One-Mile Track Walk, Gender, Age, and Body Weight. *Med Sci Sports Exer* 19: 253–259, 1987.

15. McCutcheon MC, Stricha SA, Giese MD, and Nagle FJ. A Further Analysis of the 12-Minute Run Prediction of Maximal Aerobic Power. *Res Q Exer Sport* 61: 280–283, 1990.

16. Safrit MJ, Glaucia Costa M, Hooper LM, Patterson P, and Ehlert SA. The Validity Generalization of Distance Run Tests. *Can J Sport Sci* 13: 188–196, 1988.

COOPER 1.5-MILE RUN/ WALK TEST

EQUIPMENT

- Measured 1.5-mile (2.4 km) distance, ideally on a 0.25-mile (400 m) track
- Physician's scale, stadiometer, or equivalent electronic scale
- Stopwatch
- Individual and group data sheets
- Microsoft Excel or equivalent spreadsheet program

Find the group data sheets for this laboratory online at www.HumanKinetics.com/ LaboratoryManualForExercisePhysiology.

Figure 8.2 presents a traditional 440-yard or 0.25-mile track, in which each straightaway and curve is 110 yards long and half of each length is 55 yards. Distance traveled may differ slightly depending on the track; therefore, when performing timed distances, make sure to understand the various lengths on the track. For example, because a 400 m track is 2.5 yd (2.3 m) shorter than a 440-yard track, a subject performing a 1.5-mile run on a 400 m track must go 14 m (15 yd) beyond 6 laps to complete a mile and a half. Similarly, when

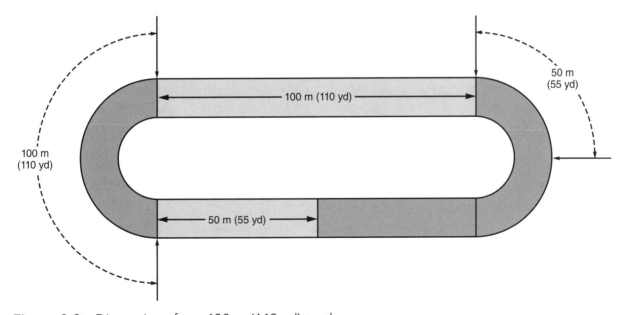

Figure 8.2 Dimensions for a 400 m (440 yd) track.

performing a 1-mile (1.6 km) Rockport test, one must go 9 m (9 yd) beyond 4 laps to complete a mile.

WARM-UP

As with any performance-based test, the subject should perform a structured warm-up to prepare for the assessment. As a rule, when working with athletes or other fit individuals, devote 5 min to general warm-up activity (e.g., jogging, cycling, jumping rope), then use 5 min for dynamic stretching (e.g., high knees, walking lunges, walking knee tucks, butt kicks, inchworms, power skips). With sedentary or untrained individuals, use less rigorous activities (e.g., leg swings, toe touches). After the warm-up, ensure that the subject clearly understands that the objective of the test is to complete the 1.5-mile (2.4 km) distance in as little time as possible. Here is a summary of steps to use when administering this test:

COOPER 1.5-MILE RUN/WALK TEST

Step 1: Measure the height and weight of each person being tested and record the results on their individual laboratory data sheets. Measure body mass to the nearest 0.01 kg and height to the nearest 0.1 cm.

Step 2: Have each subject complete a structured warm-up of about 10 min.

Step 3: Prior to starting the test, clearly explain that each individual should walk or run the 1.5-mile (2.4 km) distance as fast as possible.

Step 4: Start a stopwatch at the same time that the run/walk is initiated.

Step 5: When a subject completes the distance, his or her time should be recorded to the nearest second on the laboratory 8.1 individual data sheets.

Step 6: After completing the assessment, each tested individual should perform a cool-down consisting of slow walking followed by stretching.

Step 7: Use the equations presented on the individual data sheet to estimate each individual's $\dot{V}O_2$max, then record the result.

QUESTION SET 8.1

1. What is the underlying physiological reason for the relationship found between field tests and laboratory measurements of aerobic power?

2. Based on your results, rank your aerobic fitness in relation to the norms and percentile ranks presented in tables 7.1 (p. 171) and 8.2.

3. How did your aerobic fitness results compare with the class averages?

4. Based on the class averages, how would you rate your classmates' overall aerobic fitness?

5. What factors associated with this aerobic power test may result in variations in the values estimated for aerobic power?

Find the case studies for this laboratory online at www.HumanKinetics.com/ LaboratoryManualForExercisePhysiology.

Table 8.2 Percentile Ranks for 1.5-Mile Run/Walk Time (min:s)

Age (y)	20–29		30–39		40–49		50–59		60–69	
Percentile	Men	Women	Men	Women	Men	Women	Men	Women	Men	Women
90	9:34	10:59	9:52	11:43	10:09	12:25	11:09	13:58	12:10	15:32
80	10:08	11:56	10:38	12:53	11:09	13:38	12:08	15:14	13:25	16:46
70	10:49	12:51	11:09	13:41	11:52	14:33	12:53	16:26	14:33	18:05
60	11:27	13:25	11:49	14:33	12:25	15:17	13:53	17:19	15:20	18:52
50	11:58	14:15	12:25	15:14	13:05	16:13	14:33	18:05	16:19	20:08
40	12:29	15:05	12:53	15:56	13:50	17:11	15:14	19:10	17:19	20:55
30	13:08	15:56	13:48	16:46	14:33	18:26	16:16	20:17	18:39	22:34
20	13:58	17:11	14:33	18:18	15:32	19:43	17:30	21:57	20:13	23:55
10	15:14	18:39	15:56	20:13	17:04	21:52	19:24	23:55	23:27	26:32

Adapted, by permission, from Cooper Institute, *Physical fitness assessments and norms for adults and law enforcement* (Dallas, TX: The Cooper Institute). For more information: wwwcooperinstitute.org.

Name or ID number: _____   Tester: _____

Date: _____ Time: _____

Sex: M / F (circle one)   Age: _____ y

Height: _____ in.   _____ cm

Weight: _____ lb   _____ kg

Temperature: _____ °F   _____ °C

Relative humidity: _____ %

Barometric pressure: _____ mmHg

Location of Testing

❑ Outdoor field

❑ Indoor field

❑ Indoor track

❑ Outdoor track

❑ Gym

❑ Other

Footwear

❑ Jogging shoe

❑ Walking shoe

❑ Tennis shoe

❑ Basketball shoe

❑ Running shoe

❑ Cross-trainer

❑ Other _____

Men

$$\dot{V}O_2 \max (ml \cdot kg^{-1} \cdot min^{-1}) = 91.736 - (0.1656 \times \underset{\text{body mass (kg)}}{\underline{\hspace{1.5cm}}}) - (2.767 \times \underset{\text{time (min)}}{\underline{\hspace{1.5cm}}})$$

$$= \underline{\hspace{4cm}} ml \cdot kg^{-1} \cdot min^{-1} \text{(Equation 1)}$$

$$\dot{V}O_2 \max (ml \cdot kg^{-1} \cdot min^{-1}) = 65.404 - (7.707 \times \underset{\text{sex}}{\underline{\hspace{1.5cm}}}) - 0.159 \times \underset{\text{body mass (kg)}}{\underline{\hspace{1.5cm}}} - 0.843 \times \underset{\text{time (min)}}{\underline{\hspace{1.5cm}}}$$

$$= \underline{\hspace{4cm}} ml \cdot kg^{-1} \cdot min^{-1} \text{(Equation 2)}$$

Women

$$\dot{V}O_2 \max (ml \cdot kg^{-1} \cdot min^{-1}) = 88.020 - (0.1656 \times \underset{\text{body mass (kg)}}{\underline{\hspace{1.5cm}}}) - (2.767 \times \underset{\text{time (min)}}{\underline{\hspace{1.5cm}}})$$

$$= \underline{\hspace{4cm}} ml \cdot kg^{-1} \cdot min^{-1} \text{(Equation 1)}$$

$$\dot{V}O_2 \max (ml \cdot kg^{-1} \cdot min^{-1}) = 65.404 - (7.707 \times \underset{\text{sex}}{\underline{\hspace{1.5cm}}}) - 0.159 \times \underset{\text{body mass (kg)}}{\underline{\hspace{1.5cm}}} - 0.843 \times \underset{\text{time (min)}}{\underline{\hspace{1.5cm}}}$$

$$= \underline{\hspace{4cm}} ml \cdot kg^{-1} \cdot min^{-1} \text{(Equation 2)}$$

Percentile rank: _____

$\dot{V}O_2$max classification: _____

COOPER 12-MINUTE RUN/ WALK TEST

EQUIPMENT

- Measured distance, ideally on a 0.25-mile (400 m) track (see figure 8.2, p. 193)
- Physician's scale, stadiometer, or equivalent electronic scale
- Stopwatch
- Individual and group data sheets
- Microsoft Excel or equivalent spreadsheet program

 Find the group data sheets for this laboratory online at www.HumanKinetics.com/ LaboratoryManualForExercisePhysiology.

WARM-UP

As with any performance-based test, the subject should perform a structured warm-up to prepare for the assessment. As a rule, when working with athletes or other fit individuals, devote 5 min to general warm-up activity (e.g., jogging, cycling, jumping rope), then move on to dynamic stretching (e.g., high knees, knee tucks, walking toe touches, inchworms, skipping, and walking lunges). With sedentary or untrained individuals, use less rigorous activities (e.g., leg swings, toe touches). After the warm-up, ensure that the subject clearly understands that the objective of the test is to travel as far as possible in the allotted 12 min. Here is a summary of steps to use when administering this test:

COOPER 12-MINUTE RUN/WALK TEST

Step 1: Measure the height and weight of each person being tested and record the results on his or her individual laboratory data sheet. Measure body mass to the nearest 0.01 kg and height to the nearest 0.1 cm.

Step 2: Have each subject complete a 5 min general warm-up followed by 5 min of dynamic stretching.

Step 3: Clearly explain that the objective of the test is to cover as much distance as possible in the allotted 12 min.

Step 4: Start the stopwatch at the same time that the 12-min run/walk test is initiated.

Step 5: Provide encouragement to those undertaking the test.

Step 6: Make periodic time checks and provide feedback to those undertaking the test—for example, "5 minutes to go," "1 minute remaining," "10 seconds . . . 5 seconds . . . and stop."

Step 7: Estimate the distance covered based on the number of laps completed and the place on the track where the subject stopped. Record this distance on the individual data sheet.

Step 8: Allow adequate time for a cool-down consisting of slow walking and stretching.

Step 9: Calculate each individual's $\dot{V}O_2$max using the equations presented on the individual data sheet and record the results on the group data sheet.

QUESTION SET 8.2

1. Why is the 12-minute run/walk test not recommended for unconditioned individuals or those who have heart disease?

2. Based on your results, rank your aerobic fitness in relation to the norms and percentile ranks presented in tables 7.1 (p. 171) and 8.3?

3. How did your aerobic fitness results compare with the class averages?

4. Based on the class averages, how would you rate your classmates' overall aerobic fitness?

5. Do you find a difference between the percentile ranks determined for distance covered and estimated $\dot{V}O_2$max? If so, why do you think this is?

Find the case studies for this laboratory online at www.HumanKinetics.com/ LaboratoryManualForExercisePhysiology.

missing %

Table 8.3 Percentile Ranks for 12-Minute Run/Walk Distance

20–29				30–39				40–49				60–69				≥60			
Men		Women		Men		Women		Men		Women		Men		Women		Men		Women	
mi	km	mi	km	mi	km	mi	km	mi	km	mi	km	mi	km	mi	km	mi	km	mi	km
1.81	2.91	1.63	2.62	1.77	2.85	1.56	2.51	1.73	2.78	1.49	2.40	1.61	2.59	1.27	2.04	1.51	2.43	1.29	2.08
1.73	2.78	1.54	2.48	1.67	2.69	1.41	2.27	1.61	2.59	1.29	2.08	1.52	2.45	1.21	1.95	1.41	2.27	1.18	1.90
1.65	2.66	1.46	2.35	1.61	2.59	1.39	2.24	1.54	2.48	1.23	1.98	1.45	2.33	1.15	1.85	1.33	2.14	1.13	1.82
1.58	2.54	1.41	2.27	1.55	2.49	1.33	2.14	1.49	2.40	1.19	1.92	1.38	2.22	1.12	1.80	1.29	2.08	1.07	1.72
1.53	2.46	1.35	2.17	1.49	2.40	1.29	2.08	1.44	2.32	1.15	1.85	1.33	2.14	1.08	1.74	1.23	1.98	1.03	1.66
1.49	2.40	1.3	2.09	1.45	2.33	1.25	2.01	1.38	2.22	1.11	1.79	1.29	2.08	1.05	1.69	1.19	1.92	0.99	1.59
1.43	2.30	1.25	2.01	1.38	2.22	1.21	1.95	1.33	2.14	1.07	1.72	1.24	2.00	1.01	1.63	1.13	1.82	0.97	1.56
1.37	2.20	1.19	1.92	1.33	2.14	1.15	1.85	1.28	2.06	1.02	1.64	1.18	1.90	0.97	1.56	1.08	1.74	0.94	1.51
1.29	2.08	1.13	1.82	1.25	2.01	1.08	1.74	1.20	1.93	0.97	1.56	1.10	1.77	0.92	1.48	0.99	1.59	0.89	1.43

Adapted, by permission, from Cooper Institute, *Physical fitness assessments and norms for adults and law enforcement* (Dallas, TX: The Cooper Institute). For more information: wwwcooperinstitute.org.

Name or ID number: _____   Tester: _____

Date: _____ Time: _____

Sex: M / F (circle one)   Age: _____ y

Height: _____ in.   _____ cm

Weight: _____ lb   _____ kg

Temperature: _____ °F   _____ °C

Relative humidity: _____ %

Barometric pressure: _____ mmHg

Location of Testing

❑ Outdoor field

❑ Indoor field

❑ Indoor track

❑ Outdoor track

❑ Gym

❑ Other

Footwear

❑ Jogging shoe

❑ Walking shoe

❑ Tennis shoe

❑ Basketball shoe

❑ Running shoe

❑ Cross-trainer

❑ Other _____

Total distance: _____ yd $\times$ 0.9144 = _____ m

$\dot{V}O_2max\ (ml \cdot kg^{-1} \cdot min^{-1}) = 0.0268 \times \underset{\text{Distance (m)}}{\underline{\hspace{3cm}}} - 11.2$

$\dot{V}O_2max$ = _____ $ml \cdot kg^{-1} \cdot min^{-1}$

Percentile rank: _____

$\dot{V}O_2max$ classification: _____

ROCKPORT FITNESS WALKING TEST

EQUIPMENT

- Measured 1-mile distance, ideally on a 0.25-mile (400 m) track (see figure 8.2, p. 193)
- Physician's scale, stadiometer, or equivalent electronic scale
- Heart rate monitor (optional)
- Stopwatch
- Individual and group data sheets
- Microsoft Excel or equivalent spreadsheet program

Find the group data sheets for this laboratory online at www.HumanKinetics.com/ LaboratoryManualForExercisePhysiology.

WARM-UP

Before starting the test, you must teach the person performing the test how to palpate for his or her heart rate. Make sure that the person can find his or her heart rate and knows how to count the heartbeat. It may be prudent to have the subject practice doing so while the tester also palpates to ensure that the subject obtains accurate results. Alternatively, you could have the subject wear a heart rate monitor or have the tester measure the subject's heart rate. For the sake of this laboratory activity, however, students should also palpate their own heart rate.

After the subject learns how to palpate, be sure to measure and record his or her weight, which will be used in the estimation equations. Next, have the subject perform an appropriate warm-up consisting of general and specific activities. When working with athletes or other fit individuals, start with 5 min of general warm-up (e.g., jogging, cycling, jumping rope), then move on to 5 min of dynamic stretching (e.g., leg swings, walking toe touches, walking lunges, knee tucks, high knees). With sedentary or untrained individuals, use less rigorous activities (e.g., leg swings, toe touches). After the warm-up, make sure that the subject clearly understands that he or she is only to walk during this test. Here is a summary of basic steps to use in administering the test:

1-MILE ROCKPORT FITNESS WALKING TEST

This is an ideal test for assessing cardiorespiratory fitness. It is particularly good for individuals who are unconditioned because it uses both walking and heart rate in order to estimate aerobic power.

Step 1: Measure the height and weight of each person being tested and record the results on their individual laboratory data sheets. Measure body mass to the nearest 0.01 kg and height to the nearest 0.1 cm.

Step 2: Have each subject complete a 5-minute general warm-up followed by 5 min of dynamic stretching.

Step 3: Clearly explain that the objective of this test is to walk the 1-mile (1.6 km) distance as fast as possible and that jogging or running is not acceptable.

Step 4: Start the stopwatch at the same time that the walk test is initiated.

Step 5: Record the time taken to complete the distance on the individual data sheet. Make sure to convert the time to the nearest hundredth of a minute. For example, if the time is 12:35 min:s, convert the value by dividing 35 s by 60 s·min^{-1}, which produces a result of 0.58 min, thus making the time 12.58 min.

Step 6: Immediately after the subject passes the 1-mile (1.6 km) mark, take the heart rate. Palpate the heart rate for 15 s, then multiply the result by 4. Alternatively, have the subject wear a heart rate monitor. Record the heart rate in the appropriate location on the individual data sheet.

Step 7: Allow adequate time for a cool-down consisting of slow walking and stretching.

Step 8: Calculate each individual's $\dot{V}O_2$max using the equations presented on the individual data sheet and record the results on the group data sheet. Compare the result with the norms given in tables 8.4 and 8.5.

Table 8.4 Norms for the 1-Mile Walk Test for Subjects Aged 30–69 Years (min:s)

Ranking	Men	Women
Excellent	<10:12	<11:40
Good	10:13–11:42	11:41–13:08
High average	11:43–13:13	13:09–14:36
Low average	13:14–14:44	14:37–16:04
Fair	14:45–16:23	16:05–17:31
Poor	>16:24	>17:32

Reprinted, by permission, from J. Morrow, A. Jackson, J. Disch, and D. Mood, 2010, *Measurement and evaluation in human performance*, 4th ed. (Champaign, IL: Human Kinetics), 201.

Table 8.5 Norms for the 1-Mile Walk Test for Subjects Aged 18–30 Years (min:s)

Percentile	Men	Women
90	11:08	11:45
75	11:42	12:49
50	12:38	13:15
25	13:38	14:12
10	14:37	15:03

Reprinted, by permission, from J. Morrow, A. Jackson, J. Disch, and D. Mood, 2010, *Measurement and evaluation in human performance*, 4th ed. (Champaign, IL: Human Kinetics), 201.

QUESTION SET 8.3

1. Based on the results of your test, rank your aerobic fitness according to the data presented in tables 7.1, 8.4, and 8.5. How does your ranking compare with your class average?

2. Create a bar graph of your individual results and of the class average. How do your results differ from the class average? Are you above or below average for the class?

3. Do the values determined for the various equations produce different results? If yes, why might this be?

4. What are some limitations associated with the Rockport fitness walking test?

5. Based on the class averages, how would rate your classmates' overall fitness?

6. What are some issues related to heart rate palpation? Any procedural concerns? If so, what are they?

 Find the case studies for this laboratory online at www.HumanKinetics.com/ LaboratoryManualForExercisePhysiology.

Laboratory Activity 8.3   Individual Data Sheet

Name or ID number: _____ Tester: _____

Date: _____ Time: _____

Sex: M / F (circle one)   Age: _____ y

Height: _____ in.   _____ cm

Weight: _____ lb   _____ kg

Temperature: _____ °F   _____ °C

Relative humidity: _____ %

Barometric pressure: _____ mmHg

Location of Testing

❑ Outdoor field

❑ Indoor field

❑ Indoor track

❑ Outdoor track

❑ Gym

❑ Other

Footwear

❑ Jogging shoe

❑ Walking shoe

❑ Tennis shoe

❑ Basketball shoe

❑ Running shoe

❑ Cross-trainer

❑ Other _____

Raw Data

Raw time: _____ (min:s)   Converted time: _____ (min)

Body weight: _____ (kg) × 2.2 = _____ (lb)   Sex (circle one):   M = 1   F = 0

$\dot{V}O_2$max Determination

Men (18–29 y):

$$\dot{V}O_2\max\,(ml \cdot kg^{-1} \cdot min^{-1}) = 97.660 - (0.0957 \times \underset{\text{body weight (lb)}}{\underline{\hspace{2cm}}}) - (1.4537 \times \underset{\text{walk time (min)}}{\underline{\hspace{2cm}}}) - (0.1194 \times \underset{\text{HR}}{\underline{\hspace{1.5cm}}})$$

$\dot{V}O_2\max$ = _____ $ml \cdot kg^{-1} \cdot min^{-1}$

Ranking: _____

Percentile: _____

Women (18–29 y):

$$\dot{V}O_2 max\,(ml \cdot kg^{-1} \cdot min^{-1}) = 88.768 - (0.0957 \times \underline{\hspace{2cm}}_{\text{body weight (lb)}}) - (1.4537 \times \underline{\hspace{2cm}}_{\text{walk time (min)}}) - (0.1194 \times \underline{\hspace{2cm}}_{\text{HR}})$$

$\dot{V}O_2 max =$ _____ $ml \cdot kg^{-1} \cdot min^{-1}$

Ranking: _____

Percentile: _____

College-age students:

$$\dot{V}O_2 max\,(ml \cdot kg^{-1} \cdot min^{-1}) = 88.768 + (8.892 \times \underline{\hspace{2cm}}_{\text{sex}}) -$$

$$(0.2109 \times \underline{\hspace{2cm}}_{\text{body weight (kg)}}) - (1.4537 \times \underline{\hspace{2cm}}_{\text{walk time (min)}}) - (0.1194 \times \underline{\hspace{2cm}}_{\text{HR}})$$

$\dot{V}O_2 max =$ _____ $ml \cdot kg^{-1} \cdot min^{-1}$

Ranking: _____

Percentile: _____

Laboratory Activity 8.3

LABORATORY 9

Maximal Oxygen Consumption Measurements

Objectives

1. Understand the concept of maximal oxygen consumption as a fitness assessment and describe factors that affect $\dot{V}O_2$max.
2. Define the terms *maximal oxygen consumption* and *ventilatory threshold*.
3. Review units of measurement for work on a treadmill.
4. Understand the criteria for determining whether $\dot{V}O_2$max was attained during a graded exercise test.
5. Measure $\dot{V}O_2$max during graded treadmill exercise (Bruce treadmill protocol) using open circuit calorimetry.
6. Measure $\dot{V}O_2$max during graded exercise on a cycle ergometer using open circuit calorimetry.
7. Interpret changes in $\dot{V}O_2$, $\dot{V}CO_2$, $\dot{V}_E$, respiratory exchange ratio (RER), and heart rate as a function of power output.

Definitions

graded exercise test (GxT)—Ergometer protocol used to assess maximal oxygen consumption; increases intensity every 1- to 3-minute stage to exhaustion.

indirect calorimetry—Measurement of energy expenditure from oxygen consumption and carbon dioxide production as registered in expired gases.

maximal aerobic power—Maximal rate at which oxygen can be taken up, transported, and used during physical activity.

maximal oxygen consumption ($\dot{V}O_2$max)— Highest rate of oxygen consumption that a person can attain during exercise; com-

monly expressed in either the absolute units of $L \cdot min^{-1}$ or the relative units of $ml \cdot kg^{-1} \cdot min^{-1}$.

oxygen consumption ($\dot{V}O_2$)—Amount of oxygen taken up and used by an individual at a given time.

respiratory exchange ratio (RER or RQ)— Ratio of carbon dioxide production ($\dot{V}CO_2$) to oxygen consumption ($\dot{V}O_2$).

steady state—Plateau in metabolic parameters at a constant amount of power.

ventilatory threshold—Point during increasing exercise intensity where ventilation begins to increase disproportionately as the body tries to clear excess CO_2.

Maximal oxygen consumption (\u1E56O₂max) is a measure of the maximum rate of **oxygen consumption** in the mitochondria during oxidative phosphorylation. Thus, it is the maximal capability of the oxidative energy system to produce adenosine triphosphate (ATP) during exercise. Moreover, it is a function of the capacity of the heart, lungs, and blood to transport oxygen to the working muscle and of the ability of the muscles to use oxidative phosphorylation to create ATP aerobically. The term **maximal aerobic power** can be synonymous with $\dot{V}O_2max$, indicating the maximal rate at which oxygen can be taken up, transported, and used during physical activity. Several other terms are also used for this capacity—among them, *cardiorespiratory endurance, aerobic fitness, cardiovascular fitness,* and *cardiovascular endurance.* For the purposes of this lab, we refer to this phenomenon as maximal *oxygen uptake* or $\dot{V}O_2max$.

$\dot{V}O_2max$ is usually measured by means of a **graded exercise test (GxT)** on a treadmill or cycle ergometer using **indirect calorimetry** (see figure 9.1). The test begins at a relatively easy workload and becomes progressively more demanding due to increasing intensity. Each work stage is typically 1 to 3 min long. If an automated system ("metabolic cart") is being used, then $\dot{V}O_2$ is measured continuously; it can also be measured manually at the end of each work stage (i.e., by collecting bags of expired air to be analyzed by hand; see appendix C). The subject is encouraged to exercise as long as possible—until volitional fatigue. The highest $\dot{V}O_2$ value measured during the test is taken as $\dot{V}O_2max$ if two of the following four criteria have been met (2):

1. Plateau in $\dot{V}O_2$ (i.e., increase of ≤150 ml · min⁻¹) despite increase in work rate
2. Heart rate during the last work stage no more than 10 beats · min⁻¹ below the subject's age-predicted maximal heart rate
3. **Respiratory exchange ratio (RER)** greater than 1.10 in the final work stage
4. Rating of Perceived Exertion (RPE) greater than 17 using the original Borg scale

If these criteria are not met, then the subject probably did not attain $\dot{V}O_2max$ even with volitional fatigue.

The $\dot{V}O_2max$ test is generally considered the best noninvasive measure of cardiorespiratory fitness. It is highly correlated to maximal cardiac output and therefore provides an excellent index of the heart's capacity to pump blood. It is not, however, considered to be a good predictor of performance in endurance events.

A subject's $\dot{V}O_2max$ can be influenced by several factors:

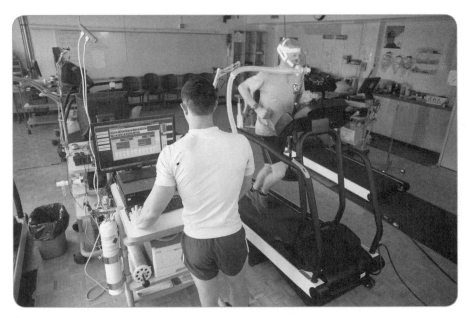

Figure 9.1 $\dot{V}O_2max$ test equipment.

- Heredity—25% to 50% genetic component
- Sex—male values 15% to 20% higher than female values for same population
- Age—peak at 15 to 18 y; gradual decline of about 8% per decade from 30 years on (can be counteracted by regular exercise)
- Training status—possible 6% to 25% improvement in values with training (higher values for aerobically trained individuals)
- Mode of exercise—outcome affected by choice of exercise

See table 7.1 (p. 171) for percentile norms according to sex and age group.

Selecting a Test Protocol

$\dot{V}O_2$max testing can be done on a number of ergometers—for example, treadmill, stair stepper, rowing machine, stationary bicycle, and even a wide treadmill that allows the subject to roller-ski or skate. $\dot{V}O_2$max is generally higher in modalities that recruit greater amounts of muscle. The highest $\dot{V}O_2$maxes recorded for a male (94 ml·kg^{-1}·min^{-1}) and for a female (77 ml·kg^{-1}·min^{-1}) were in cross-country skiers (10). Of course, laboratories may be limited by the available ergometers, but insofar as possible when testing an athlete you should choose the ergometer that most closely resembles his or her sport. When this is not possible—or when testing an untrained subject—the most popular modes of testing are treadmills and cycle ergometers.

$\dot{V}O_2$max may also be dependent on the protocol. Many treadmill and cycle ergometer protocols are available, and one area of skill for an exercise physiologist lies in choosing a testing protocol that will give interpretable results. When selecting a protocol, try to stay within the following guidelines:

- The test should be progressive, and each increase in workload should be equal to the previous one. You should not, for example, use small increases in the early stages, then switch to larger increases later in the protocol. Equal stage increments can range from 0.5 to 3 METs depending on the subject's anticipated fitness.
- Each stage should be 1 to 3 min long, and the goal is to reach **steady state** by the end of each stage.

- Total testing should last from 8 to 15 min. If the duration is too short, it is unlikely that the subject will reach steady state, and the anaerobic component is typically too large. If the duration is too long, other mechanisms of fatigue may cause the subject to quit before $\dot{V}O_2$max is reached.

The maximum measured $\dot{V}O_2$ may depend on the protocol or modality chosen or on the effort put forth by the subject. Because the measured $\dot{V}O_2$ is dependent on protocol, modality, or effort, the maximum measured $\dot{V}O_2$ may be either less than or the same as $\dot{V}O_2$max and is often referred to as $\dot{V}O_2peak$. As an example, $\dot{V}O_2$ measured during arm cranking would produce a $\dot{V}O_2peak$ that is less than the whole-body $\dot{V}O_2$max elicited during treadmill testing.

Not all subjects or clients should be tested for $\dot{V}O_2$max. Exercising to exhaustion involves inherent risks including light-headedness, pallor, angina (chest pain), nausea, dyspnea (undue shortness of breath), arrhythmia, myocardial infarction, and death. Because of this risk, the American College of Sports Medicine (ACSM) has produced contraindications to maximal exercise testing; for a complete list, see page 43 in chapter 2.

Participation is voluntary. Subjects should be informed that they can quit at any time before or during the test. For a list of indications for terminating a test, see page 167 in chapter 7.

Monitoring Progress With RPE Scales

In performing a $\dot{V}O_2$max test, a subject is of course unable to talk while wearing a mouthpiece or mask. It is useful to testers, however, to know how the subject is feeling, and thus it is common to use scales for rating of perceived exertion (RPE). This type of measure can be used to monitor progress toward maximal exertion. Two subjective scales have been developed (6), and both are appropriate for use during maximal exercise testing (see lab 7). The RPE scales include recommended pretest instructions that help clarify their use to the subject being tested (8). Read these instructions to the subject prior to the start of the test.

During the test, we want you to pay close attention to how hard you feel the exercise work rate is. This feeling should reflect your total amount of exertion and fatigue, combining all sensations and feelings of physical stress, effort, and fatigue. Don't concern yourself with any one factor, such as leg pain, shortness of breath, or exercise intensity, but try to concentrate on your total, inner feeling of exertion. Try not to underestimate or overestimate your feelings of exertion; be as accurate as you can.

Anecdotally, fit subjects are better than untrained subjects at rating their exertion; in fact, you may find that unfit subjects go quickly from moderate RPE to having to quit the test. Even so, the validity of RPE has been tested in both healthy and diseased populations (8). Because RPE can also be affected by external information (e.g., heart rate or work rate), it is important to keep the subject blind to his or her own metabolic data, since such knowledge could lead the subject to alter his or her effort.

Estimating Fuel Usage With RER

As discussed in laboratory 5, the ratio of oxygen consumed to carbon dioxide produced is referred to as the *respiratory exchange ratio*. For the sake of review, and to elaborate, RER (the volume of carbon dioxide expelled divided by the volume of oxygen consumed) can be calculated by measuring $\dot{V}O_2$ and $\dot{V}CO_2$. RER can be used to determine the fuel mixture during exercise, which is typically somewhere between 0.70 (pure fat) and 1.0 (pure carbohydrate), thus indicating that a mixture of these two fuels is being used.

In order to use RER to estimate the fuels being used during exercise, we have to make a few assumptions. First, the breakdown of protein contributes very little to the overall energy expenditure during exercise (typically <5%); as a result, the RER that we measure in the laboratory is sometimes referred to as "nonprotein RER." Also, the subject must be in steady state, meaning that the $\dot{V}O_2$ and RER should not change if there is no change in workrate. RER is affected by the production of CO_2 during exercise. As intensity increases during the $\dot{V}O_2$max test, RER increases as a result of a shift from more fat use to greater reliance on carbohydrate. With this in mind, RER

during submaximal steady state work can be used to determine accurate training zones for clients interested in maximizing absolute fat oxidation. We discussed in laboratory 5 how RER can be converted to $kcal \cdot min^{-1}$ when $\dot{V}O_2$ in $L \cdot min^{-1}$ is known. RER is also useful as subjects approach $\dot{V}O_2$max. As we will see shortly, CO_2 production can come from other sources besides macronutrient (fat and carbohydrate) oxidation. This information helps us understand how RER can exceed 1.0; indeed, this is one of the criteria for justifying a valid $\dot{V}O_2$max test (RER > 1.1).

Ventilatory Threshold Signals

$\dot{V}O_2$ increases as a linear function of power output during a graded exercise test up to $\dot{V}O_2$max. On the other hand, ventilation ($\dot{V}_E$) increases linearly as a function of power output only during exercise requiring up to about 50% to 75% $\dot{V}O_2$max, above which $\dot{V}_E$ increases at a steeper rate (see figure 9.2).

The greater rate of increase in $\dot{V}_E$ above this **ventilatory threshold** coincides with a similar rise in $\dot{V}CO_2$ and a sharp increase in blood lactate.

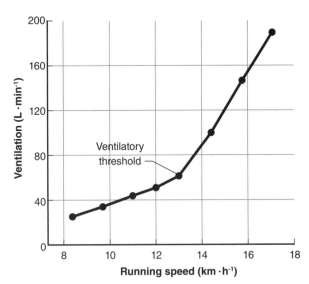

Figure 9.2   Ventilation during exercise of increasing intensity.

Reprinted, by permission, from W. L. Kenney, J.H. Wilmore, and D.L. Costill, 2012, *Physiology of sport and exercise*, 5th ed. (Champaign, IL: Human Kinetics), 198.

Remember that lactic acid dissociates into a lactate ion and H^+ in the blood, and the H^+ combines with bicarbonate ion to form carbonic acid, which in turn dissociates into H_2O and CO_2.

$$La^-H^+ + HCO_3^- \leftrightarrow H_2CO_3 \leftrightarrow H_2O + CO_2$$

The accumulation of CO_2 in the blood leads to an increase in CO_2 in the expired air ($\dot{V}CO_2$). The increase in blood CO_2 stimulates an increase in ventilation. Thus, the ventilatory threshold is indicative of an abrupt increase in blood lactate and a subsequent decrease in blood pH. This "nonmetabolic" or "excess" CO_2 contributes to the $\dot{V}CO_2$ measured during the test, which is why at high intensities RER can exceed 1.0. Many people misunderstand the heavy breathing at maximum exercise as an attempt to "get more oxygen." In fact, at altitudes of less than 2,000 m, ventilation is driven more by CO_2 production than by O_2 consumption. This concept is discussed further in laboratory 10. Ventilation can be monitored during a $\dot{V}O_2$max test to gauge intensity and anticipate volitional fatigue.

References

1. American College of Sports Medicine. *ACSM's Guidelines for Exercise Testing and Prescription.* 8th ed. Philadelphia: Lippincott Williams & Wilkins, 2010.
2. Borg GA. Perceived Exertion. *Exer Sport Sci Rev* 2: 131–153, 1974.
3. Bruce RA, Kusumi F, and Hosmer D. Maximal Oxygen Intake and Nomographic Assessment of Functional Aerobic Impairment in Cardiovascular Disease. *Am Heart J* 85: 546–562, 1973.
4. Foster C, Jackson AS, Pollock ML, Taylor MM, Hare J, Sennett SM, Rod JL, Sarwar M, and Schmidt DH. Generalized Equations for Predicting Functional Capacity From Treadmill Performance. *Am Heart J* 107: 1229–1234, 1984.
5. Kenney WL, Wilmore JH, and Costill DL. *Physiology of Sport and Exercise.* 5th ed. Champaign, IL: Human Kinetics, 2012.
6. Mcardle WD, Katch FI, and Pechar GS. Comparison of Continuous and Discontinuous Treadmill and Bicycle Tests For Max $\dot{V}O_2$. *Med Sci Sports* 5: 156–160, 1973.
7. Noble BJ, Borg GA, Jacobs I, Ceci R, and Kaiser P. A Category-Ratio Perceived Exertion Scale: Relationship to Blood and Muscle Lactates and Heart Rate. *Med Sci Sports Exer* 15: 523–528, 1983.
8. Pollock ML, Foster C, Schmidt D, Hellman C, Linnerud AC, and Ward A. Comparative Analysis of Physiologic Responses to Three Different Maximal Graded Exercise Test Protocols in Healthy Women. *Am Heart J* 103: 363–373, 1982.
9. Robertson RJ and Noble BJ. Perception of Physical Exertion: Methods, Mediators, and Applications. *Exer Sport Sci Rev* 25: 407–452, 1997.
10. Whaley MH, Brubaker PH, Kaminsky LA, and Miller CR. Validity of Rating of Perceived Exertion During Graded Exercise Testing in Apparently Healthy Adults and Cardiac Patients. *J Cardiopulm Rehabil* 17: 261–267, 1997.

GRADED TREADMILL
$\dot{V}O_2$MAX TEST

EQUIPMENT

- Running treadmill
- Online $\dot{V}O_2$/$\dot{V}CO_2$ collection system (metabolic cart)
- Breathing valve assembly, mouthpiece (or mask), hoses, nose clip
- Heart rate monitor
- Stopwatch
- RPE scale
- Individual data sheets

BRUCE TREADMILL PROTOCOL FOR $\dot{V}O_2$MAX

Step 1: At least one subject from each laboratory section will perform the max test on the treadmill. The subject should come prepared to exercise in appropriate shoes and attire, should be well rested, and should have fasted for at least 2 h.

Step 2: Record the subject's descriptive characteristics—height, weight, and age.

Step 3: After the lab instructor explains the procedures for using the metabolic cart, proceed to calibrate and operate the cart, then enter the subject's data into the cart.

Step 4: Choose one person to acting as a "spotter" and have him or her stand behind the treadmill. Others will monitor the subject's heart rate, RPE, and gas values, as well as the overall and stage time, and record the information on the data collection sheet.

Step 5: Prepare the subject for testing with the HR monitor and headgear.

Step 6: Before the test begins, read the subject the RPE instructions presented earlier in this chapter. Allow the subject to ask questions prior to being connected to the metabolic equipment.

Step 7: Connect the subject to the mouthpiece or mask apparatus while the subject stands on the treadmill.

Step 8: Measure the subject's resting heart rate and metabolic data.

Step 9: Allow a 3- to 5-minute warm-up at an intensity no greater than that of the first stage.

Step 10: As soon as the subject begins the first stage, start a stopwatch; since overall time in the Bruce protocol can be used as an accurate estimate of $\dot{V}O_2$max.

Step 11: Follow the Bruce treadmill protocol presented in figure 9.3 to determine the work output at each stage. Increase speed and percent grade at the end of every 3-minute stage until your subject can go no longer.

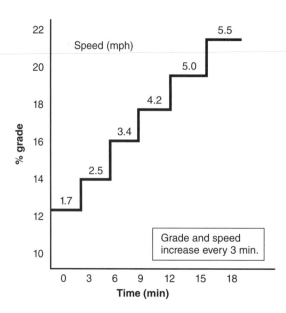

Figure 9.3 Bruce treadmill protocol.

Reprinted, by permission, from J. Hoffman, 2006, *Norms for fitness, performance, and health* (Champaign, IL: Human Kinetics), 24. Adapted from *American Heart Journal*, Vol. 85, R.A. Bruce, F. Kusumi and D. Hosmer, "Maximal oxygen uptake and nomographic assessment of functional aerobic impairment in cardiovascular disease," 546-562, Copyright 1973, with permission of Elsevier.

Step 12: Collect data during the last 30 s of each stage, specifically, record HR, $\dot{V}O_2$, RER, $\dot{V}_E$, and RPE for each stage on the individual data collection sheet.

Step 13: Be observant of the exercising individual to monitor for signs and symptoms that call for terminating the test. Encouragement is effective in motivating a subject to attain maximal effort.

Step 14: Since the subject is unable to communicate through the mask or mouthpiece, several common hand signals can be used during max testing. For example, a thumbs-up indicates that the subject is doing well and is willing to continue. A raised index finger indicates that the subject will need to stop in about 1 minute, and knowing this allows the testers to prepare for the end of the test and to collect maximum data for all of the outcome measures.

Step 15: Stop the test when $\dot{V}O_2$max test criteria are reached, or when indicators for stopping the test arise, or when the subject stops the test by grabbing the handrail. Slow the treadmill to 3 mph and reduce the grade to 0%. Have the subject keep the mouthpiece in his or her mouth for a 5 min cool-down period (record the RER during the cool-down). Immediately record the total time in the protocol, which will be used to predict $\dot{V}O_2$max.

Step 16: The cool-down is important in order to keep the subject moving to avoid blood pooling in the lower extremities and potential syncope. During the cool-down, which should last at least 5 min, monitor the subject's heart rate. An appropriate cool-down would have the subject attain an HR of at most 120 beats·min^{-1} before he or she gets off the ergometer. Continue to monitor the subject's physical appearance and symptoms during the cool-down period.

Laboratory Activity 9.1

BRUCE PROTOCOL FORMULAS

Because the standard Bruce protocol (2) has been used for so long, estimations have been developed from the amount of time spent in the protocol. Thus, if protocol stage times are strictly adhered to, the following equations can be used to estimate $\dot{V}O_2max$ (3):

Men

$$\dot{V}O_2max \ (ml \cdot kg^{-1} \cdot min^{-1}) = 14.76 - 1.379 \times (time) + 0.451 \times (time^2) - 0.012 \times (time^3)$$

Women

$$\dot{V}O_2max \ (ml \cdot kg^{-1} \cdot min^{-1}) = 4.38 \times (time) - 3.90 \ (6)$$

These predictions correlate significantly ($r = 0.98$ and 0.91 for men and women, respectively) with measured $\dot{V}O_2 max$ and therefore may be useful if respiratory gas analysis with a metabolic cart is not possible.

Calculate the power output (or work rate) for each work stage on a treadmill using the following equation:

$$\text{Power output } (kg \cdot m \cdot min^{-1}) = \text{subject's body weight (kg)} \times \text{treadmill speed} \ (m \cdot min^{-1}) \times \text{fractional angle of the treadmill}$$

where fractional angle = percent grade divided by 100. For example, the fractional angle at a 2% grade = 2% / 100 = 0.02. In order to convert mph to $m \cdot min^{-1}$, multiply by $26.8 \ m \cdot min^{-1}$.

QUESTION SET 9.1

1. Prepare a data table showing values for HR, $\dot{V}O_2$, $\dot{V}_E$, RPE, and RER for each stage that the subject completed.
2. Plot $\dot{V}O_2$, HR, and RER (y-axis) for your subject as a function of the calculated power output ($kg \cdot m \cdot min^{-1}$ or watts).
3. Did your subjects attain $\dot{V}O_2max$? Justify your answer.
4. Determine relative ($ml \cdot kg^{-1} \cdot min^{-1}$) and absolute ($L \cdot min^{-1}$) $\dot{V}O_2max$ values for your subject. Which is a better index of cardiovascular fitness? Why?
5. For the subjects for which you collected data, plot $\dot{V}_E$ (y-axis) as a function of HR. Did any of your subjects have a distinct ventilatory threshold? Why does $\dot{V}_E$ typically increase more sharply at heavy workloads?
6. RER at $\dot{V}O_2max$ is usually greater than 1.0. If an RER of 1.0 is 100% carbohydrate oxidation, how can RER exceed 1.0?
7. Compare the collected data with the normative values in table 7.1. How does your subject rank according to the information given?
8. Using the prediction equations provided for the Bruce protocol, how well does the time in the protocol compare with the measured $\dot{V}O_2max$?
9. How can a person increase his or her $\dot{V}O_2max$? Be specific as possible.

Find the case studies for this laboratory online at www.HumanKinetics.com/ LaboratoryManualForExercisePhysiology.

Laboratory Activity 9.1 Individual Data Sheet

Name or ID number: _____ Date: _____

Tester: _____ Time: _____

Sex: M / F (circle one) Age: _____ y Height: _____ in. _____ cm

Temperature: _____°F _____°C Weight: _____ lb _____ kg

Barometric pressure: _____ mmHg Relative humidity: _____ %

RPE											
RER											
$\dot{V}CO_2$ (L · min^{-1})											
METs											
$\dot{V}O_2$ (ml · kg^{-1} · min^{-1})											
$\dot{V}O_2$ (L · min^{-1})											
$\dot{V}_E$ (L · min^{-1})											
HR (bpm)											
Power (kg · m · min^{-1})	Rest										
Grade (%)	0										
Speed (mph)	0										
Exercise duration (min)	0										

CYCLE ERGOMETER $\dot{V}O_2$MAX TEST

EQUIPMENT

- Cycle ergometer
- Online $\dot{V}O_2/\dot{V}CO_2$ collection system (metabolic cart)
- Breathing valve assembly, mouthpiece (or mask), hoses, nose clip
- Heart rate monitor
- Stopwatch
- RPE scale
- Individual data sheets

CYCLE ERGOMETER PROTOCOLS

The graded protocol on a bike is relatively easy. If your lab has an electronically braked bike, then the subject can use a self-selected cadence since the ergometer will adjust resistance accordingly. If you are using a mechanically braked cycle ergometer, such as a Monark, you need to have the subject cycle at a fixed cadence, since $rev \cdot min^{-1}$ is a component of power on a bike (see below). We recommend using a metronome to keep the subject at the determined cadence (between 60 and 100 $rev \cdot min^{-1}$). Stage time should be either 2 or 3 min.

The main considerations are initial work output, time in each stage, and amount of increase between stages. Initial work output is a function of fitness and the subject's familiarity with cycling. Less fit subjects should start between 25 and 75 W (about 150 and 450 $kg \cdot m \cdot min^{-1}$), whereas more fit subjects who are familiar with cycling can start between 75 and 150 W (about 450 and 920 $kg \cdot m \cdot min^{-1}$). Similar, when working with less fit subjects, the amount of increase between stages should be smaller (e.g., 10–15 W, or about 60–90 $kg \cdot m \cdot min^{-1}$); when working with fitter subjects who are experienced in cycling, the increase can be as much as 25-50 W (about 150-300 $kg \cdot m \cdot min^{-1}$).

In order to serve as an accurate $\dot{V}O_2$max test, the session should last 8–15 min. As with the Bruce treadmill test, several continuous cycle ergometer tests have been published, and two of the most popular are the Åstrand and the McArdle protocols (4). However, there is nothing particularly special about these protocols, and in fact their required cadences are lower than that preferred by many subjects (50 and 60 $rev \cdot min^{-1}$, respectively). The Åstrand test for men starts at 600 $kg \cdot m \cdot min^{-1}$ (about 100 W) and increases by 300 $kg \cdot m \cdot min^{-1}$ (about 50 W) every 2 min while pedaling at 50 $rev \cdot min^{-1}$; for women, it starts at 300 $kg \cdot m \cdot min^{-1}$ and increases by 150 $kg \cdot m \cdot min^{-1}$ (about 25 W) every 2 min at 50 $rev \cdot min^{-1}$. The McArdle protocol, in contrast, is not gender specific. It starts at 900 $kg \cdot m \cdot min^{-1}$ (about 150 W) and increases by 240 $kg \cdot m \cdot min^{-1}$ (about 40 W) every 2 min at 60 $rev \cdot min^{-1}$ (4). Thus this protocol should be used only for fit individuals who have experience in cycling. These two protocols can be used literally or as a gauge for students in designing their own protocol to individualize to their subject.

CYCLE ERGOMETER $\dot{V}O_2$MAX TEST

Step 1: At least one subject from each laboratory section will perform the max test on the cycle ergometer. The subject should come prepared to exercise in appropriate shoes and attire, should be well rested, and should have fasted for at least 2 hours.

Step 2: Record the subject's descriptive characteristics—height, weight, and age.

Step 3: After the lab instructor explains the procedures for using the metabolic cart, proceed to calibrate and operate the cart, then enter the subject's data into the cart.

Step 4: Prepare subject for testing with the HR monitor, chest strap, and head-gear.

Step 5: Before the test begins, read the subject the RPE instructions presented earlier in this chapter. Allow the subject to ask questions prior to being connected to the metabolic equipment.

Step 6: Connect the subject to the mouthpiece or mask apparatus while the subject sits on the cycle ergometer.

Step 7: Measure the subject's resting heart rate and metabolic data.

Step 8: Allow a 3- to 5-minute warm-up at an intensity no greater than that of the first stage.

Step 9: Follow one of the cycle ergometer protocols (Åstrand, McArdle, or self-created) to determine the work output at each stage.

Step 10: Collect data during the last 30 s of each stage. Record HR, $\dot{V}O_2$, RER, $\dot{V}_E$, and RPE at each stage.

Step 11: Be observant of the exercising individual to monitor for signs and symptoms that call for terminating the test. Encouragement is effective in motivating a subject to attain maximal effort.

Step 12: Since the subject is unable to communicate through the mask or mouthpiece, several common hand signals can be used during max testing. For example, a thumbs-up indicates that the subject is doing well and is willing to continue. A raised index finger indicates that the subject will need to stop in about 1 minute, and knowing this allows the testers to prepare for the end of the test and to collect maximum data for all of the outcome measures.

Step 13: Stop the test when $\dot{V}O_2$max test criteria are reached, or when indicators for stopping the test arise, or when the subject stops the test by ceasing pedaling. Reduce the workload immediately (to the $kg \cdot m \cdot min^{-1}$ of the first stage) so that the subject can begin cool-down, during which the subject should keep the mouthpiece in his or her mouth (notice RER during the cool-down period).

Step 14: The cool-down is important in order to keep the subject moving to avoid blood pooling in the lower extremities and potential syncope. During the cool-down, which should last at least 5 minutes, monitor the subject's heart rate. An appropriate cool-down would have the subject attain an HR of at most 120 beats·min^{-1} before he or she gets off the ergometer. Continue to monitor the subject's physical appearance and symptoms during the cool-down period.

CYCLE ERGOMETER POWER OUTPUT FORMULAS

Calculate the **power output** (or work rate) for each work stage on a stationary bicycle using the following equation:

$$\text{Power output (kg·m·min}^{-1}) = \text{pedal cadence (rev·min}^{-1}) \times$$
$$\text{flywheel distance (m·rev}^{-1}) \times \text{resistance on the flywheel (kg)}$$

Power on a bike is often expressed in watts. To convert kg·m·min^{-1} to watts, divide by 6.12 (1 watt = 6.12 kg·m·min^{-1}). Flywheel distance is the distance the bike would travel for one pedal revolution *if* it could travel. For a multigear bicycle, this would change with every gear; for a stationary bicycle, it is fixed. To measure flywheel distance, first measure the circumference of the flywheel, then count the number of revolutions of the flywheel for one complete pedal stroke (a piece of tape on the flywheel can be helpful here). Many commonly used laboratory cycle ergometers such as the Monark have a flywheel distance of 6 m·rev^{-1}. For more information on calibrating your cycle ergometer, see appendix E.

QUESTION SET 9.2

1. Prepare a data table showing values for HR, $\dot{V}O_2$, $\dot{V}_E$, RPE, and RER for each stage that the subject completed.
2. Plot $\dot{V}O_2$, HR, and RER (y-axis) for your subject as a function of the calculated power output (kg·m·min^{-1} or watts).
3. Did your subject(s) attain $\dot{V}O_2$max? Justify your answer.
4. Determine relative (ml·kg^{-1}·min^{-1}) and absolute (L·min^{-1}) $\dot{V}O_2$max values for your subject. Which is a better index of cardiovascular fitness? Why?
5. For the subjects that you have collected data for, plot $\dot{V}_E$ (y-axis) as a function of HR. Did any of your subjects have a distinct ventilatory threshold? Why does $\dot{V}_E$ typically increase more sharply at heavy workloads?
6. RER at $\dot{V}O_2$max is usually greater than 1.0. If an RER of 1.0 is 100% carbohydrate oxidation, how can RER exceed 1.0?
7. Compare the collected data with the normative values in table 7.1. How does your subject rank according to the information given?
8. How does the measured $\dot{V}O_2$max on the cycle compare with the predicted $\dot{V}O_2$ from the ACSM prediction equations (see appendix B)? Why might they differ?
9. How can a person increase his or her $\dot{V}O_2$max? Be specific.
10. Why are males expected to have higher $\dot{V}O_2$max values? What does this suggest is important for attaining a high $\dot{V}O_2$max?

Find the case studies for this laboratory online at www.HumanKinetics.com/ LaboratoryManualForExercisePhysiology.

Name or ID number: _____ Date: _____

Tester: _____ Time: _____

Sex: M / F (circle one)   Age: _____ y Height: _____ in. _____ cm

Temperature: _____ °F _____ °C Weight: _____ lb _____ kg

Barometric pressure: _____ mmHg Relative humidity: _____ %

RPE											
RER											
$\dot{V}CO_2$ (L · min⁻¹)											
METs											
$\dot{V}O_2$ (ml · kg⁻¹ · min⁻¹)											
$\dot{V}O_2$ (L · min⁻¹)											
$\dot{V}_E$ (L · min⁻¹)											
HR (bpm)											
Power (kg · m · min⁻¹)	Rest										
Grade (%)	0										
Speed (mph)	0										
Exercise duration (min)	0										

Blood Lactate Threshold Assessment

Objectives

1. Understand the concept of lactate threshold and its importance in predicting endurance performance.

2. Learn the process of measuring lactate threshold by lactate analysis and noninvasive respiratory gas analysis.

3. Compare methods of assessing the lactate threshold.

Definitions

fast (or anaerobic) glycolysis—Hydrolysis of six-carbon glucose to two three-carbon lactate molecules.

lactate—Anion that remains after lactic acid disassociates a proton.

lactate threshold (LT)—Point during exercise of increasing intensity at which the rate of lactate production exceeds the rate of lactate clearance; defined as the point preceding an increase in lactate of >1 mM with increases in intensity.

lactic acid—Three-carbon molecule $(C_3H_6O_3)$ formed from pyruvate; as a carboxylic acid, disassociates a proton under most conditions (remaining anion is called *lactate*).

metabolic acidosis—State of lowered pH resulting from acids produced in metabolism; thought to be the result during exercise of lactic acid production from fast glycolysis.

onset of blood lactate accumulation (OBLA)—Point at which blood lactate levels reach 4.0 mM during an exercise bout with increasing intensity.

Lactic acid is the product of **fast (or anaerobic) glycolysis.** It is formed from pyruvate catalyzed by the enzyme lactate dehydrogenase (LDH) through the following reaction:

$$\text{pyruvate } (C_3H_4O_3) + \text{NADH} + H^+ \leftrightarrow \text{lactic acid } (C_3H_6O_3) + \text{NAD}^+$$

The end point of glycolysis, pyruvate, has two potential fates—lactic acid, as just illustrated, and acetyl CoA, which is found in the mitochondria and is formed through a reaction catalyzed by the enzyme pyruvate dehydrogenase (PDH). The fate of pyruvate is important because it involves a com-mitment to either anaerobic or aerobic metabolism; as a result, exercise physiologists are interested in measuring lactate production.

Lactic acid and **lactate** are often used synonymously—and incorrectly. Lactate is the anion of the lactic acid molecule once it has released its proton. Within the muscle-cell cytoplasm and blood, lactic acid typically takes the form of lactate and H^+ (thus we use the term *lactate* in this lab). Because of this loss of a proton, lactic acid has been thought to contribute to changes in pH in muscle and blood during exercise and therefore to **metabolic acidosis**.

However, the magnitude of lactic acid's contribution to metabolic acidosis in muscle is disputed (for a discussion see sources 11 and 12 in this chapter's reference list).

Lactate is constantly produced, even in the condition of rest. For many years, it was thought to be a waste product of glycolysis that caused muscle fatigue and soreness. In fact, lactate can be used as a fuel source like many other carbon-containing compounds. However, as exercise intensity increases, blood lactate concentrations increase due to a number of factors, including both increased production and reduced removal. As exercise intensity increases, the demand for production of adenosine triphosphate (ATP) also increases, which can overwhelm the ability of the mitochondria to meet the need aerobically. In addition, the recruitment of fast-twitch fibers (with less mitochondria) increases, which facilitates greater lactate production. At these higher intensities, less fat and more carbohydrate are being used for fuel, and this "glycolytic flux" contributes to greater lactate production. Hormones (epinephrine and norepinephrine) increase glycolytic flux at higher intensities by stimulating glycogen breakdown. Many exercise physiology texts also report reduced oxygen as a reason for greater lactate production, but this is not the best explanation when at normal altitudes (1). In reality, increasing lactate production can be attributed to the inability of oxidative phosphorylation in meeting the energy demand of exercise. As the rate of ATP demand exceeds the capacity of the mitochondria to produce ATP, greater anaerobic contribution to ATP synthesis is required.

Removal of blood lactate (and its subsequent oxidation) can be performed by almost any metabolically active tissue (e.g., muscle, heart, liver, kidneys). Blood lactate removal is reduced at high exercise intensities due to the reduction in blood flow away from lactate-clearing tissues (e.g., non-working muscle, liver, kidneys). Therefore, when lactate production exceeds removal, this is typically an indication that exercise intensity is above that where the ATP requirement can be met aerobically. This fact makes blood lactate a useful measure in assessing an individual's ability to produce energy aerobically without significant lactate accumulation in the blood, which can be best illustrated by performing repeated blood draws during an incremental test to volitional fatigue. The point where lactate production exceeds clearance is referred to as the *lactate deflection point*. Blood lactate concentrations (measured in $mmol \cdot L^{-1}$ or mM) at a given absolute intensity can therefore be used as an indicator of an individual's ability to produce energy aerobically. Thus it is an effective measure of fitness for aerobic athletes. Table 10.1 illustrates the effects of a 400 m run on blood lactate levels.

Determining the Deflection Point

Over the years, a number of methods have been developed to define and describe the lactate deflection point from an incremental test that best corresponds to endurance performance (1). This point can be crucial because it indicates when one switches from mostly aerobic metabolism—and therefore an effort that should be sustainable for some time—to mostly anaerobic metabolism, which may hasten fatigue. At submaximal exercise intensities, blood lactate values are similar to resting values (0.9–2.0 $mmol \cdot L^{-1}$). As intensity continues to increase, blood lactate concentrations demonstrate an accelerated increase above

Table 10.1 Blood and Muscle pH and Lactate Concentration 5 min After a 400 m Run

| Runner | Time (s) | Muscle | | Blood | |
		pH	Lactate (mmol · kg⁻¹)	pH	Lactate (mmol · L⁻¹)
1	61.0	6.68	19.7	7.12	12.6
2	57.1	6.59	20.5	7.14	13.4
3	65.0	6.59	20.2	7.02	13.1
4	58.5	6.68	18.2	7.10	10.1
Average	**60.4**	**6.64**	**19.7**	**7.10**	**12.3**

Reprinted, by permission, from W. L. Kenney, J.H. Wilmore, and D.L. Costill, 2012, *Physiology of sport and exercise*, 5th ed. (Champaign, IL: Human Kinetics), 202.

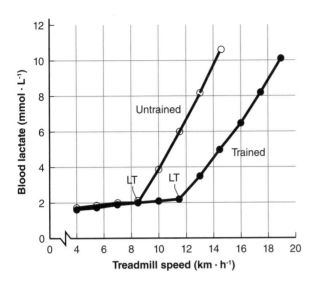

Figure 10.1   Blood lactate concentrations increase with exercise of increasing intensity.

Reprinted, by permission, from W. L. Kenney, J.H. Wilmore, and D.L. Costill, 2012, *Physiology of sport and exercise*, 5th ed. (Champaign, IL: Human Kinetics), 263.

resting values (see figure 10.1). The point at which blood lactate concentrations increase non-linearly is known as the **lactate threshold (LT)**.

Endurance performance depends on an athlete's ability to perform for extended periods of time at the highest possible intensity without experiencing the effects of fatigue and lactate accumulation (2, 5, 10). These factors are dependent on LT because they occur at a certain percentage of $\dot{V}O_2$max and determine the point at which lactate concentrations begin to rise. The LT occurs at different percentages of $\dot{V}O_2$max for different groups of individuals:

Untrained, sedentary individuals	~50%–60% of $\dot{V}O_2$max
Well-trained distance runners	>75% of $\dot{V}O_2$max

Studies show that actual endurance race performance correlates more closely with LT than with $\dot{V}O_2$max (4, 10, 13), which helps explain why athletes with the same $\dot{V}O_2$max often perform differently during endurance events. It also helps explain continued performance improvement without increases in $\dot{V}O_2$max. Determining LT can aid in the design of training programs to improve race times for endurance athletes. If an athlete knows the percentage of $\dot{V}O_2$max at which he or she can train before experiencing lactate accumulation, then he or she can improve the threshold through interval training. Training at or near the LT shifts the curve to the right; in order to do so, of course, it is necessary to know when LT occurs and then to train at that percentage of $\dot{V}O_2$max and heart rate. More information on training associated with lactate threshold can be found in *Lactate Threshold Training* by Peter Janssen (9).

The most clearly defined methods for determining the lactate deflection point are the **onset of blood lactate accumulation (OBLA)** and the anaerobic or lactate threshold (LT). OBLA is defined as the workload associated with 4 mmol·L⁻¹ during an incremental exercise test (13). The lactate threshold is considered to be the workload where a nonlinear increase of >1 mmol·L⁻¹ is observed in successive workloads (6, 8). It is thought that OBLA typically occurs at a higher intensity and percentage of $\dot{V}O_2$max than LT (see figures 10.1 and 10.2) and thus that LT better predicts performance lasting 60–75 min whereas OBLA better predicts those lasting 20–40 min (7).

Selecting a Test Method

Several methods are available for measuring blood lactate. Of course, a small sample of blood needs to be drawn, but this rarely involves more than a drop from a finger stick. Lactate and glucose can be measured in a matter of minutes by automated machines such as the YSI 2300 STAT (Yellow Springs Incorporated, Yellow Springs, OH). Because of the interest of endurance coaches, portable analyzers are also commercially available (for examples, see www.lactate.com). These units work like glucose analyzers, where a drop of blood is applied to a strip that engages the machine, which produces results in a matter of seconds. Lactate can also be measured by other automated machines, such as the i-stat (see www.abbottpointofcare.com/Products-and-Services/iSTAT-Handheld.aspx). See figure 10.3 for an example of a portable lactate analyzer. Finally, lactate can be analyzed using a spectrophotometer, which is considered the most accurate method but requires more blood and takes longer. More than likely, your lab has a portable analyzer that your lab instructor will show you how to use.

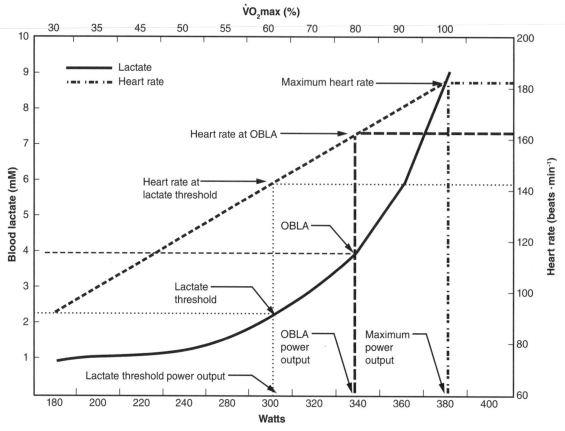

Figure 10.2 Definitions associated with lactate threshold and OBLA.

Adapted, by permission, from T. Bompa and G. Haff, 2009, *Periodization: Theory and methodology of training,* 5th ed. (Champaign, IL: Human Kinetics), 295.

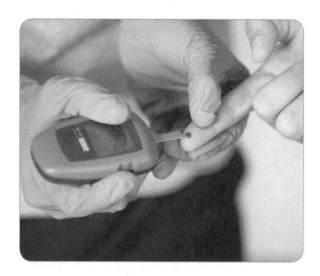

Figure 10.3 Portable lactate analyzer.

The Role of the Ventilatory Threshold

As mentioned in laboratory 9, $\dot{V}O_2$ increases as a linear function of power output during a graded exercise test up to $\dot{V}O_2$max. On the other hand, ventilation ($\dot{V}_E$) increases linearly as a function of power output only during exercise requiring up to about 50% to 75% $\dot{V}O_2$max, above which $\dot{V}_E$ increases at a steeper rate (see figure 9.2, p. 212).

The greater rate of increase in $\dot{V}_E$ above this ventilatory threshold coincides with a similar rise in $\dot{V}CO_2$ and a sharp increase in blood lactate. Remember that lactic acid dissociates into a lactate ion and H^+ in the blood, and the H^+ combines with bicarbonate ion to form carbonic acid, which in turn dissociates into H_2O and CO_2.

$$La^-H^+ + HCO_3^- \leftrightarrow H_2CO_3 \leftrightarrow H_2O + CO_2$$

The accumulation of CO_2 in the blood leads to an increase in CO_2 in the expired air ($\dot{V}CO_2$). The increase in blood CO_2 stimulates an increase in ventilation. Thus, the ventilatory threshold is indicative of an abrupt increase in blood lactate and a subsequent decrease in blood pH. This "nonmetabolic" or "excess" CO_2 contributes to the $\dot{V}CO_2$ measured during the test, which is why at high intensities RER can exceed 1.0.

The bicarbonate is not clearing the lactate ion; it is simply buffering the proton. As mentioned above, the lactate must be cleared by being taken up by tissues such as muscle, liver, or kidneys.

As you will learn in this lab, measurement of blood lactate complicates an incremental test of increasing intensity to exhaustion. The necessity of taking even a small drop of blood during every stage increases the number of people needed to carry out the test. More supplies are also needed (e.g., lactate analyzer, gloves, gauze, alcohol swabs, bandages). For this reason, the ventilatory threshold is often used to approximate the lactate threshold. In this lab, you will relate the ventilatory threshold to the lactate threshold.

References

1. Australian Sports Commission. *Physiological Tests for Elite Athletes*. Champaign, IL: Human Kinetics, 2000.

2. Bentley DJ, McNaughton LR, Thompson D, Vleck VE, and Batterham AM. Peak Power Output, the Lactate Threshold, and Time Trial Performance in Cyclists. *Med Sci Sports Exer* 33: 2077–2081, 2001.

3. Brooks GA, Fahey TD, and Baldwin KM. *Exercise Physiology: Human Bioenergetics and Its Applications*. McGraw Hill, 2005.

4. Coyle EF, Coggan AR, Hopper MK, and Walters TJ. Determinants of Endurance in Well-Trained Cyclists. *J Appl Physiol* 64: 2622–2630, 1988.

5. Coyle EF, Feltner ME, Kautz SA, Hamilton MT, Montain SJ, Baylor AM, Abraham LD, and Petrek GW. Physiological and Biomechanical Factors Associated With Elite Endurance Cycling Performance. *Med Sci Sports Exer* 23: 93–107, 1991.

6. Coyle EF, Martin WH, Ehsani AA, Hagberg JM, Bloomfield SA, Sinacore DR, and Holloszy JO. Blood Lactate Threshold in Some Well-Trained Ischemic Heart Disease Patients. *J Appl Physiol* 54: 18–23, 1983.

7. Dumke CL, Brock DW, Helms BH, and Haff GG. Heart Rate at Lactate Threshold and Cycling Time Trials. *J Strength Cond Res* 20: 601–607, 2006.

8. Hagberg JM and Coyle EF. Physiological Determinants of Endurance Performance as Studied in Competitive Racewalkers. *Med Sci Sports Exer* 15: 287–289, 1983.

9. Janssen P. *Lactate Threshold Training*. Champaign, IL: Human Kinetics, 2001.

10. Kindermann W, Simon G, and Keul J. The Significance of the Aerobic-Anaerobic Transition for the Determination of Work Load Intensities During Endurance Training. *Eur J Appl Physiol Occup Physiol* 42: 25–34, 1979.

11. Lindinger MI, Kowalchuk JM, and Heigenhauser GJ. Applying Physicochemical Principles to Skeletal Muscle Acid-Base Status. *Am J Physiol Regul Integr Comp Physiol* 289: R891–R894 (author reply R904–R910), 2005.

12. Robergs RA, Ghiasvand F, and Parker D. Biochemistry of Exercise-Induced Metabolic Acidosis. *Am J Physiol Regul Integr Comp Physiol* 287: R502–R516, 2004.

13. Sjodin B and Jacobs I. Onset of Blood Lactate Accumulation and Marathon Running Performance. *Int J Sports Med* 2: 23–26, 1981.

BLOOD LACTATE MEASUREMENT AT REST

EQUIPMENT

- Lactate analyzer with strips
- Lancets
- Exam gloves
- Alcohol swabs
- Sterile gauze, bandages
- Group data sheet

 Find the group data sheets for this laboratory online at www.HumanKinetics.com/ LaboratoryManualForExercisePhysiology.

LABORATORY ORIENTATION

Collecting a drop of blood is a simple, relatively noninvasive measure. However, because blood is a bodily fluid, you must take care in handling it. You should wear gloves when handling blood and be careful to dispose of blood-stained materials into bins designated for biological waste. It is a common mistake for students to spread blood contamination onto other equipment or elsewhere in lab because of residual blood on their gloves. Be conscious of potential contamination.

As mentioned at the beginning of this lab, lactate can be measured in blood samples by means of several methods. Because of the availability and affordability of portable lactate analyzers, the following steps assume that you will be using one. Note that calibration of portable analyzers may require that a calibration strip be used for each batch of lactate test strips. Follow any manufacturer's guidelines for calibration of your lactate monitor. It is possible to collect blood from earlobes or fingertips; because fingertips are both more common and easier for students, we will assume that you are using fingertip blood draws in this lab. Your laboratory instructor will clarify the methods to use for your particular lactate analyzer.

LACTATE ANALYZER TEST

Step 1: Your laboratory instructor will demonstrate the procedure for taking blood samples for the lactate analyzer in your lab. Assuming that your lab uses a portable analyzer, calibration is relatively easy. Each batch of lactate strips comes with a code that needs to be confirmed in the analyzer, either by inserting a code strip or by manual entry. The machine should then recognize a lactate strip once it is inserted, and be ready for the application of a drop of blood.

Step 2: Have subject sit for about 3 min prior to taking a resting blood sample.

Step 3: While wearing exam gloves, clean the subject's fingertip with an alcohol swab. Wait for the alcohol to evaporate so that the blood is not mixed with the alcohol during sampling.

Step 4: With a lancet, prick the swabbed finger and pulse-squeeze it to obtain a drop of blood.

Step 5: Place the lactate analyzer strip in contact with the drop of blood to begin measurement. Capillary action should draw a blood sample into a reservoir in the strip. Be careful not to introduce bubbles of air into the strip reservoir.

Step 6: Wipe any remaining blood off of the finger with sterile gauze and apply a bandage if necessary.

Step 7: Record the data on the group data sheet.

QUESTION SET 10.1

1. Prepare a data table showing resting lactate values for your class.
2. Did resting lactate differ by sex? Why might this be the case?
3. What do these values say about lactate production at rest?
4. What could explain a resting lactate value of >2.0 mM?

 Find the case studies for this laboratory online at www.HumanKinetics.com/ LaboratoryManualForExercisePhysiology.

Laboratory Activity 10.1

LACTATE THRESHOLD DURING AN INCREMENTAL CYCLE TEST

EQUIPMENT

- Cycle ergometer
- Metabolic cart, including gas flow meter and O_2 and CO_2 analyzers
- Heart rate monitor
- Lactate analyzer with strips
- Exam gloves
- Lancets
- Alcohol swabs
- Sterile gauze and bandages
- Individual data sheets

LABORATORY ORIENTATION

Collecting a drop of blood is a simple, relatively noninvasive measure. However, because blood is a bodily fluid, you must take care in handling it. You should wear gloves when handling blood and be careful to dispose of blood-stained materials into bins designated for biological waste. It is a common mistake for students to spread blood contamination onto other equipment or elsewhere in lab because of residual blood on their gloves. Be conscious of potential contamination.

As mentioned at the beginning of this lab, lactate can be measured in blood samples by means of several methods. Because of the availability and affordability of portable lactate analyzers, the following steps assume that you will be using one. Note that calibration of portable analyzers may require that a calibration strip be used for each batch of lactate test strips. Follow any manufacturer's guidelines for calibration of your lactate monitor. It is possible to collect blood from earlobes or fingertips; because fingertips are both more common and easier for students, we will assume that you are using fingertip blood draws in this lab. Your laboratory instructor will clarify the methods to use for your particular lactate analyzer.

INCREMENTAL CYCLE TEST FOR LACTATE THRESHOLD

Step 1: One student will serve as the subject for determining lactate threshold during an incremental test. Other students will facilitate the testing by performing such tasks as blood sampling, controlling workload on the cycle ergometer, taking RPE and HR measurements, and controlling the metabolic cart.

Step 2: Record the subject's descriptive characteristics—height, weight, and age.

Step 3: Fit the subject on the cycle ergometer comfortably with a slight knee angle (5–15°) when the leg is fully extended.

Step 4: Your lab instructor will explain procedures for using the metabolic cart, but you and other students will operate the cart and conduct the test, including calibration and data recording. Enter the subject's data into the metabolic cart.

Step 5: Prepare subject for testing by helping him or her with the HR monitor chest strap and headgear.

Step 6: Prior to the test, read the subject the RPE instructions (see lab 9). Allow the subject to ask questions prior to being connected to the metabolic equipment.

Step 7: Connect the subject to the mouthpiece or mask apparatus while the subject sits on the bike.

Step 8: Allow the subject to sit for about 3 min to attain resting steady state.

Step 9: Your laboratory instructor will demonstrate the procedure for taking blood samples for the lactate analyzer. While wearing exam gloves, clean the subject's fingertip with an alcohol swab. Wait for alcohol to evaporate so that the blood is not mixed with the alcohol during sampling. With a lancet, prick the swabbed finger and pulse-squeeze it to obtain a drop of blood. Place the lactate analyzer strip in contact with the drop of blood to begin measurement. Wipe any remaining blood off of the finger with sterile gauze and apply a bandage if necessary.

Step 10: Measure the subject's resting heart rate, RPE, and metabolic data.

Step 11: Allow a 3 to 5 min warm-up at an intensity no greater than that of the first stage.

Step 12: From your knowledge of cycle protocols discussed in laboratory 9, you can select a protocol such as the Åstrand or McArdle. However, these protocols have 2 min stages, which can be challenging for blood draws. Thus we suggest that you have noncyclists pedal at a cadence of 75 rev · min^{-1} at an initial resistance of 0.5 kg and then increase it by 0.5 kg every 3 min. If a fit cyclist is the volunteer subject, have him or her start at 1 or 1.5 kg.

Step 13: At each stage, collect the following data during the last 30 s of the stage: power, HR, $\dot{V}O_2$, RER, $\dot{V}_E$, and RPE. Record the results on the individual data sheet.

Step 14: During the last minute of each stage, collect a lactate sample. Be prepared with alcohol, gauze, strips, and analyzer to obtain the blood sample as quickly as possible. Doing so may take several students. It is ideal not to interrupt the stage timing, but if necessary for the purposes of this lab you may prolong the stage until a good blood sample is obtained for lactate analysis. Do not attempt to get a drop of blood from a previous stage lancet site—it can delay the blood draw and possibly result in coagulated blood being placed on the strip. Use a fresh lancet and new blood draw site for each stage.

Step 15: Be observant of the exercising individual to monitor for signs and symptoms indicating that the test should be terminated. Encouragement is effective in motivating a subject to attain maximal effort.

Step 16: Since the subject is unable to communicate through the mask or mouthpiece, several common hand signals can be used during maximal testing. For example, a thumbs-up indicates that the subject is doing well and is willing to continue. A raised index finger indicates that the subject will need to stop in about 1 minute, and knowing this allows the testers to prepare for the end of the test and to collect maximum data for all of the outcome measures.

Step 17: Stop the test when $\dot{V}O_2$max test criteria are reached, or when indicators for stopping the test arise (see laboratory 7, p. 167, for criteria for terminating a test), or when the subject stops the test by volitional fatigue. At this point, reduce the workload on the bike to that of the initial stage of the protocol. Have the subject keep the mouthpiece in his or her mouth for a 5-minute cool-down period (record RER during the cool-down).

Step 18: The cool-down is important in order to keep the subject moving to avoid blood pooling in the lower extremities and potential syncope. During the cool-down, which should last at least 5 min, monitor the subject's heart rate. An appropriate cool-down would have the subject attain an HR of at 120 beats $\cdot$ min^{-1} before he or she gets off the ergometer.

Step 19: Continue to monitor the subject's physical appearance and symptoms during the cool-down period.

Step 20: Record the data from each stage and complete the individual data sheet for this subject.

CYCLE ERGOMETER WORK AND POWER FORMULAS

The design of the Monark cycle ergometer allows you to accurately set the work rate by adjusting the tension on a belt around the flywheel and having the subject pedal at a constant rev $\cdot$ min^{-1}. The circumference of the flywheel is such that for each pedal revolution, a given spot on the flywheel would travel 6 m. The flywheel velocity is, therefore, equal to the pedal rev $\cdot$ min^{-1} $\times$ 6 m $\cdot$ rev^{-1}. As with the treadmill, no calculated work is done on a cycle ergometer when pedaling at no resistance. Work (force × distance) and power (work / time or force × velocity) can be calculated on a Monark cycle ergometer by using the following formulas:

$$\text{Work} = \text{force} \times \text{distance}$$

$$\text{Work on bike} = \text{resistance (kg)} \times (\text{pedal rev} \cdot \text{min}^{-1} \times 6 \text{ m} \cdot \text{rev}^{-1} \times \text{total min of exercise})$$

$$\text{Power} = \text{force} \times \text{velocity}$$

$$\text{Power on bike} = \text{resistance (kg)} \times (\text{pedal rev} \cdot \text{min}^{-1} \times 6 \text{ m} \cdot \text{rev}^{-1})$$

Note that the result of the power equation will be expressed in $\text{kg} \cdot \text{m} \cdot \text{min}^{-1}$, which means that you can also divide it by 6.12 to express it in watts.

QUESTION SET 10.2

1. Prepare a graph illustrating the relationship between blood lactate concentration and work rate.

2. Prepare a graph illustrating the relationship between HR and work rate.

3. Prepare a graph illustrating the relationship between $\dot{V}_E$ and work rate.

4. Recall the laboratory 9 discussion. Did your subject meet the criteria for having achieved a true $\dot{V}O_2\text{max}$?

5. At what workload, HR, and %HRmax does the subject reach his or her lactate threshold (LT)? OBLA? Ventilatory threshold? (Hint: a table would work best.)

6. Does the LT occur at the same workload as the ventilatory threshold? Why or why not?

7. What might explain a subject achieving different $\dot{V}O_2\text{max}$ values on the cycle ergometer and the treadmill?

8. What causes the abrupt increase in blood lactate accumulation, indicated by LT, that occurs during dynamic exercise in an incremental test?

9. Explain the importance of LT for performance prediction.

Find the case studies for this laboratory online at www.HumanKinetics.com/ LaboratoryManualForExercisePhysiology.

Laboratory Activity 10.2

Name or ID number: _____ Date: _____

Tester: _____ Time: _____

Sex: M / F (circle one)   Age: _____ y Height: _____ in. _____ cm

Weight: _____ lb _____ kg

Mode and protocol: _____

Metabolic cart: _____

Stage time	Resistance (kg)	Power (kg · m · min⁻¹)	HR (bpm)	$\dot{V}_E$ (L· min⁻¹)	$\dot{V}O_2$ (ml· min⁻¹)	$\dot{V}O_2$ (ml· kg⁻¹·min⁻¹)	METs	$\dot{V}CO_2$ (ml· kg⁻¹·min⁻¹)	RER	RPE	Lactate
	Rest	Rest									

BLOOD LACTATE AFTER ANAEROBIC EXERCISE

EQUIPMENT

- Monark cycle or equivalent ergometer equipped to perform a Wingate anaerobic test (e.g., Monark 814E, 824E, or 834E, all of which contain a peg or basket upon which the external load can be applied, as shown in figure 12.7 on p. 347)
- Revolutions counter that counts number of pedal revolutions performed in the test
- Timing apparatus (e.g., countdown set-timer, stopwatch, or timer attached to the cycle ergometer)s
- Heart rate monitor
- Lactate analyzer with strips
- Exam gloves
- Lancets
- Alcohol swabs
- Sterile gauze and bandages
- Individual data sheets

LABORATORY ORIENTATION

Lactate is typically cleared within minutes following exercise. The removal of lactate from the blood following exercise depends on blood flow to tissues such as muscle, liver, heart, and kidneys, which is typically facilitated by an exercise cool-down (see figure 10.4). This process is particularly important following a bout of exercise that is more reliant on fast glycolysis—the more anaerobic the bout of exercise, the greater the lactate production, and thus the more time needed to clear lactate from the blood. In this

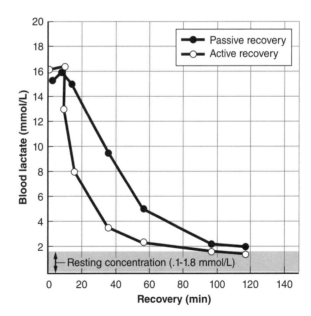

Figure 10.4 Active versus passive recovery on blood lactate clearance post exercise.

Reprinted, by permission, from W.L. Kenney, J.H. Wilmore, and D.L. Costill, 2012, *Physiology sport and exercise*, 4th ed. (Champaign, IL: Human Kinetics), 202.

lab activity, you will use a Wingate test to demonstrate the contribution to energy from fast glycolysis and the ability to clear lactate post-exercise.

BLOOD LACTATE LEVELS FOLLOWING A WINGATE TEST

Step 1: Two similarly sized students will perform Wingate testing. Following the test, one subject (referred to as SS for sitting subject) will sit quietly without a cool-down, while the other student (referred to as CDS for cool-down subject) will continue pedaling at a low workload.

Step 2: Set up the cycle ergometer and check to see if it is in working order.

Step 3: Measure both subjects' height and weight without shoes and record the data.

Step 4: Calculate the prescribed load for each subject based on the subject's training status (see table 12.2, p. 315). Record the load on the individual data sheet.

Step 5: Fit the cycle ergometer to the subjects so that while seated on the bike their extended legs have a slight bend (5–15°) (see figure 4.3a, p. 128).

Step 6: Completely explain the test protocol to the subjects and emphasize the fact that this is an all-out effort that lasts 30 s.

Step 7: Have the subjects perform the standardized warm-up and recovery periods outlined in table 12.10 (p. 348).

Step 8: Instruct the subjects to increase their pedaling rate to maximum. Instruct the force setter to apply the load, tell the timer to start the timer, and yell "go" when the subjects reach a maximal pedaling rate.

Step 9: The counter should begin counting pedal revolutions when he or she hears the "go" command. (If you are using a computerized system, this step need not be performed.)

Step 10: The counter should tell the recorder the number of revolutions completed at the end of each 5 s time interval during the test (i.e., 5, 10, 15, 20, 25, and stop).

Step 11: At the 30-second mark, the timer yells "stop" and the force setter reduces the load so that the subject can pedal for 2 to 5 min or until he or she has recovered. At this time, the counter tells the recorder the revolutions completed for the final 5 s interval.

Step 12: Following the test, collect blood samples in both the CDS and SS subjects (see details in this chapter's preceding lab activities for blood sampling procedures) every 2 min for the first 10 min, then once every 5 min for the next 30 min. Record blood lactate results in the data table for both subjects.

Step 13: To convert the prescribed load to newtons, multiply the load by 9.80665. Calculate the power output from the total number of revolutions, the distance the flywheel travels per rotation (6 m), and the duration of the interval. Calculate the power output for each time interval and record it in the appropriate location on the individual data sheet.

Step 14: Divide the power output calculated for each time interval by body weight in order to determine a relative power output. Record these values on the individual data sheet.

Step 15: Calculate the absolute and relative work accomplished during the 30 s time interval and record these values on the individual data sheet for both subjects.

Step 16: Calculate the absolute and relative mean power outputs accomplished during the 30 s test and record these data on the individual data sheet for both subjects.

Step 17: Record the data from each test stage and complete the data table for this subject.

QUESTION SET 10.3

1. What are the physiological reasons for the large amount of lactate produced after a Wingate test?

2. Prepare a graph illustrating the relationship between blood lactate concentration and time for both subjects.

3. At what time point did peak lactate occur following the Wingate test? What is the reason for this delayed response?

4. What do your data indicate about the importance of an active cool-down? For which sports is this particularly important?

Find the case studies for this laboratory online at www.HumanKinetics.com/ LaboratoryManualForExercisePhysiology.

Date: _____ Time: _____

Tester: _____

Temperature: _____°F _____°C Barometric pressure: _____mmHg

Relative humidity: _____ %

Wingate Test Subject Data: Cool-Down Subject (CDS)

Name or ID number: _____

Sex: M / F (circle one)   Age: _____ y Height: _____ in. _____ cm

Weight: _____ lb _____ kg

Prescribed load = _____ (body weight [kg]) × _____ (kp/kg body mass*) = _____ (kp)

*Note: In most instances, the multiplier should be 0.075 kp/kg body mass.

Force (N) = _____ (force setting in kp) × 9.80665 = _____ (N)

Wingate Test Subject Data: Sitting Subject (SS)

Name or ID number: _____

Sex: M / F (circle one)   Age: _____ y Height: _____ in. _____ cm

Weight: _____ lb _____ kg

Prescribed load = _____ (body weight [kg]) × _____ (kp/kg body mass*) = _____ (kp)

*Note: In most instances, the multiplier should be 0.075 kp/kg body mass.

Force (N) = _____ (force setting in kp) × 9.80665 = _____ (N)

Quantification of Pedal Revolutions

Time interval (s)	Revolutions (CDS)	Revolutions (SS)
0–5		
5–10		
10–15		
15–20		
20–25		
25–30		

Absolute Power Data for CDS

Time interval (s)	Peak power (W)
0–5	Peak power (W) = (_____ $\times$ _____ $\times$ 6 m) / 5 s = _____ W force (N) / revolutions
5–10	Peak power (W) = (_____ $\times$ _____ $\times$ 6 m) / 5 s = _____ W force (N) / revolutions
10–15	Peak power (W) = (_____ $\times$ _____ $\times$ 6 m) / 5 s = _____ W force (N) / revolutions
15–20	Peak power (W) = (_____ $\times$ _____ $\times$ 6 m) / 5 s = _____ W force (N) / revolutions
20–25	Peak power (W) = (_____ $\times$ _____ $\times$ 6 m) / 5 s = _____ W force (N) / revolutions
25–30	Peak power (W) = (_____ $\times$ _____ $\times$ 6 m) / 5 s = _____ W force (N) / revolutions

Absolute Power Data for SS

Time interval (s)	Peak power (W)
0–5	Peak power (W) = (_____ $\times$ _____ $\times$ 6 m) / 5 s = _____ W force (N) / revolutions
5–10	Peak power (W) = (_____ $\times$ _____ $\times$ 6 m) / 5 s = _____ W force (N) / revolutions
10–15	Peak power (W) = (_____ $\times$ _____ $\times$ 6 m) / 5 s = _____ W force (N) / revolutions
15–20	Peak power (W) = (_____ $\times$ _____ $\times$ 6 m) / 5 s = _____ W force (N) / revolutions
20–25	Peak power (W) = (_____ $\times$ _____ $\times$ 6 m) / 5 s = _____ W force (N) / revolutions
25–30	Peak power (W) = (_____ $\times$ _____ $\times$ 6 m) / 5 s = _____ W force (N) / revolutions

Relative Power Output for CDS

Time interval (s)	Relative power (W $\cdot$ kg^{-1})
0–5	Relative peak power = _____ / _____ = _____ W $\cdot$ kg^{-1} peak power (W) / body mass (kg)
5–10	Relative peak power = _____ / _____ = _____ W $\cdot$ kg^{-1} peak power (W) / body mass (kg)
10–15	Relative peak power = _____ / _____ = _____ W $\cdot$ kg^{-1} peak power (W) / body mass (kg)
15–20	Relative peak power = _____ / _____ = _____ W $\cdot$ kg^{-1} peak power (W) / body mass (kg)
20–25	Relative peak power = _____ / _____ = _____ W $\cdot$ kg^{-1} peak power (W) / body mass (kg)
25–30	Relative peak power = _____ / _____ = _____ W $\cdot$ kg^{-1} peak power (W) / body mass (kg)

Laboratory Activity 10.3

Relative Power for SS

Time interval (s)	Relative power (W · kg⁻¹)
0–5	Relative peak power = _____ / _____ = _____ W · kg⁻¹ peak power (W) body mass (kg)
5–10	Relative peak power = _____ / _____ = _____ W · kg⁻¹ peak power (W) body mass (kg)
10–15	Relative peak power = _____ / _____ = _____ W · kg⁻¹ peak power (W) body mass (kg)
15–20	Relative peak power = _____ / _____ = _____ W · kg⁻¹ peak power (W) body mass (kg)
20–25	Relative peak power = _____ / _____ = _____ W · kg⁻¹ peak power (W) body mass (kg)
25–30	Relative peak power = _____ / _____ = _____ W · kg⁻¹ peak power (W) body mass (kg)

Work Data for CDS

Time interval (s)	Work
0–30	Total work = (_____ × _____ × 6 m) = _____ J force (N) revolutions
0–30	Relative work = (_____ / _____ × 6 m) = _____ J · kg⁻¹ total work (J) body mass (kg)

Work Data for SS

Time interval (s)	Work
0–30	Total work = (_____ × _____ × 6 m) = _____ J force (N) revolutions
0–30	Relative work = (_____ / _____ × 6 m) = _____ J · kg⁻¹ total work (J) body mass (kg)

Mean Power Output for CDS

Time interval (s)	Mean power output
0–30	Mean power output = _____ / 30 s = _____ W total work (J)
0–30	Relative mean power output = _____ / _____ = _____ W · kg⁻¹ mean power (W) body mass (kg)

Mean Power Output for SS

Time interval (s)	Mean power output
0–30	Mean power output = _____ / 30 s = _____ W total work (J)
0–30	Relative mean power output = _____ / _____ = _____ W · kg⁻¹ mean power (W) body mass (kg)

Calculating the Fatigue Index

Fatigue index (CDS) = [(_____ – _____) / _____] × 100 = _____%

highest power (W)   lowest power (W)   highest power (W)

Fatigue index (SS) = [(_____ – _____) / _____] × 100 = _____%

highest power (W)   lowest power (W)   highest power (W)

Lactate Results Following Wingate Test

	CDS	SS
Peak power (W)		
Mean power (W)		
Lactate 2 min post (mM)		
Lactate 4 min post (mM)		
Lactate 6 min post (mM)		
Lactate 8 min post (mM)		
Lactate 10 min post (mM)		
Lactate 15 min post (mM)		
Lactate 20 min post (mM)		
Lactate 25 min post (mM)		
Lactate 30 min post (mM)		

Musculoskeletal Fitness Measurements

Objectives

1. Become familiar with various methods of evaluating muscular strength and endurance.
2. Differentiate between direct and indirect methods of determining muscular strength.
3. Understand the proper methods for performing a 1-repetition maximum bench press and leg press test.
4. Introduce various prediction equations that are useful when attempting to estimate muscular strength.

Definitions

1-repetition maximum (1RM)—Maximum amount of weight or load that can be lifted one time.

alternated grip—Grip in which the dominant hand is supinated and the other hand is pronated

muscular endurance—Ability to exert submaximal forces repetitively (2, 4).

muscular fitness—Combination of muscular endurance and maximal strength.

muscular strength—Highest amount of force that can be generated by a muscle or group of muscles during a single contraction (11, 21).

prediction equation—Equation established to predict one variable from one or a series of other variables. (In this laboratory, equations are used that predict 1RM strength from the load lifted and the number of repetitions performed.)

pronated grip—Overhand grip.

repetition maximum (RM)—Maximal amount of weight that can be lifted for a prescribed repetition range (e.g., 1RM is the heaviest weight that can be lifted one time, whereas 10RM is the heaviest weight that can be lifted 10 times).

supinated grip—Underhanded grip.

Musculoskeletal or **muscular fitness** is associated with numerous health benefits, including a reduced risk of coronary heart disease, osteoporosis, glucose intolerance, and musculoskeletal injuries (2, 23). A healthy musculoskeletal system is also associated with improved ability to complete activities of daily living and is directly related to quality of life (23, 46). Ultimately, high levels of musculoskeletal fitness are associated with positive health status, whereas lower levels of musculoskeletal fitness are associated with lower health status (46).

Since musculoskeletal fitness is such an important contributor to overall health and wellness, it should be evaluated as part of a comprehensive health and wellness screening process (21, 41). The

American College of Sports Medicine (ACSM) defines muscular or musculoskeletal fitness as the combination of muscular strength and muscular endurance (2). **Muscular strength** refers to the largest amount of force that a muscle or group of muscles can generate during a single contraction (11, 21), whereas **muscular endurance** refers to the ability to exert submaximal forces repetitively (2, 4). Muscular strength and endurance can be evaluated by means of numerous methods, and individual tests are generally selected based upon the muscle group being tested, the equipment available, and the individual subject's capabilities.

Assessments of Muscular Strength

No single assessment evaluates total body muscular strength or endurance; rather, musculoskeletal fitness testing is very specific to the muscle groups tested, the velocity of movement employed, the type of contraction and range of motion used, and the type of equipment used to perform the assessment (2). These issues sometimes make it difficult to compare individual results with those reported in the literature.

When conducting assessments of musculoskeletal or muscular fitness, you should familiarize clients or subjects with the equipment and protocol being used, since taking these steps ensures that the testing process is as reliable as possible (2). You should also perform all protocols under stan-

dardized conditions. The ACSM recommends six steps—including using proper lifting technique and familiarizing clients and subjects with the equipment—to increase the accuracy and reliability of any musculoskeletal fitness assessment (table 11.1).

Muscular strength can be assessed with either dynamic or static methods; specifically, dynamic assessments involve moving an external load or body part, whereas static methods exhibit no overt muscular or limb movement (2). Muscular strength assessment can also be divided into two categories in another fashion: (1) the **1-repetition maximum (1RM)** test and (2) static or isometric tests.

1-Repetition Maximum Testing

Traditionally, the 1RM test has been considered the gold standard when evaluating dynamic muscular strength. It requires the client to exert maximal force dynamically through a range of motion in a controlled manner while maintaining proper technique (2, 5). When performing this assessment with a client for the first time, it is often difficult to obtain an accurate and reliable 1RM because of the number of attempts that may be used or the fact that the client may have unstable technique (5, 29). However, if the client is properly familiarized with the exercise and protocol used in the assessment, the 1RM test is highly reliable, and 1RM tests overall exhibit high test–retest reliabilities as indicated by intraclass correlations ranging between ICC = 0.79 and ICC = 0.99 (26).

Table 11.1 ACSM Recommendations for Standardized Conditions for Muscular Strength and Endurance Tests

Guideline	Rationale
Maintenance of proper technique	Maintaining proper technique ensures that the exercise is performed correctly and safely.
Consistent repetition duration or movement speed	Maintaining consistent repetition speed allows for easier interpretation of results.
Full range of motion	Using a full range of motion allows for easier interpretation of results and maximizes safety.
Use of spotters	In exercises such as the bench press or back squat, spotters must be used in order to maximize safety.
Familiarization with equipment	All subjects should be familiarized with the equipment and protocols used in order to maximize the reliability of the assessment.
Proper warm-up	As with all testing, employ a proper warm-up in order to maximize performance capacity while minimizing potential risks that may be inherent in the testing process.

Adapted from American College of Sports Medicine (2).

The basic 1RM procedure can be applied to both free weight and machine-based exercises and can be completed through the following methods (2, 26, 42):

1. Have the subject estimate his or her maximum. This is easier to do with trained individuals because you can use the loads that they train with in order to form the estimation. With untrained or novice individuals, it is more difficult to determine the perceived maximum, and the process is rather exploratory. In most cases, body weight is used as the perceived 1RM in order to get these subjects testing loads.

2. The subject should warm up with 5 to 10 repetitions at 40% to 60% of perceived maximum.

3. After 1 min of rest, the subject should perform 3 to 5 repetitions with a resistance between 60% and 80% of perceived maximum.

4. After 3 min rest, the subject should perform 1 repetition equal to ~90% of perceived maximum.

5. If done correctly, step 4 of this procedure should take the subject close to the 1RM. At this point, make conservative increases in resistance and have the subject perform 1 repetition. If the lift is completed, the subject should rest for 3 min, then make another attempt with an increased load. Continue this process (resting 3 min, then increasing the load) until the subject is unable to complete the repetition with good technique. To maximize the reliability of the process, the 1RM should be achieved within 3 to 5 sets.

6. The 1RM value is reported as the heaviest weight that the client successfully completed.

1RM testing is considered safe for most populations (26); specifically, investigations looking at 1RM testing with children, older adults, and athletes have found few or no injuries when testing is conducted according to standard procedures and is properly supervised (8, 11, 20, 26, 39, 40). Collectively, the literature highlights the fact that 1RM testing is safe for clinical clients and athletes, but it also stresses the importance of familiarizing clients with the exercise being tested and of ensuring proper supervision by qualified testers. Ultimately, the tester supervising the test must assess the client's technique and determine whether or not the client should continue in the testing process.

Once the 1RM is established, you can represent the results in several ways. The first method is to simply report the heaviest load (in kg) that the client or subject lifted. This method is acceptable when attempting to track a client's musculoskeletal fitness over time, but it is not as useful when making comparisons between individuals (6). The most common method for evaluating or comparing individuals is to simply divide the 1RM by body mass, then compare the results with data from norm tables (e.g., tables 11.2 and 11.3) (21).

Table 11.2 Age- and Sex-Based Norms for the 1RM Bench Press

Age (y)		20–29		30–39		40–49		50–59		≥60	
Descriptors	**% rank**	**M**	**F**	**M**	**F**	**M**	**F**	**M**	**F**	**M**	**F**
Well above average	90	1.48	0.54	1.24	0.49	1.10	0.46	0.97	0.40	0.89	0.41
	80	1.32	0.49	1.12	0.45	1.00	0.40	0.90	0.37	0.82	0.38
Above average	70	1.22	0.42	1.04	0.42	0.93	0.38	0.84	0.35	0.77	0.36
	60	1.14	0.41	0.98	0.41	0.88	0.37	0.79	0.33	0.72	0.32
Average	50	1.06	0.40	0.93	0.38	0.84	0.34	0.75	0.31	0.68	0.30
	40	0.99	0.37	0.88	0.37	0.80	0.32	0.71	0.28	0.66	0.29
Below average	30	0.93	0.35	0.83	0.34	0.76	0.30	0.68	0.26	0.63	0.28
	20	0.88	0.33	0.78	0.32	0.72	0.27	0.63	0.23	0.57	0.26
Well below average	10	0.80	0.30	0.71	0.27	0.65	0.23	0.57	0.19	0.53	0.25

Values represented in table are 1RM (kg) / body mass (kg). M = male and F = female.

Adapted, by permission, from V. Heyward, 2010, *Advanced fitness assessment and exercise prescription*, 6th ed. (Champaign, IL; Human Kinetics), 136. Data for women provided by the Women's Exercise Research Center, The George Washington University Medical Center, Washington, D.C., 1998. Data for men provided by The Cooper Institute for Aerobics Research, The Physical Fitness Specialist Manual, The Cooper Institute, Dallas, TX, 2005.

Table 11.3 Age- and Sex-Based Norms for the 1RM Leg Press

Age (y)		20–29		30–39		40–49		50–59		≥60	
Descriptors	% rank	M	F	M	F	M	F	M	F	M	F
Well above average	90	2.27	2.05	2.07	1.73	1.92	1.63	1.80	1.51	1.73	1.40
	80	2.13	1.66	1.93	1.50	1.82	1.46	1.71	1.30	1.62	1.25
Above average	70	2.05	1.42	1.85	1.47	1.74	1.35	1.64	1.24	1.56	1.18
	60	1.97	1.36	1.77	1.32	1.68	1.26	1.58	1.18	1.49	1.15
Average	50	1.91	1.32	1.71	1.26	1.62	1.19	1.52	1.09	1.43	1.08
	40	1.83	1.25	1.65	1.21	1.57	1.12	1.46	1.03	138	1.04
Below average	30	1.74	1.23	1.59	1.16	1.51	1.03	1.39	0.95	1.30	0.98
	20	1.63	1.13	1.52	1.09	1.44	0.94	1.32	0.86	1.25	0.94
Well below average	10	1.51	1.02	1.43	0.94	1.35	0.76	1.22	0.75	1.16	0.84

Values represented in table are 1RM (kg) / body mass (kg). M = male and F= female.

Adapted, by permission, from V. Heyward, 2010, *Advanced fitness assessment and exercise prescription*, 6th ed. (Champaign, IL; Human Kinetics), 136. Data for women provided by the Women's Exercise Research Center, The George Washington University Medical Center, Washington, D.C., 1998. Data for men provided by The Cooper Institute for Aerobics Research, The Physical Fitness Specialist Manual, The Cooper Institute, Dallas, TX, 2005.

The ACSM recommends using the bench press and leg press for assessing upper and lower body strength when conducting a health and wellness screening (2). However, you can also assess overall strength fitness by examining the 1RM values for six specific exercises: bench press, arm curl, latissimus pull-down, leg press, leg extension, and leg curl. Each 1RM is divided by body mass to determine relative strength levels that are converted to points and summed in order to determine a total strength fitness score, which is then classified (see table 11.4).

Predicting 1RM

In some scenarios, the use of a direct 1RM test may not be feasible or safe, and in these cases a strength endurance test may be recommended as an indirect method of determining muscular strength (8). Prediction or indirect determination of 1RM is based upon a negative linear relationship between the percent of 1RM and the number of repetitions that can be performed (5). This linear relationship suggests that the percent of 1RM load decreases by about 2% to 2.5% per maximal repetition performed. Thus the 1RM load represents the maximal resistance that can be lifted 1 time and represents a 100% 1RM load, while lifting a load 5 times would be associated with approximately 90% of 1RM load (5).

There are two indirect methods for determining 1RM—the repetition maximum test (25) and the use of **prediction equations** (1, 13, 17, 28, 30, 33, 34, 37).

Estimating 1RM

One method for estimating the individual's 1RM is to use a **repetition maximum** test or strength endurance test. When such a test is used in this capacity, it should be performed with a load that ranges between 5 and 10 repetitions (i.e., the heaviest load that allows the client to complete between 5 and 10 repetitions with proper technique). Once the maximum number of repetitions that can be performed with a specific load is established, it can then be used in conjunction with a 1RM estimation table (e.g., table 11.5) to predict the client's maximal muscular strength with a specific exercise (8).

Table 11.4 Relative Strength Ratios* for Selected 1RM Tests in College-Age Men and Women

Points	Bench press		Lat pull-down		Leg press		Leg extension		Leg curl		Arm curl	
	M	F	M	F	M	F	M	F	M	F	M	F
10	1.50	0.90	1.20	0.85	3.00	2.70	0.80	0.70	0.70	0.60	0.70	0.50
9	1.40	0.85	1.15	0.80	2.80	2.50	0.75	0.65	0.65	0.55	0.65	0.45
8	1.30	0.80	1.10	0.75	2.60	2.30	0.70	0.60	0.60	0.52	0.60	0.42
7	1.20	0.70	1.05	0.73	2.40	2.10	0.65	0.55	0.55	0.50	0.55	0.38
6	1.10	0.65	1.00	0.70	2.20	2.00	0.60	0.52	0.50	0.45	0.50	0.35
5	1.00	0.60	0.95	0.65	2.00	1.80	0.55	0.50	0.45	0.40	0.45	0.32
4	0.90	0.55	0.90	0.63	1.80	1.60	0.50	0.45	0.40	0.35	0.40	0.28
3	0.80	0.50	0.85	0.60	1.60	1.40	0.45	0.40	0.35	0.30	0.35	0.25
2	0.70	0.45	0.80	0.55	1.40	1.20	0.40	0.35	0.30	0.25	0.30	0.21
1	0.60	0.35	0.75	0.50	1.20	1.00	0.35	0.30	0.25	0.20	0.25	0.18

Interpretation

To determine strength fitness category, sum the points for the 6 exercises tested:

Total points	Strength fitness classification
48–60	Excellent
37–47	Good
25–36	Average
13–24	Fair
0–12	Poor

* Relative values are determined by dividing the 1RM in kg by the subject's body mass in kg.

M = male and F = female.

Adapted, by permission, from V. Heyward, 2010, *Advanced fitness assessment and exercise prescription*, 6th ed. (Champaign, IL: Human Kinetics), 138.

Using 1RM Prediction Equations

The second indirect method is to conduct a strength endurance test that allows the client to perform fewer than 10 repetitions (1, 13, 15, 35, 47). In taking this approach, you can choose from numerous prediction equations (see table 11.6) (1, 13, 17, 28, 30, 33, 35, 37). The predictions are generally more accurate with heavier loads and fewer repetitions (1).

Isometric Strength Tests

You can also assess strength by means of static or isometric tests, in which there is no visible movement and the length of the active muscle is constant. Isometric tests are usually specific to the muscle group and joint angle being assessed and thus are limited in their ability to describe an individual's overall strength (2). The results of any isometric test are affected by factors including joint

Table 11.5 1-RM Estimation Table

Max reps (RM)	1	2	3	4	5	6	7	8	9	10	12	15
%1RM	100	95	93	90	87	85	83	80	77	75	67	65
Load (lb or kg)	10	10	9	9	9	9	8	8	8	8	7	7
	20	19	19	18	17	17	17	16	15	15	13	13
	30	29	28	27	26	26	25	24	23	23	20	20
	40	38	37	36	35	34	33	32	31	30	27	26
	50	48	47	45	44	43	42	40	39	38	34	33
	60	57	56	54	52	51	50	48	46	45	40	39
	70	67	65	63	61	60	58	56	54	53	47	46
	80	76	74	72	70	68	66	64	62	60	54	52
	90	86	84	81	78	77	75	72	69	68	60	59
	100	95	93	90	87	85	83	80	77	75	67	65
	110	105	102	99	96	94	91	88	85	83	74	72
	120	114	112	108	104	102	100	96	92	90	80	78
	130	124	121	117	113	111	108	104	100	98	87	85
	140	133	130	126	122	119	116	112	108	105	94	91
	150	143	140	135	131	128	125	120	116	113	101	98
	160	152	149	144	139	136	133	128	123	120	107	104
	170	162	158	153	148	145	141	136	131	128	114	111
	180	171	167	162	157	153	149	144	139	135	121	117
	190	181	177	171	165	162	158	152	146	143	127	124
	200	190	186	180	174	170	166	160	154	150	134	130
	210	200	195	189	183	179	174	168	162	158	141	137
	220	209	205	198	191	187	183	176	169	165	147	143
	230	219	214	207	200	196	191	184	177	173	154	150
	240	228	223	216	209	204	199	192	185	180	161	156
	250	238	233	225	218	213	208	200	193	188	168	163
	260	247	242	234	226	221	206	208	200	195	174	169
	270	257	251	243	235	230	224	216	208	203	181	176
	280	266	260	252	244	238	232	224	216	210	188	182
	290	276	270	261	252	247	241	232	223	218	194	189
	300	285	279	270	261	255	249	240	231	225	201	195
	310	295	288	279	270	264	257	248	239	233	208	202
	320	304	298	288	278	272	266	256	246	240	214	208
	330	314	307	297	287	281	274	264	254	248	221	215
	340	323	316	306	296	289	282	272	262	255	228	221
	350	333	326	315	305	298	291	280	270	263	235	228
	360	342	335	324	313	306	299	288	277	270	241	234
	370	352	344	333	322	315	307	296	285	278	248	241
	380	361	353	342	331	323	315	304	293	285	255	247
	390	371	363	351	339	332	324	312	300	293	261	254
	400	380	372	360	348	340	332	320	308	300	268	260

Max reps (RM)	1	2	3	4	5	6	7	8	9	10	12	15
%1RM	100	95	93	90	87	85	83	80	77	75	67	65
Load (lb or kg) *(continued)*	410	390	381	369	357	349	340	328	316	308	274	267
	420	399	391	378	365	357	349	336	323	315	281	273
	430	409	400	387	374	366	357	344	331	323	288	280
	440	418	409	396	383	374	365	352	339	330	295	286
	450	428	419	405	392	383	374	360	347	338	302	293
	460	437	428	414	400	391	382	368	354	345	308	299
	470	447	437	423	409	400	390	376	362	353	315	306
	480	456	446	432	418	408	398	384	370	360	322	312
	490	466	456	441	426	417	407	392	377	368	328	319
	500	475	465	450	435	425	415	400	385	375	335	325
	510	485	474	459	444	434	423	408	393	383	342	332
	520	494	484	468	452	442	432	416	400	390	348	338
	530	504	493	477	461	451	440	424	408	398	355	345
	540	513	502	486	470	459	448	432	416	405	362	351
	550	523	512	495	479	468	457	440	424	413	369	358
	560	532	521	504	487	476	465	448	431	420	375	364
	570	542	530	513	496	485	473	456	439	428	382	371
	580	551	539	522	505	493	481	464	447	435	389	377
	590	561	549	531	513	502	490	472	454	443	395	384
	600	570	558	540	522	510	498	480	462	450	402	390

Reprinted, by permission, from National Strength and Conditioning Association, 2008, Resistance training, by T.R. Baechle, R.W. Earle, and D. Wathen. In *Essentials of strength training and conditioning*, 3rd ed., edited by T.R. Baechle and R.W. Earle (Champaign, IL: Human Kinetics), 398.

Table 11.6 1RM Prediction Equations

Reference	Equation
Adams (5)	$1RM = RepWt / (1 - 0.02 \times RTF)$
Berger (10)	$1RM = RepWt / (1.0261 - 0.00262 \times RTF)$
Brown (12)	$1RM = (RTF \times 0.0338 + 0.9849) \times RepWt$
Cummings and Finn (16)	$1RM = 1.175 \times RepWt + 0.839 \times RTF - 4.29787$
Mayhew et al. (32)	$1RM = RepWt / (0.522 + 0.419\, e^{-0.055 \times RTF})$
O'Conner et al. (37)	$1RM = 0.025\, (RepWt \times RTF) + RepWt$

1RM = 1-repetition maximum; RepWT = repetition weight, load < 1RM to perform repetitions; and RTF = repetitions to failure.

Adapted from Mayhew et al. (33).

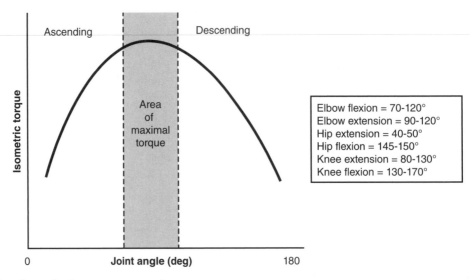

Figure 11.1 Sample human strength curve.

Reprinted, by permission, from P.J. Maud and C. Foster, 2008, *Physiological assessment of human fitness*, 2nd ed. (Champaign, IL: Human Kinetics), 131.

angle, testing protocol, feedback, and individual motivation (20, 26) Since force varies across any joint's range of motion, it is important that the joint angle assessed with this type of testing is standardized (figure 11.1).

Isometric strength and endurance can be assessed by using an isometric dynamometer, which can assess strength in the grip (figure 11.2*a*), the back and legs (figure 11.2*b*), and the upper body. The most simplistic isometric assessment is done with a handgrip dynamometer, and results obtained this way have been related to muscle mass (22) and to the function of the forearm muscles. As a whole, handgrip tests are very reliable (ICC > 0.90) (31) and constitute valid methods for assessing strength (5). The strength levels determined for each hand are then summed to create a composite score that represents a level of strength (see table 11.7).

Figure 11.2 Isometric dynamometers can be used to assess *(a)* grip strength and *(b)* back and leg strength. Force-power racks *(c)* can be used to test a variety of isometric movements.

Figure 11.2*c*: Reprinted, by permission, from M. Stone, M. Stone, and W. Sands, 2007, *Principles and practice of resistance training* (Champaign, IL: Human Kinetics), 174.

Table 11.7 Age- and Sex-Based Norms for Combined Handgrip Test

Age (y)	15–19		20–29		30–39		40–49		50–59		60–69	
Classification	M	F	M	F	M	F	M	F	M	F	M	F
Excellent	≥108	≥68	≥115	≥70	≥115	≥71	≥108	≥69	≥101	≥61	≥100	≥54
Very good	98–107	60–67	104–114	63–69	104–114	63–70	97–107	61–68	92–100	54–60	91–99	48–53
Good	90–97	53–59	95–103	58–62	95–103	58–62	88–96	54–60	84–91	49–53	84–90	45–47
Fair	79–89	48–52	84–94	52–59	84–94	51–57	80–87	49–53	76–83	45–48	73–83	41–44
Needs improvement	≤78	≤47	≤83	≤51	≤83	≤50	≤79	≤48	≤75	≤44	≤72	≤40

All values represent combined right and left handgrip strength in kilograms (kg). M = male, and F = female.

Source: The Canadian Physical Activity, Fitness & Lifestyle Approach: CSEP-Health & Fitness Program's Health-Related Appraisal and Counselling Strategy, 3rd Edition © 2003. Reprinted with permission of the Canadian Society for Exercise Physiology.

More complex multijoint isometric tests (e.g., mid-thigh pull, back squat, bench press) can be performed with the combination of a force plate and a custom isometric rack (26, 45) (see figure 11.2c). The reliability and validity of this approach is generally related to the standardization of both position and protocol (44). When standardized positions are used, these testing methods have been shown to be very reliable (19, 43, 45).

The most basic measurement of isometric strength reveals the peak force that can be generated and the time to achieve peak force (5). More complex isometric testing procedures yield a detailed force-time curve that can be analyzed (see figure 11.3) (20) in terms of three consider-ations: (1) overall peak force, sometimes termed *maximal voluntary contraction (MVC)*; (2) rate of force development, sometimes referred to as *explosive strength;* and (3) starting strength (44).

Assessments of Muscular Endurance

Muscular endurance is the ability of the neuromuscular system to perform one of two types of action: (1) repeated contractions over a period of time until fatigue occurs with dynamic exercises and (2) maintenance of a specific percentage of a maximal voluntary contraction for a period of time with static exercises (2). Generally, an individual's

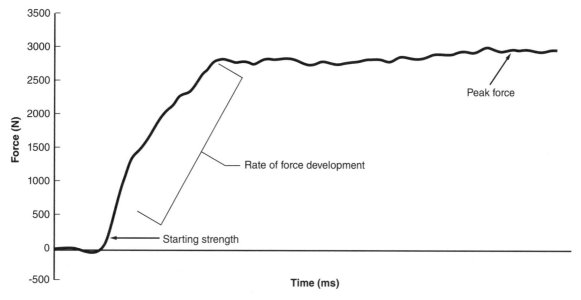

Figure 11.3 Isometric force-time curve.

muscular endurance is represented by the total number of repetitions that he or she can perform with a percentage of 1RM. Methods for evaluating a client's muscular endurance include the YMCA bench press test, the push-up test, and the curl-up test (2, 4, 21).

YMCA Bench Press Protocol

The YMCA bench press test is a classic protocol for muscular endurance. It assesses the muscular endurance of the upper body, specifically the pectoralis major, anterior deltoid, and triceps. This widely used assessment of muscular endurance requires the client to lie on his or her back on a bench and press a constant load at a cadence of 30 reps · min^{-1} until he or she experiences volitional fatigue or an inability to maintain the prescribed cadence (24). Kim et al. (24) have recently reported that the YMCA bench press can also be used to estimate maximal bench press strength with the following prediction equations:

$$\text{Men 1RM (kg)} = 1.55 \times$$
$$(\text{repetitions performed at } 30 \text{ reps} \cdot \text{min}^{-1}) + 37.9$$

$$\text{Women 1RM (kg)} = 0.31 \times$$
$$(\text{repetitions performed at } 30 \text{ reps} \cdot \text{min}^{-1}) + 19.2$$

Once a maximum number of repetitions has been established, you can compare it with data from a norm table (e.g., table 11.8) in order to classify the client's upper body muscular endurance.

Push-Up Test

This test of upper body (pectoralis major, anterior deltoid, and triceps) muscular endurance has no time limit and requires only minimal equipment (21). It is performed in a continuous fashion and requires the client to maintain a straight back at all times. The test is stopped when proper technique cannot be maintained for two consecutive repetitions. The total number of repetitions is then used to tabulate the score, which is compared with data from age- and sex-based norm tables (e.g., table 11.9).

Table 11.8 Muscular Endurance Norms for the Bench Press

Age (y)	18–25		26–35		36–45		46–55		56–65		>65	
% rank	M	F	M	F	M	F	M	F	M	F	M	F
90	49	49	48	46	41	41	33	33	28	29	22	22
80	34	30	30	29	26	26	21	20	17	17	12	12
70	26	21	22	21	20	17	13	12	10	9	8	6
60	17	13	16	16	12	10	8	6	4	4	3	2
50	5	2	4	2	2	2	1	0	0	0	0	0

Score is based upon the number of repetitions completed in 1 min with an 80 lb (36.3 kg) bar for men and a 35 lb (15.9 kg) bar for women. M = male and F = female.

Reprinted, by permission, from V. Heyward, 2010, *Advanced fitness assessment and exercise prescription*, 6th ed. (Champaign, IL: Human Kinetics), 138.

Table 11.9 Age- and Sex-Based Norms for the Push-Up Test

Age (y)	15–19		20–29		30–39		40–49		50–59		60–69	
Classification	M	F	M	F	M	F	M	F	M	F	M	F
Excellent	≥39	≥33	≥36	≥30	≥30	≥27	≥25	≥24	≥21	≥21	≥18	≥17
Very good	29–38	25–32	29–35	21–29	22–29	20–26	17–24	15–23	13–20	11–20	11–17	12–16
Good	23–28	18–24	22–28	15–20	17–21	13–19	13–16	11–14	10–12	7–10	8–10	5–11
Fair	18–22	12–17	17–21	10–14	12–16	8–12	10–12	5–10	7–9	2–6	5–7	2–4
Needs improvement	≤17	≤11	≤16	≤9	≤11	≤7	≤9	≤4	≤6	≤1	≤4	≤1

Source: The Canadian Physical Activity, Fitness & Lifestyle Approach: CSEP-Health & Fitness Program's Health-Related Appraisal and Counselling Strategy, 3rd Edition © 2003. Adapted with permission of the Canadian Society for Exercise Physiology.

Curl-Up Test

This test, which assesses abdominal muscular endurance, is commonly conducted in the exercise physiology laboratory. It comes in several variations that range in duration from 60 s to 120 s and require a cadence of 20 to 25 reps·min⁻¹ (21). The

ACSM recommends using a 1 min timed curl-up test with a cadence of 25 reps·min⁻¹(2). Repetitions of the curl-up are counted only if proper technique is used, and the total number of completed repetitions is used to score the test. Results can then be compared with data from age- and sex-based norm tables (e.g., table 11.10).

Table 11.10 Age- and Sex-Based Norms for the Curl-Up Test

Age (y)	15–19		20–29		30–39		40–49		50–59		60–69	
Classification	M	F	M	F	M	F	M	F	M	F	M	F
Excellent	25	25	25	25	25	25	25	25	25	25	25	25
Very good	23–24	22–24	21–24	18–24	18–24	19–24	18–24	19–24	17–24	19–24	16–24	17–24
Good	21–22	17–21	16–20	14–17	15–17	10–18	13–17	11–18	11–16	10–18	11–15	8–16
Fair	16–20	12–16	11–15	5–13	11–14	6–9	6–12	4–10	8–10	6–9	6–10	3–7
Needs improvement	≤15	≤11	≤10	≤4	≤10	≤5	≤5	≤3	≤7	≤5	≤5	≤2

Source: The Canadian Physical Activity, Fitness & Lifestyle Approach: CSEP-Health & Fitness Program's Health-Related Appraisal and Counselling Strategy, 3rd Edition © 2003. Adapted with permission of the Canadian Society for Exercise Physiology.

References

1. Abadie BR and Wentworth M. Prediction of 1RM Strength From a 5-10 Repetition Submaximal Test in College Aged Females. *J Exer Physiol* (online) 3: 1–5, 2000.

2. American College of Sports Medicine. *ACSM's Guidelines for Exercise Testing and Prescription*. 8th ed., Philadelphia: Lippincott Williams & Wilkins, 2010.

3. American College of Sports Medicine. *ACSM's Health-Related Physical Fitness Assessment Manual*. 3rd ed. Baltimore: Lippincott Williams & Wilkins, 2010.

4. Acevedo EO and MA Starks. *Exercise Testing and Prescription Lab Manual*. Champaign, IL: Human Kinetics, 2003, p. 165.

5. Adams GM. *Exercise Physiology Laboratory Manual*. 3rd ed. Boston: McGraw-Hill, 1998.

6. Adams KJ, Swank AM, Berning JM, Sevene-Adams PG, Barnard KL, and Shimp-Bowerman J. Progressive Strength Training in Sedentary, Older African American Women. *Med Sci Sports Exer* 33: 1567–1576, 2001.

7. Adams V, Jiang H, Yu J, Mobius-Winkler S, Fiehn E, Linke A, Weigl C, Schuler G, and Hambrecht R. Apoptosis in Skeletal Myocytes of Patients With Chronic Heart Failure Is Associated With Exercise Intolerance. *J Am Coll Cardiol* 33: 959–965, 1999.

8. Baechle TR, Earle RW, and Wathen D. Resistance Training. In: Baechle TR and Earle RW, eds., *Essentials of Strength Training and Conditioning*. 3rd ed. National Strength and Conditioning Association. Champaign, IL: Human Kinetics, 2008, pp. 381–412.

9. Barnard KL, Adams KJ, Swank AM, Mann E, and Denny DM. Injuries and Muscle Soreness During the One Repetition Maximum Assessment in a Cardiac Rehabilitation Population. *J Cardiopulm Rehabil* 19: 52–58, 1999.

10. Berger RA. Relationship Between Dynamic Strength and Dynamic Endurance. *Res Q* 41: 115–116, 1970.

11. Bompa TO and Haff GG. *Periodization: Theory and Methodology of Training*. 5th ed. Champaign, IL: Human Kinetics, 2009.

12. Brown HL. *Lifetime Fitness*. 3rd ed. Scottsdale, AZ: Gorsuch Scarisbrick, 1992.

13. Bryzychi M. Assessing Strength. *Fitness Manage* June: 34–37, 2000.

14. Bryzychi M. Strength Testing: Predicting a One Rep Max From Reps to Fatigue. *J Phys Educ Rec Dance* 64: 88–90, 1993.

15. Chapman P, Whitehead JR, and Binkert RH. The 225-lb Reps-to-Failure Test as a Submaximal Estimation of 1RM Bench Press Performance in

College Football Players. *J Strength Cond Res* 12: 258–261, 1998.

16. Cummings B and Finn KJ. Estimation of a One Repetition Maximum Bench Press for Untrained Women. *J Strength Cond Res* 12: 262–265, 1998.

17. Epley B. Poundage Chart. In: *Boyd Epley Workout*. Lincoln, NE: University of Nebraska, 1985.

18. Faigenbaum AD, Westcott WL, Loud RL, and Long C. The Effects of Different Resistance Training Protocols on Muscular Strength and Endurance Development in Children. *Pediatrics* 104: 1–7, 1999.

19. Haff GG, Jackson JR, Kawamori N, Carlock JM, Hartman MJ, Kilgore JL, Morris RT, Ramsey MW, Sands WA, and Stone MH. Force-Time Curve Characteristics and Hormonal Alterations During an Eleven-Week Training Period in Elite Women Weightlifters. *J Strength Cond Res* 22: 433–446, 2008.

20. Haff GG, Stone MH, O'bryant HS, Harman E, Dinan CN, Johnson R, and Han KH. Force-Time Dependent Characteristics of Dynamic and Isometric Muscle Actions. *J Strength Cond Res* 11: 269–272, 1997.

21. Heyward VH. *Advanced Fitness Assessment and Exercise Prescription*. 6th ed. Champaign, IL: Human Kinetics, 2010.

22. Kallman DA, Plato CC, and Tobin JD. The Role of Muscle Loss in the Age-Related Decline of Grip Strength: Cross-Sectional and Longitudinal Perspectives. *J Gerontol* 45: M82–M88, 1990.

23. Kell RT, Bell G, and Quinney A. Musculoskeletal Fitness, Health Outcomes and Quality of Life. *Sports Med* 31: 863–873, 2001.

24. Kim PS, Mayhew JL, and Peterson DF. A Modified YMCA Bench Press Test as a Predictor of 1 Repetition Maximum Bench Press Strength. *J Strength Cond Res* 16: 440–445, 2002.

25. Kraemer WJ and Fry AC. Strength Testing: Development and Evaluation of Methodology. In: Maud PJ and Foster C, eds., *Physiological Assesment of Human Fitness*. 2nd ed. Champaign, IL: Human Kinetics, 1995, pp. 115–138.

26. Kraemer WJ, Ratamess NA, Fry AC, and French DN. Strength Training: Development and Evaluation of Methodology. In: Maud PJ and Foster C, eds., *Physiological Assessment of Human Fitness*. 2nd ed. Champaign, IL: Human Kinetics, 2006, pp. 119–150.

27. Kraemer WJ, Vingren JL, Hatfield DL, Spiering BA, and Fragala MS. Resistance Training Programs. In: *ACSM's Resources for the Personal Trainer*. Baltimore: Lippincott Williams & Wilkins, 2007, pp. 372–403.

28. Kravitz L, Akalan C, Nowicki K, and Kinzey SJ. Prediction of 1 Repetition Maximum in High-School Power Lifters. *J Strength Cond Res* 17: 167–172, 2003.

29. Kuramoto AK and Payne VG. Predicting Muscular Strength in Women: A Preliminary Study. *Res Q Exer Sport* 66: 168–172, 1995.

30. Lander JE. Maximum Based on Reps. *Natl Strength Cond Assoc* 6: 60–61, 1985.

31. Mathiowetz V, Weber K, Volland G, and Kashman N. Reliability and Validity of Grip and Pinch Strength Evaluations. *J Hand Surg Am* 9: 222–226, 1984.

32. Mayhew JL, Ball TE, Arnold ME, and Bowen JC. Relative Muscular Endurance Performance as a Predictor of Bench Press Strength in College Men and Women. *J Appl Sport Sci Res* 6: 200–206, 1992.

33. Mayhew JL, Johnson BD, Lamonte MJ, Lauber D, and Kemmler W. Accuracy of Prediction Equations for Determining One Repetition Maximum Bench Press in Women Before and After Resistance Training. *J Strength Cond Res* 22: 1570–1577, 2008.

34. Mayhew JL, Prinster JL, Ware JS, Zimmer DL, Arabas JR, and Bemben MG. Muscular Endurance Repetitions to Predict Bench Press Strength in Men of Different Training Levels. *J Sports Med Phys Fitness* 35: 108–113, 1995.

35. Mayhew JL, Ware JS, Cannon K, Corbett S, Chapman PP, Bemben MG, Ward TE, Farris B, Juraszek J, and Slovak JP. Validation of the NFL-225 Test for Predicting 1RM Bench Press Performance in College Football Players. *J Sports Med Phys Fitness* 42: 304–308, 2002.

36. Montoye HJ and Lamphiear DE. Grip and Arm Strength in Males and Females, Age 10 to 69. *Res Q* 48: 109–120, 1977.

37. O'Conner B, Simmons J, and O'Shea P. *Weight Training Today*. St. Paul, MN: West, 1989, pp. 26–33.

38. Sale DG. Testing Strength and Power. In: MacDougall JD, Wenger HA, and Green HJ, eds., *Physiological Testing of the High-Performance Athlete*. Champaign, IL: Human Kinetics, 1991, pp. 21–106.

39. Schlicht J, Camaione DN, and Owen SV. Effect of Intense Strength Training on Standing Balance, Walking Speed, and Sit-to-Stand Performance in Older Adults. *J Gerontol A Biol Sci Med Sci* 56: M281–M286, 2001.

40. Shaw CE, Mccully KK, and Posner JD. Injuries During the One Repetition Maximum Assessment in the Elderly. *J Cardiopulm Rehab* 15: 283–287, 1995.

41. Spring T, Franklin B, and Dejong A. Muscular Fitness and Assessment. In: *ACSM's Resource Manual for Guidelines for Exercise Testing and Prescription.* 8th ed. Baltimore: Lippincott Williams & Wilkins, 2010, pp. 332–348.

42. Stone MH and O'Bryant HS. *Weight Training: A Scientific Approach.* Edina, MN: Burgess, 1987.

43. Stone MH, Sands WA, Pierce KC, Carlock J, Cardinale M, and Newton RU. Relationship of Maximum Strength to Weightlifting Performance. *Med Sci Sports Exer* 37: 1037–1043, 2005.

44. Stone MH, Sands WA, Pierce KC, Ramsey MW, and Haff GG. Power and Power Potentiation Among Strength-Power Athletes: Preliminary Study. *Int J Sports Physiol Perf* 3: 55–67, 2008.

45. Stone MH, Stone ME, and Sands WA. *Principles and Practice of Resistance Training.* Champaign, IL: Human Kinetics, 2007, p. 376.

46. Warburton DE, Gledhill N, and Quinney A. Musculoskeletal Fitness and Health. *Can J Appl Physiol* 26: 217–237, 2001.

47. Ware JS, Clemens CT, Mayhew JL, and Johnston TJ. Muscular Endurance Repetitions to Predict Bench Press and Squat Strength in College Football Players. *J Strength Cond Res* 9: 99–103, 1995.

48. Wathen D. Load Assignment. In: *Essentials of Strenght Training and Conditioning.* T.R. Baechle, ed. Champaigne, IL: Human Kinetics, 1994. pp. 435–446.

MAXIMAL UPPER BODY STRENGTH

EQUIPMENT

- Bench press and appropriate free weights
- Physician's scale or equivalent electronic scale
- Stadiometer
- Stopwatch
- Individual and group data sheets
- Microsoft Excel or equivalent spreadsheet program

Find the group data sheets for this laboratory online at www.HumanKinetics.com/ LaboratoryManualForExercisePhysiology.

WARM-UP

Regardless of which method is used to assess muscular strength, each subject should perform a structured warm-up designed to prepare him or her for the tests that will be performed. The warm-up should contain a 5 min general portion and a 5 min dynamic stretching portion (8). Sample general warm-up exercises include light jogging, cycling, and jumping rope. Dynamic stretching activities might include push-ups, arm circles and swings, as well as other activities target- ing the upper body. After completing the warm-up protocol, the subject can begin the testing process.

1RM BENCH PRESS TEST

The bench press test requires the use of a flat bench and free weights that pro- vide a range of possible loads. For example, various weight bars may be needed (e.g., 5 kg, 15 kg, 20 kg), depending on the population being tested. Ideally, you will have a barbell set with loads ranging from 0.5 kg to 25 kg to allow for a wide variety of loads. When performing the bench press test, it is essential to use appropriate spotting techniques (see figure 11.4 on p. 265). Briefly, the spotter should stand behind the bench press and use an **alternated grip** (i.e., one hand using a **supinated grip** and the other hand using a **pronated grip**) when lifting the bar out of the racked position on the bench and when helping rerack the barbell. It is also important for the subject to maintain five points of contact with the floor or bench—head, shoulders and upper back, right foot, left foot, and buttocks. Before starting the 1RM test, the tester should consult with the subject to estimate his or her perceived maximum. This process can be challenging with untrained individuals because it is likely to be speculative at best. For males, use the subject's body weight as an estimated 1RM; for example, if the subject weighs 60 kg, then his perceived 1RM would be about 60 kg. With athletes, in contrast, it is relatively easy to determine the perceived maximum, since you can likely use the actual 1RM or estimates based on the individual's training weights. This value should be noted on the individual data sheet for laboratory activity 11.1 and used to calculate the 40% to 60% load, 60% to 80% load, and 90% load needed to conduct the test. Once the loads have been determined, use the following steps to conduct the assessment.

Step 1: Set up the bench press area with all equipment needed to conduct the test.

Step 2: Have the subject remove his or her shoes, then measure his or her height and weight. Record these numbers on the individual data sheet. Have the subject put on his or her shoes.

Step 3: Lead the subject through a 10 min warm-up that includes dynamic warm-up activities such as arms swings in order to prepare the upper body for the bench press test.

Step 4: Load the bar with the weight corresponding to the subject's warm-up weight (40%–60% of perceived maximum), then move into appropriate spotting position behind the bench press (figure 11.4a).

Step 5: Instruct the subject to lie on his or her back on the bench in a five-point contact position (figure 11.4a). The subject's body should be in a position on the bench such that the racked bar is directly above his or her eyes. All repetitions should begin in this position.

Step 6: Lift the bar off of the rack using an alternated grip upon the command of the subject. Guide the bar to a position over the subject's chest, then release the bar to the subject (figure 11.4b).

Step 7: Have the subject lower the bar in a controlled manner to touch his or her chest at approximately nipple height while maintaining the five points of contact and inhaling (figure 11.4c). During this movement, keep your hands in the alternated position near the bar but not touching the bar.

Step 8: The subject should push the bar upward until his or her elbows are fully extended while maintaining the five points of contact and exhaling (figure 11.4d). At the end of the prescribed number of repetitions—or if maximum is achieved—grasp the bar with an alternated grip and place it back on the rack (figure 11.4e).

Step 9: After the subject completes 5 to 10 repetitions with the 40% to 60% load, quickly change the load to the predetermined second-set load. The subject should have only 1 min to rest during this change.

Step 10: After the 1 min rest interval, conduct steps 4 through 7 and have the subject perform 3 to 5 repetitions with a load that is 60% to 80% of perceived maximum.

Figure 11.4 Bench press series: *(a)* lift-off, *(b)* starting position, *(c)* downward position, *(d)* upward movement, and *(e)* racking the bar.

Step 11: After the completion of set 2, the subject rests for 3 min while you change the resistance to a load that is 90% of perceived maximum.

Step 12: After the 3 min rest, conduct steps 4 through 7 and have the subject perform 1 repetition with the 90% load.

Step 13: While the subject rests for 3 min, increase the load, depending upon how well the subject performed the previous attempt. If the attempt appeared relatively easy, increase the load by 5 to 10 kg; if, however, the attempt was more difficult, increase the load by 1 to 5 kg, depending upon the weights available.

Step 14: Repeat steps 12 and 13, continuing to have the subject perform only 1 repetition until a 1RM is achieved.

Step 15: If an attempt is unsuccessful, reduce the load (by 1 to 10 kg) but keep it above the last successful attempt and have the subject rest for 3 min.

Step 16: After the 3 min rest, have the subject perform another attempt with the basic procedures outlined in steps 4 to 7.

Step 17: If the attempt is successful, increase the load by a small amount (1–5 kg) and repeat step 16. If the attempt is unsuccessful, the 1RM load is determined as the heaviest load successfully completed. Record the maximum weight lifted in the appropriate location on the individual data sheet.

Step 18: Divide the maximal weight lifted by the subject's body mass and record the result in the appropriate location on the individual data sheet. This value should then be compared with the data in the norm table for the bench press (table 11.2). Based upon the results, note the subject's strength classification and percent rank on the individual data sheet.

QUESTION SET 11.1

1. Why is it important to test muscular strength?

2. When performing the 1RM bench press tests, what are some important factors to remember as a tester? As a subject?

3. Based upon your individual data, how do you compare with the class averages? What is the class's overall ranking?

4. Using Excel or an equivalent spreadsheet, graph the class, male, and female averages for the bench press. Include standard deviations on your graph.

5. How do the male and female results differ when examining absolute weight lifted? Relative weight lifted?

Find the case studies for this laboratory online at www.HumanKinetics.com/ LaboratoryManualForExercisePhysiology.

Laboratory Activity 11.1   Individual Data Sheet

Name or ID number: _____   Date: _____

Tester: _____   Time: _____

Sex: M / F (circle one)   Age: _____ y   Height: _____ in. _____ cm

Temperature: _____ °F _____ °C   Weight: _____ lb _____ kg

Barometric pressure: _____ mmHg   Relative humidity: _____ %

1RM Bench Press

Perceived maximum		Load	kg	Completed
Set 1	5–10 repetitions at 40%–60% of perceived maximum			❏
Set 2	3–5 repetitions at 60%–80% of perceived maximum			❏
Set 3	1 repetition at 90% of perceived maximum			❏
Set 4	1RM attempt			❏
Set 5	1RM attempt			❏

Note: 1 min rest between sets 1 and 2; 3 min rest preceding sets 3, 4, and 5.

Absolute 1-repetition maximum = _____ kg

Relative 1-repetition maximum = Load (kg) / body weight (kg) = _____

Strength classification = % rank = _____

Notes:

MAXIMAL LOWER BODY STRENGTH

EQUIPMENT

- Leg press machine
- Physician's scale or equivalent electronic scale
- Stadiometer
- Stopwatch
- Individual and group data sheets
- Microsoft Excel or equivalent spreadsheet program

Find the group data sheets for this laboratory online at www.HumanKinetics.com/ LaboratoryManualForExercisePhysiology.

WARM-UP

Regardless of which method is used to assess muscular strength, each subject should perform a structured warm-up designed to prepare him or her for the tests that will be performed. The warm-up should contain a 5 min general portion and a 5 min dynamic stretching portion (8). Sample general warm-up exercises include light jogging, cycling, and jumping rope. For this test, dynamic stretching activities might include leg swings, walking lunges, high knees, and trunk circles. After completing the warm-up protocol, the subject can begin the testing process.

1RM LEG PRESS TEST

The 1-repetition maximum leg press test is performed in a fashion very similar to that of the bench press test. The subject sits in a leg press machine with his or her feet hip-width apart on the machine's foot platform (figure 11.5a). Make sure that the subject is properly fitted to the machine. The foot and chair placement should allow the subject to full extend his or her legs (figure 11.5b) and to achieve a bottom position where the knee is bent to about 90° (figure 11.5c). The tester must monitor the test to ensure that the subject's hips and back remain in contact with the back pad throughout the lift. The tester should consult with the subject to determine a perceived maximum, and this load should be recorded in the appropriate location on the individual data sheet and then used to calculate the loads needed for the various sets contained in the 1RM testing protocol.

Step 1: Prepare the leg press machine to perform the test.

Step 2: Have the subject remove his or her shoes and then measure his or her height and weight. Record these numbers on the individual data sheet for laboratory 11.2. Have the subject put on his or her shoes.

Step 3: Direct the subject through a 10 min warm-up that includes dynamic exercises specifically targeting the lower body (e.g., walking knee hugs, toy soldiers, walking lunges).

Step 4: Following the warm-up, have the subject get into the leg press machine. Adjust the machine as needed to ensure that the subject is properly placed and that his or her feet are in the appropriate position (figure 11.5a). Load the leg press with the warm-up load (40%–60% of perceived maximum).

Step 5: Have the subject initiate the leg press (figure 11.5b) by removing the support mechanism from the foot platform and then grasping the handles or seat.

Figure 11.5 Leg press series: (a) foot position, (b) starting position, and (c) downward and upward movements.

Laboratory Activity 11.2

Step 6: Instruct the subject to allow the hips and knees to slowly flex, effectively lowering the foot platform. The knees should flex until the top of the thighs are parallel to the foot platform (figure 11.5c). While the subject is performing this motion, remind him or her to inhale during this portion of the lift.

Step 7: Having achieved the bottom position, the subject should extend the legs, which will effectively move the foot platform upward (figure 11.5c). Encourage the subject to exhale and fully extend his or her legs during this portion of the lift (figure 11.5b).

Step 8: Following the warm-up, increase the load to 60% to 80% of the subject's perceived maximum. After 1 min of rest, have the subject repeat steps 5 through 7.

Step 9: After completion of the second set, instruct the subject to rest for 3 min while you increase the load to correspond to about 90% of the subject's perceived maximum. After the recovery period, have the subject repeat steps 5 through 7, but perform only 1 repetition with an increased load.

Step 10: Based upon the how the subject's form looked, increase the load for the next set. For example, if it appeared difficult for the subject to complete the prescribed repetitions, increase the resistance by a smaller amount; if, on the other hand, it seemed relatively easy, then increase the weight by more. Be careful neither to increase the load too rapidly or to be overly conservative. After a 3 min rest, the subject will repeat steps 5 through 7 while performing only 1 repetition with an increased load.

Step 11: Repeat step 10 until the subject reaches the heaviest weight that he or she can lift with appropriate technique.

Step 12: Record the heaviest load properly lifted in the appropriate location on the individual data sheet as the subject's 1RM for the leg press.

Step 13: Divide the maximal weight lifted by the subject's body mass and record the result in the appropriate location on the individual data sheet. Compare this value with the data in the norm table for the leg press (table 11.3). Based upon the results, note the subject's strength classification and percent rank on the individual data sheet.

QUESTION SET 11.2

1. What is the basic procedure for conducting a 1RM test?
2. Based on your individual results, how do you compare with the class average and with the norms presented in this laboratory activity?
3. Graph the class, male, and female results using Microsoft Excel. Based on this graph, what can you say about the class's data?
4. Based on the class results, perform a t-test comparing the absolute and relative male data to the absolute and relative female data.

Find the case studies for this laboratory online at www.HumanKinetics.com/LaboratoryManualForExercisePhysiology.

Name or ID number: _____   Date: _____

Tester: _____   Time: _____

Sex: M / F (circle one)   Age: _____ y   Height: _____ in. _____ cm

Temperature: _____ °F _____ °C   Weight: _____ lb _____ kg

Barometric pressure: _____ mmHg   Relative humidity: _____ %

1RM Leg Press

Perceived maximum		Load	kg	Completed
Set 1	5–10 repetitions at 40%–60% of perceived maximum			❏
Set 2	3–5 repetitions at 60%–80% of perceived maximum			❏
Set 3	1 repetition at 90% of perceived maximum			❏
Set 4	1RM attempt			❏
Set 5	1RM attempt			❏

Note: 1 min rest between sets 1 and 2; 3 min rest preceding sets 3, 4, and 5.

Absolute 1-repetition maximum = _____ kg

Relative 1-repetition maximum = Load (kg) / body weight (kg) = _____

Strength classification = % rank = _____

Notes:

MAXIMAL HANDGRIP STRENGTH

EQUIPMENT

- Handgrip dynamometer (hydraulic, spring, other)
- Physician's scale or equivalent electronic scale
- Stadiometer
- Stopwatch
- Individual and group data sheets
- Microsoft Excel or equivalent spreadsheet program

 Find the group data sheets for this laboratory online at www.HumanKinetics.com/ LaboratoryManualForExercisePhysiology.

WARM-UP

Regardless of which method is used to assess muscular strength, each subject should perform a structured warm-up designed to prepare him or her for the tests that will be performed. The warm-up should contain a 5 min general portion and a 5 min dynamic stretching portion (8). Sample general warm-up exercises include light jogging, cycling, and jumping rope. Dynamic stretching activities might include wrist, finger, and elbow rotations. After completing the warm-up protocol, the subject can begin the testing process.

ISOMETRIC HANDGRIP STRENGTH TEST

The assessment of handgrip strength is generally completed with the use of a handgrip dynamometer. Of the many types, the hydraulic (e.g., Jamar) and spring (e.g., Stoelting, Smedley, Lafayette; figure 11.2a) are the most prevalent. Regardless of type, the basic testing procedures are the same.

Step 1: Gather all the equipment needed to perform the test and set up a testing area. Ensure that the dynamometer's pointer is set at zero.

Step 2: Direct the subject to remove his or her shoes and determine the subject's height and weight. Record the results on the individual data sheet for laboratory activity 11.3.

Step 3: Once the subject has put his or her shoes on, direct him or her through a 10 min warm-up.

Step 4: Have the subject stand and face straight ahead.

Step 5: Adjust the dynamometer grip so that the middle portion of the middle finger is at a right angle. The method for adjusting the grip is instrument dependent, and you should refer to the equipment manufacturer's instructions in order to understand how to properly adjust the dynamometer.

Step 6: Record the grip setting in the appropriate box on the individual data sheet. Use this setting for all subsequent tests.

Step 7: Have the subject place his or her forearm between an angle of 90-180° to the upper arm, which is kept in a vertical position. Effectively, this puts the arm in a position that is either bent at a right angle or straight.

Step 8: Ensure that the subject keeps his or her wrist and forearm in a midprone position.

Step 9: Give the subject the following commands (7, 31):

    a. *"Are you ready?"* Ask just prior to having the subject initiate the test; when the subject answers yes, move to the next command.

    b. *"Squeeze as hard as possible."* Say this as the subject initiates the test.

    c. *"Harder! . . . Harder! . . . Relax."* Say this as the subject undertakes and then finishes the test.

Step 10: Have the subject undertake two or three (36, 38) trials for each hand (and alternating between hands). Each test should be separated by 1 min (21).

Step 11: After each test, record the maximum force generated in kilograms in the appropriate location on the individual data sheet.

Step 12: After each test, reset the dynamometer's pointer to zero.

QUESTION SET 11.3

1. What are some important considerations when performing a handgrip dynamometer test?

2. Why is it important to reset the pointer after each test? If this was not done, how would the test be affected?

3. How do your results compare with those of the class?

4. Using Excel or an equivalent spreadsheet program, graph the class, male, and female results. Make sure to place standard deviations on your graph. Based on the graph, what can you say about these data?

5. Graph the relative and absolute strength of the five strongest men and the five strongest women. Based on this graph, what can you say about the class?

 Find the case studies for this laboratory online at www.HumanKinetics.com/ LaboratoryManualForExercisePhysiology.

Laboratory Activity 11.3

Laboratory Activity 11.3 Individual Data Sheet

Name or ID number: _____ Date: _____

Tester: _____ Time: _____

Sex: M / F (circle one)   Age: _____ y Height: _____ in. _____ cm

Temperature: _____ °F _____ °C Weight: _____ lb _____ kg

Barometric pressure: _____ mmHg Relative humidity: _____ %

Handgrip Strength

Circle the best of first 2 trials; underline the best of the 3 trials.

Instrument	Grip setting position		Trial 1 R	Trial 1 L	Trial 2 R	Trial 2 L	Trial 3 R	Trial 3 L
Hydraulic	1 2 3 4 5 (circle appropriate number)							
Spring	mm	Setting on handle						
Other	Setting							

Best of two trials	Right hand	+	Left hand	=	Sum		Rating
Hydraulic		+		=		kg	
Spring		+		=		kg	
Other		+		=		kg	

Ratio of grip strength to body mass	Grip sum = _____ kg	÷	Body mass = _____	=		Ratio
Ratio of dominant to nondominant hand	Dominant = _____ kg	÷	Nondominant = _____ kg	=		Ratio

Notes:

UPPER BODY MUSCULAR ENDURANCE

EQUIPMENT

- Bench press with appropriate free weights
- Floor mats (for push-up stations)
- Metronome
- Individual and group data sheets
- Microsoft Excel or equivalent spreadsheet program

Find the group data sheets for this laboratory online at www.HumanKinetics.com/ LaboratoryManualForExercisePhysiology.

WARM-UP

Regardless of which method is used to assess muscular endurance, each subject should perform a structured warm-up designed to prepare him or her for the tests that will be performed. The warm-up should contain a 5 min general portion and a 5 min dynamic stretching portion (8). Sample general warm-up exercises include light jogging, cycling, and jumping rope. Dynamic stretching activities might include arm circles and swings, as well as other activities targeting the upper body. After completing the warm-up protocol, the subject can begin the testing process.

YMCA BENCH PRESS TEST

This test requires a station with a bench press, barbell (with appropriate weights), and metronome. Men use an 80 lb (36.3 kg) barbell, and women use a 35 lb (15.9 kg) bar. During the test, a metronome counts 60 beats·min⁻¹ in order to establish a rate of 30 reps·min⁻¹. If the subject cannot maintain this rate, the test is ceased.

Step 1: Set up the testing area with the appropriate weight, as well as a bench and a metronome.

Step 2: Set the metronome to 60 beats·min⁻¹ to establish the necessary cadence (30 reps·min⁻¹).

Step 3: Have the subject remove his or her shoes, then determine his or her weight and height. Record these measures on the individual data sheet for laboratory activity 11.4. Have the subject put his or her shoes back on and move to the bench press area.

Step 4: Assume a spotting position while directing the subject to lie on his or her back on the bench. Explain to the subject that he or she must maintain all 5 points of contact (e.g., head, shoulders, buttocks, and right and left feet) during the test (figure 11.6a).

Step 5: Once the subject has been positioned properly on the bench, lift the bar off of the rack with an alternated grip (figure 11.6b) while the subject holds the bar in a pronated grip with the hands slightly farther than shoulder-width apart. Throughout the test, spot the subject with an alternated grip in order to assist when he or she reaches muscular failure.

Step 6: Have the subject lower the bar to the starting position with the bar resting on his or her chest and the elbows in a flexed position (figure 11.6c).

Step 7: Instruct the subject to press the bar up to a position in which the arms are fully extended (figure 11.6d) and then return bar to the chest at a

Figure 11.6 Bench press series: (a) lift-off, (b) starting position, (c) downward position, (d) upward movement, and (e) racking the bar.

Laboratory Activity 11.4

cadence of 30 reps · min⁻¹. The up-and-down movement will continue as long as the subject can maintain the prescribed cadence. During each repetition, ensure that a full range of motion is performed. Count each successful repetition.

Step 8: When the subject can no longer maintain the 30 reps · min⁻¹ cadence, help him or her return the barbell to the rack (figure 11.6e).

Step 9: Record the total number of repetitions completed and determine the subject's rating by comparing his or her results with the data presented in table 11.8. Record the results in the appropriate locations on the individual data sheet.

Step 10: Using the prediction equation provided on the individual data sheet, estimate the subject's 1RM strength.

PUSH-UP TEST

The push-up test is recommended by ACSM as a tool for evaluating upper body muscular endurance (2, 3). It is an ideal field test because it requires no special equipment. When conducting the push-up test, make sure that the subject maintains a flat back at all times. For men, the subject takes the standard push-up position with his hands about shoulder-width apart, back straight, head up, and toes being used as a pivot point (figure 11.7a). Women use a modified knee push-up position in which the legs are held together, the lower leg is in contact with a mat or the floor with the ankles plantar-flexed, the back is held straight, the head is up, and the hands are approximately shoulder-width apart (figure 11.8a).

Step 1: Set up the testing area and explain the basics of the test to the subject.

Step 2: Have the subject remove his or her shoes, then determine his or her weight and height. Record these measures on the individual data sheet for laboratory activity 11.4. Have the subject put his or her shoes back on and move to the test area.

Step 3: Lead the subject through a 10 min warm-up that includes both a general warm-up and a dynamic stretching routine targeting the musculature of the upper body.

Step 4: Instruct the subject to assume the proper starting position (for men, use the standard position in figure 11.8a; for women, use the modified position in figure 11.8a).

Step 5: Have the subject lower his or her body while maintaining a straight back until his or her chin touches the mat (figures 11.7b and 11.8b). The stomach should not touch the mat at any time during the assessment.

Step 6: Once the chin touches the mat, the subject should push up by extending his or her arms until achieving a straight-arm position. Record each successful push-up as it is performed.

Figure 11.7 Push-up test: *(a)* standard start position and *(b)* standard lowered position.

Figure 11.8 Push-up test: *(a)* modified start position and *(b)* modified lowered position.

Step 7: Without resting, the subject should repeat steps 5 and 6 until he or she can no longer maintain a straight-back position or complete any more push-ups.

Step 8: Record the total number of push-ups completed in the appropriate location on the individual data sheet. Compare the subject's results with the data presented in table 11.9 and record the appropriate classification.

QUESTION SET 11.4

1. What were your individual upper body muscular endurance ratings for the YMCA bench press test and the push-up test? How did these compare with the results for the class and with the values presented in the norm tables?

2. Did the men and women in the class differ in upper body muscular endurance? Did you expect this result? Why or why not?

Laboratory Activity 11.4

3. How did your push-up test results compare with the normative values? Was the muscular endurance rating similar to the YMCA bench press test results?

4. Using Excel or an equivalent spreadsheet program, perform a correlation analysis between the number of push-ups performed and the number of repetitions performed in the YMCA bench press test. Are these related? If so, how might they be related?

5. Using Excel or an equivalent spreadsheet program, perform a correlation analysis between the number of push-ups performed and the estimated maximal weight determined. Are these related? If so, how might they be related?

Find the case studies for this laboratory online at www.HumanKinetics.com/ LaboratoryManualForExercisePhysiology.

Laboratory Activity 11.4

Name or ID number: _____   Date: _____

Tester: _____   Time: _____

Sex: M / F (circle one)   Age: _____ y   Height: _____ in. _____ cm

Temperature: _____ °F _____ °C   Weight: _____ lb _____ kg

Barometric pressure: _____ mmHg   Relative humidity: _____ %

YMCA bench press test		Comments
Number of repetitions		
Rating		
Percent ranking		
1RM prediction	Men	$1.55 \times \underset{\text{\# reps}}{\underline{\hspace{2cm}}} + 37.9 = \underline{\hspace{2cm}}$ kg
	Women	$0.31 \times \underset{\text{\# reps}}{\underline{\hspace{2cm}}} + 19.2 = \underline{\hspace{2cm}}$ kg

Push-up test		Comments
Number of repetitions		
Classification		

MUSCULAR ENDURANCE OF THE TRUNK

EQUIPMENT

- Floor mats (for curl-up test)
- Masking tape
- Metronome
- Stopwatch
- Measuring tape
- Individual and group data sheets
- Microsoft Excel or equivalent spreadsheet program

 Find the group data sheets for this laboratory online at www.HumanKinetics.com/ LaboratoryManualForExercisePhysiology.

WARM-UP

Regardless of which method is used to assess muscular endurance, each subject should perform a structured warm-up designed to prepare him or her for the tests that will be performed. The warm-up should contain a 5 min general portion and a 5 min dynamic stretching portion (8). Sample general warm-up exercises include light jogging, cycling, and jumping rope. Dynamic stretching activities might include trunk swings, leg swings, and knee tucks. After completing the warm-up protocol, the subject can begin the testing process.

CURL-UP TEST

This is a simple test that is often performed as part of a health and fitness testing battery. It requires minimal equipment—a floor mat, a metronome, masking tape, and a measuring tape.

Step 1: Set up the testing area by placing a piece of masking tape across a mat. Explain the basics of the test to the subject.

Step 2: Have the subject remove his or her shoes, then determine his or her weight and height. Record these measures on the individual data sheet for laboratory activity 11.5. Have the subject put his or her shoes back on and move to the test area.

Step 3: Lead the subject through a 10 min warm-up that includes both a general warm-up and a dynamic stretching routine targeting the trunk musculature.

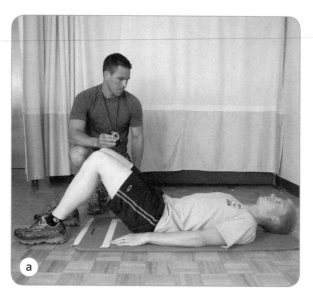

Figure 11.9 Curl-up test: *(a)* start and *(b)* end position.

Step 4: Instruct the subject to assume a supine position with his or her legs bent at 90° on a mat (figure 11.9a). Have the subject place his or her arms at his or her sides with the palms facing downward and the middle fingers touching the piece of tape. Place a second piece of tape 10 in. (3.94 cm) beyond the first piece. An alternative position would be to place the hands on the thighs and curl up until the hands reach the kneecaps.

Step 5: Set the metronome to 50 beats · min⁻¹. The subject should then perform slow, controlled curl-ups to take the shoulder blades off of the mat and slide his or her middle fingers to the second line (8 or 12 cm beyond the first) in time with the metronome at a rate of 25 per minute. For each curl-up, the lower back should be flattened before curling up (figure 11.9b).

Step 6: Instruct the subject to perform as many curl-ups as possible, without pausing, to a maximum of 25.

Step 7: Record the total number of curl-ups performed in the appropriate location on the individual data sheet, then compare this number with the norms presented in table 11.10.

ALTERNATIVE CURL-UP TEST

This simple test is often performed as part of a health and fitness testing battery (3). It is a basic modification of the traditional curl-up test that is limited to 1 min and requires the same minimal equipment as the standard version plus a stopwatch.

Step 1: Set up the testing area by placing a piece of masking tape across a mat. Explain the basics of the test to the subject.

Laboratory Activity 11.5

Step 2: Have the subject remove his or her shoes, then determine his or her weight and height. Record these measures on the individual data sheet for laboratory activity 11.5. Have the subject put his or her shoes back on and move to the test area.

Step 3: Lead the subject through a 10 min warm-up that includes both a general warm-up and a dynamic stretching routine targeting the trunk musculature.

Step 4: Instruct the subject to assume a supine position with his or her legs bent at 90° on a mat. Have the subject position his or her arms so that the palms are facing down on the thighs, then curl up until the hands reach the kneecaps.

Step 5: The subject should perform as many curl-ups as possible in 1 min without pausing. For each curl-up, make sure that the subject takes his or her shoulder blades off the mat and slides his or her hands until they reach his or her kneecap. For each curl-up, the lower back should be flattened before curling up.

Step 6: Instruct the subject to perform as many curl-ups as possible without pausing in 1 min.

Step 7: Start the stopwatch and encourage the subject. Give the subject a countdown, such as *"10 seconds to go . . . 5 seconds, 4, 3, 2, 1, stop."*

Step 8: Stop the stopwatch at 1 min and record the total number of curl-ups performed in the appropriate location on the individual data sheet. This number should then be compared with the norms presented in table 11.10.

QUESTION SET 11.5

1. How did your curl-up test result compare with the normative values and the class average?

2. Using Excel or an equivalent spreadsheet program, create a figure comparing the total class results for both versions of the curl-up test.

3. Which sex demonstrated the highest muscular endurance in this test? What might explain any differences between the sexes?

4. Based on your knowledge of exercise physiology, what factors contribute to muscular endurance? Is muscular endurance primarily impacted by neuromuscular factors, metabolic factors, or both? Why?

5. What muscle fiber type predominates in the trunk muscles? How might this impact one's muscular endurance?

6. How does strength relate to muscular endurance?

Find the case studies for this laboratory online at www.HumanKinetics.com/ LaboratoryManualForExercisePhysiology.

Name or ID number: _____ Date: _____

Tester: _____ Time: _____

Sex: M / F (circle one) Age: _____ y Height: _____ in. _____ cm

Temperature: _____ °F _____ °C Weight: _____ lb _____ kg

Barometric pressure: _____ mmHg Relative humidity: _____ %

Curl-up test		Comments
Number of repetitions		
Classification		

Alternative curl-up test		Comments
Number of repetitions		
Classification		

ESTIMATED BENCH PRESS 1RM

EQUIPMENT

- Bench press with appropriate free weights
- Calculator
- Stopwatch
- Individual and group data sheets
- Microsoft Excel or equivalent spreadsheet program

 Find the group data sheets for this laboratory online at www.HumanKinetics.com/ LaboratoryManualForExercisePhysiology.

LABORATORY REQUIREMENTS

This laboratory requires the 1RM results of the bench press determined in laboratory activity 11.1. If the subject has not completed that laboratory activity, then he or she will need to perform a 1RM assessment for the bench press prior to performing this activity. If that is not possible in the available time, use a best guess estimate for the subject's 1RM.

WARM-UP

Regardless of which method is used to assess muscular strength, each subject should perform a structured warm-up designed to prepare him or her for the tests that will be performed. The warm-up should contain a 5 min general portion and a 5 min dynamic stretching portion (8). For the bench press test, specific warm-up procedures might include doing push-ups or light bench pressing. After completing the warm-up protocol, the subject can begin the testing process.

BENCH PRESS 1RM TEST

When performing a repetition maximum test as a tool for predicting the 1RM, use a load that allows fewer than 10 repetitions to be performed (27). This repetition range is selected because it has previously been established that using heavier loads in the prediction process results in greater accuracy in predicting the 1RM (1). In this lab, you will use the 1RM to calculate a load that is 87% of 1RM, which is chosen as the target intensity because it corresponds to previously established 5-repetition maximum loads for most individuals (8, 48). Once the target load has been established, initiate the testing process with the following steps.

Step 1: Set up the bench press area with the appropriate weights.

Step 2: Have the subject remove his or her shoes, then measure his or her height and weight. Record these data on the individual data sheet for this laboratory activity. Instruct the subject to put his or her shoes back on and move to the test area.

Step 3: Explain the test to the subject, then begin a 10 min warm-up protocol that contains a 5 min general warm-up (e.g., cycling, jogging, rope jumping) followed by 5 min of dynamic stretching activities targeting the upper body.

Step 4: Use the subject's 1RM to determine the test set load.

Step 5: Load the barbell with 50% of the calculated test set load.

Step 6: Instruct the subject to lie on the bench on his or her back with five points of contact (head, shoulders and upper back, buttocks, left foot, and right foot).

Step 7: Assume the spotting position depicted in figure 11.6a.

Step 8: Instruct the subject to grasp the bar using a pronated grip with hands approximately shoulder-width apart.

Step 9: Lift the barbell off of the rack with an alternated grip while the subject is grasping the bar (figure 11.6b). The subject should hold the barbell at arm's length.

Step 10: Have the subject lower the barbell in a controlled fashion while inhaling until it touches his or her chest by moving the elbows down past the torso and slightly away from the body (figure 11.6c). Once the barbell touches the chest, instruct the subject to press the bar upward while exhaling until his or her arms are fully extended (figure 11.6d). This pattern should be completed six times. While the subject is performing each repetition, follow the bar with an alternated grip in order to assist the subject if needed.

Step 11: Once the last repetition has been completed for the set, help the subject place the bar onto the rack (figure 11.6e).

Step 12: Adjust the weight to a load corresponding to 70% of the test weight previously calculated. During this time, the subject will rest for 3 min.

Step 13: Repeat steps 7 through 11.

Step 14: While the subject rests for 3 min, change the weight to a load corresponding to 90% of the predetermined test weight.

Step 15: Repeat steps 7 through 11.

Step 16: Adjust the weight on the bar to the test set weight (87% of estimated 1RM) while the subject rests for 3 min. After the 3 min, complete steps 7 through 11 but have the subject perform as many repetitions as possible.

Step 17: If the subject completes fewer than 10 repetitions with the predetermined load, the test is complete, and the tester should record the total number of repetitions completed, as well as the load used, in the appropriate location on the individual data sheet. Skip to step 21.

Step 18: If the subject performs more than 10 repetitions, increase the load by 2.5% to 5.0% while the subject rests for 3 min.

Step 19: Repeat steps 7 through 11 but have the subject perform as many repetitions as possible.

Step 20: After the subject completes step 19, the tester can then complete step 17. Continue this process until the subject reaches a weight that he or she can lift fewer than 10 times.

Step 21: Using the Adams (5) equation, calculate the subject's predicted 1RM bench press.

$$\text{Predicted 1RM (kg)} = \frac{\text{rep weight (kg)}}{1 - (.02 \times \text{\#reps})}$$

Place the weight used for the repetition maximum test in the rep weight (kg) location and place the number of repetitions in the appropriate location. Using a calculator, perform the calculation and record the result on the individual data sheet.

Step 22: Using the Berger (10) equation, calculate the subject's predicted 1RM bench press.

$$\text{Predicted 1RM (kg)} = \frac{\text{rep weight (kg)}}{1.0261 - (.00262 \times \text{\#reps})}$$

Place the weight used for the repetition maximum test in the rep weight (kg) location and place the number of repetitions in the appropriate location. Using a calculator, perform the calculation, then record the predicted 1RM in the appropriate location on the laboratory data sheet.

Step 23: Using the Cummings and Finn (16) equation, calculate the subject's predicted 1RM bench press.

$$\text{Predicted 1RM (kg)} = \text{rep weight (kg)} + 0.839 \times \text{\#reps} - 4.29787$$

Place the weight used for the repetition maximum test in the rep weight (kg) location and place the number of repetitions in the appropriate location. Using a calculator, perform the calculation, then record the predicted 1RM in the appropriate location on the laboratory data sheet.

Step 24: Using the Brown (12) equation, calculate the subject's predicted 1RM bench press.

$$\text{Predicted 1RM (kg)} = (\text{\#reps} \times 0.0338 + 0.9849) \times \text{rep weight (kg)}$$

Place the weight used for the repetition maximum test in the rep weight (kg) location and place the number of repetitions in the appropriate location. Using a calculator, perform the calculation, then record the predicted 1RM in the appropriate location on the laboratory data sheet.

Step 25: Using the Mayhew et al. (32) equation, calculate the subject's predicted 1RM bench press.

$$\text{Predicted 1RM (kg)} = \frac{\text{rep weight (kg)}}{0.522 + 0.419 e^{-0.055 \times \text{\#reps}}}$$

Laboratory Activity 11.6

Place the weight used for the repetition maximum test in the rep weight (kg) location and place the number of repetitions in the appropriate location. Using a calculator, perform the calculation, then record the predicted 1RM in the appropriate location on the laboratory data sheet.

Step 26: Using the O'Conner et al. (37) equation, calculate the subject's predicted 1RM bench press.

$$\text{Predicted 1RM (kg)} = 0.025 \times \text{rep weight (kg)} \times \text{\#reps} + \text{rep weight (kg)}$$

Place the weight used for the repetition maximum test in the rep weight (kg) locations and place the number of repetitions in the appropriate location. Using a calculator, perform the calculation, then record the predicted 1RM in the appropriate location on the laboratory data sheet.

QUESTION SET 11.6

1. What was your predicted 1RM for the bench press? Did any of the formulas differ from your previously established 1RM? Was this pattern similar for the class as a whole?

2. Why might a 1RM prediction differ from an actual 1RM? (Hint: Is there a physiological reason?)

3. Were there any sex-specific differences in the ability of specific prediction equations to estimate the 1RM bench press? If there are what might contribute to this difference? If there is no difference, why might be the reason for this?

4. Using Excel, create a graph of each of the estimated maximum results and the actual 1-repetition maximum. Which formula best estimates the 1RM? Which formula is the worst? What might explain the differences between the regression equations used to predict 1RM?

5. Explain the pros and cons of 1RM and repetition maximum testing. Which test is safer and why?

6. Why is important to use fewer than 10 repetitions when predicting 1RM values? What the physiological reason for performing fewer than 10 repetitions?

Find the case studies for this laboratory online at www.HumanKinetics.com/ LaboratoryManualForExercisePhysiology.

Laboratory Activity 11.6

Laboratory Activity 11.6 Individual Data Sheet

Name or ID number: _____ Date: _____

Tester: _____ Time: _____

Sex: M / F (circle one)   Age: _____ y Height: _____ in. _____ cm

Temperature: _____ °F _____ °C Weight: _____ lb _____ kg

Barometric pressure: _____ mmHg Relative humidity: _____ %

1RM test results	Absolute		Relative		Load calculations		
Bench press		kg	Load (kg) / body weight (kg)		Test set	$\dfrac{\rule{2cm}{0.4pt}}{\text{1RM}} \times 0.87 =$ _____ kg	
					Warm-up set 3	$\dfrac{\rule{2cm}{0.4pt}}{\text{Test set (kg)}} \times 0.90 =$ _____ kg	
					Warm-up set 2	$\dfrac{\rule{2cm}{0.4pt}}{\text{Test set (kg)}} \times 0.70 =$ _____ kg	
					Warm-up set 1	$\dfrac{\rule{2cm}{0.4pt}}{\text{Test set (kg)}} \times 0.50 =$ _____ kg	

Note: If the 1RM leg press or bench press has not been completed, refer to laboratory 11.1 for methods and data sheets.

Repetition Maximum Tests

Bench press			
Set	**Load**	**Repetitions**	**Completed**
Warm-up set 1		6	❑
3 min rest			
Warm-up set 2		6	❑
3 min rest			
Warm-up set 3		6	❑
3 min rest			
Test set 1		To failure	❑
3 min rest			
Test set 2*		To failure	❑

*If test set 1 results in more than 10 repetitions, then repeat the test after adding weight.

Laboratory Activity 11.6

Bench press 1RM prediction equations		
Resistance	kg	Note: This resistance is 87% of 1RM.
Number of repetitions		

1RM prediction	Adams (5)	$\underline{\hspace{2cm}}_{\text{Rep weight (kg)}} / (1 - 0.02 \times \underline{\hspace{2cm}}_{\text{number of reps}}) = \underline{\hspace{1.5cm}}$ kg
	Berger (10)	$\underline{\hspace{2cm}}_{\text{Rep weight (kg)}} / 1.0261 - 0.00262 \times \underline{\hspace{2cm}}_{\text{number of reps}} = \underline{\hspace{1.5cm}}$ kg
	Cummings and Finn (16)	$1.175 \times \underline{\hspace{2cm}}_{\text{rep weight (kg)}} + 0.839 \times \underline{\hspace{2cm}}_{\text{number of reps}} - 4.29787 = \underline{\hspace{1.5cm}}$ kg
	Brown (12)	$(\underline{\hspace{2cm}}_{\text{Number of reps}} \times 0.0338 + 9.849) \times \underline{\hspace{2cm}}_{\text{rep weight (kg)}} = \underline{\hspace{1.5cm}}$ kg
	Mayhew et al. (32)	$\underline{\hspace{2cm}}_{\text{Rep weight (kg)}} / (0.522 + 0.419 e^{-0.055 \times \underline{\hspace{1cm}}_{\text{number of reps}}}) = \underline{\hspace{1.5cm}}$ kg
	O'Conner et al. (37)	$0.025 \times (\underline{\hspace{2cm}}_{\text{number of reps}} \times \underline{\hspace{2cm}}_{\text{rep weight (kg)}}) + \underline{\hspace{2cm}}_{\text{rep weight (kg)}} = \underline{\hspace{1.5cm}}$ kg

ESTIMATED LEG PRESS 1RM

EQUIPMENT

- Leg press with appropriate free weights
- Calculator
- Stopwatch
- Individual and group data sheets
- Microsoft Excel or equivalent spreadsheet program

 Find the group data sheets for this laboratory online at www.HumanKinetics.com/ LaboratoryManualForExercisePhysiology.

LABORATORY REQUIREMENTS

This laboratory requires the 1RM results of the leg press determined in laboratory activity 11.2. If the subject has not completed that laboratory activity, he or she will need to perform a 1RM assessment for the leg press prior to performing this activity. If that is not possible in the available time, use a best guess estimate for the subject's 1RM.

WARM-UP

Regardless of which method is used to assess muscular strength, each subject should perform a structured warm-up designed to prepare him or her for the tests that will be performed. The warm-up should contain a 5 min general portion and a 5 min dynamic stretching portion (8). For the leg press, the dynamic stretching warm-up might include body-weight squats, leg swings, and light leg pressing. After completing the warm-up protocol, the subject can begin the testing process.

LEG PRESS 1RM TEST

The leg press repetition maximum test should be performed with an estimated load that allows fewer than 10 repetitions to be performed. In this lab, you will use the 1RM to calculate a load that is 87% of 1RM, which is chosen as the target intensity because it corresponds to previously established 5-repetition maximum loads for most individuals (8, 48). Once the target load has been established, initiate the testing process with the following steps.

Step 1: Set up the leg press area with the appropriate weights.

Step 2: Have the subject remove his or her shoes, then measure his or her height and weight. Record these data on the individual data sheet for this laboratory activity. Instruct the subject to put his or her shoes back on and move to the test area.

Step 3: Explain the test to the subject, then begin a 10 min warm-up protocol that contains a 5 min general warm-up (e.g., cycling, jogging, or rope jumping) followed by 5 min of dynamic stretching activities targeting the lower body.

Step 4: Load the leg press machine with 50% of the calculated test set load. The test set load can be calculated using an actual or predicted 1RM.

Step 5: Instruct the subject to get into the leg press machine with his or her feet on the footpad (figure 11.5a). Adjust the leg press machine to ensure that the subject is properly placed in it.

Step 6: To initiate the lift, have the subject engage the leg press machine (figure 11.5b) by removing the support mechanism from the foot platform and then grasping the handles or seat.

Step 7: Have the subject slowly flex his or her hips and knees to lower the foot platform. The knees should flex until the tops of the thighs are parallel to the foot platform (figure 11.5c). During this portion of the lift, ensure that the subject inhales. After achieving the bottom position, the subject should extend his or her legs, while exhaling, to move the foot platform upward (figure 11.5c). Once the knees are fully extended, the subject should repeat this step for a total of 6 repetitions.

Step 8: Adjust the leg press weight to a load corresponding to 70% of the test weight previously calculated. During this time, the subject will rest for 3 min.

Step 9: Have the subject repeat steps 6 and 7.

Step 10: While the subject rests for 3 min, change the weight to 80% of the predetermined test weight.

Step 11: Have the subject repeat steps 6 and 7.

Step 12: Adjust the leg press weight on the leg press load to the test set weight (87% of 1RM) while the subject rests for 3 min. After the 3 min, have the subject repeat steps 6 and 7 but complete as many repetitions as possible.

Step 13: If the subject completes fewer than 10 repetitions with the predetermined load, the test is complete, and the tester should record the total number of repetitions completed, as well as the load used, in the appropriate location on the individual data sheet. Skip to step 17.

Step 14: If the subject performs more than 10 repetitions, increase the load by 2.5% to 5.0% while the subject rests for 3 min.

Step 15: Have the subject repeat steps 6 and 7 but complete as many repetitions as possible.

Step 16: After the subject completes step 15, the tester can then complete step 13. Continue this process until the subject reaches a weight that he or she can lift fewer than 10 times.

Laboratory Activity 11.7

Step 17: Using the Adams (5) equation, calculate the subject's predicted 1RM leg press.

$$\text{Predicted 1RM (kg)} = \frac{\text{rep weight (kg)}}{1 - (.02 \times \#\text{reps})}$$

Place the weight used for the repetition maximum test in the rep weight (kg) location and place the number of repetitions in the appropriate location. Using a calculator, perform the calculation and record the result in the appropriate location on the individual data sheet.

Step 18: Using the Berger (10) equation, calculate the subject's predicted 1RM leg press.

$$\text{Predicted 1RM (kg)} = \frac{\text{rep weight (kg)}}{1.0261 - (.00262 \times \#\text{reps})}$$

Place the weight used for the repetition maximum test in the rep weight (kg) location and place the number of repetitions in the appropriate location. Using a calculator, perform the calculation, then record the predicted 1RM in the appropriate location on the laboratory data sheet.

Step 19: Using the Cummings and Finn (16) equation, calculate the subject's predicted 1RM leg press.

$$\text{Predicted 1RM (kg)} = \text{rep weight (kg)} + (0.839 \times \#\text{reps}) - 4.29787$$

Place the weight used for the repetition maximum test in the rep weight (kg) location and place the number of repetitions in the appropriate location. Using a calculator, perform the calculation, then record the predicted 1RM in the appropriate location on the laboratory data sheet.

Step 20: Using the Brown (12) equation, calculate the subject's predicted 1RM leg press.

$$\text{Predicted 1RM (kg)} = (\#\text{reps} \times 0.0338 + 0.9849) \times \text{rep weight (kg)}$$

Place the weight used for the repetition maximum test in the rep weight (kg) location and place the number of repetitions in the appropriate location. Using a calculator, perform the calculation, then record the predicted 1RM in the appropriate location on the laboratory data sheet.

Step 21: Using the Mayhew et al. (32) equation, calculate the subject's predicted 1RM leg press.

$$\text{Predicted 1RM (kg)} = \frac{\text{rep weight (kg)}}{0.522 + 0.419e^{-0.055 \times \#\text{reps}}}$$

Place the weight used for the repetition maximum test in the rep weight (kg) location and place the number of repetitions in the appropriate location. Using a calculator, perform the calculation, then record the predicted 1RM in the appropriate location on the laboratory data sheet.

Step 22: Using the O'Conner et al. (37) equation, calculate the subject's predicted 1RM leg press.

$$\text{Predicted 1RM (kg)} = 0.025 \times \text{rep weight (kg)} \times \#\text{reps} + \text{rep weight (kg)}$$

Place the weight used for the repetition maximum test in the rep weight (kg) locations and place the number of repetitions in the appropriate location. Using a calculator, perform the calculation, then record the predicted 1RM in the appropriate location on the laboratory data sheet.

QUESTION SET 11.7

1. What was your predicted 1RM for the leg press? Did any of the formulas differ from your previously established 1RM? Was this difference consistent for the class as a whole? Was there a difference between the male and female data?

2. Why might a 1RM prediction differ from an actual 1RM? (Hint: Is there a physiological reason?)

3. Why might you want to perform fewer than 10 repetitions when conducting this type of test? How would performing more than 10 repetitions affect the 1RM?

4. Using Excel, create a graph of each of the estimated maximum results and the actual 1-repetition maximum. Which formula best estimates the 1RM? Which formula is the worst? Is this consistent between males and females?

Find the case studies for this laboratory online at www.HumanKinetics.com/ LaboratoryManualForExercisePhysiology.

Laboratory Activity 11.7

Laboratory Activity 11.7 Individual Data Sheet

Name or ID number: _____ Date: _____

Tester: _____ Time: _____

Sex: M / F (circle one)   Age: _____ y   Height: _____ in. _____ cm

Temperature: _____ °F _____ °C   Weight: _____ lb _____ kg

Barometric pressure: _____ mmHg   Relative humidity: _____ %

1RM test results	Absolute		Relative		Load calculations		
Leg press		kg		Load (kg) / body weight (kg)	Test set	$\dfrac{\rule{2cm}{0.4pt}}{1RM} \times 0.87 = $ _____ kg	
					Warm-up set 3	$\dfrac{\rule{2cm}{0.4pt}}{\text{Test set (kg)}} \times 0.90 = $ _____ kg	
					Warm-up set 2	$\dfrac{\rule{2cm}{0.4pt}}{\text{Test set (kg)}} \times 0.70 = $ _____ kg	
					Warm-up set 1	$\dfrac{\rule{2cm}{0.4pt}}{\text{Test set (kg)}} \times 0.50 = $ _____ kg	

Note: If the 1RM leg press or bench press has not been completed, refer to laboratory 11.1 for methods and data sheets.

Repetition Maximum Tests

Leg press			
Set	Load	Repetitions	Completed
Warm-up set 1		6	❏
3 min rest			
Warm-up set 2		6	❏
3 min rest			
Warm-up set 3		6	❏
3 min rest			
Test set 1		To failure	❏
3 min rest			
Test set 2*		To failure	❏

*If test set 1 results in more than 10 repetitions, then repeat the test after adding weight.

Laboratory Activity 11.7

303

Leg press 1RM prediction equations			
Resistance		kg	Note: This resistance is 87% of 1RM.
Number of repetitions			

1RM prediction

Adams (5)

$$\underline{\hspace{2cm}}_{\text{Rep weight (kg)}} / (1 - 0.02 \times \underline{\hspace{2cm}}_{\text{number of reps}}) = \underline{\hspace{1.5cm}} \text{ kg}$$

Berger (10)

$$\underline{\hspace{2cm}}_{\text{Rep weight (kg)}} / 1.0261 - 0.00262 \times \underline{\hspace{2cm}}_{\text{number of reps}} = \underline{\hspace{1.5cm}} \text{ kg}$$

Cummings and Finn (16)

$$1.175 \times \underline{\hspace{2cm}}_{\text{rep weight (kg)}} + 0.839 \times \underline{\hspace{2cm}}_{\text{number of reps}} - 4.29787 = \underline{\hspace{1.5cm}} \text{ kg}$$

Brown (12)

$$(\underline{\hspace{2cm}}_{\text{Number of reps}} \times 0.0338 + 9.849) \times \underline{\hspace{2cm}}_{\text{rep weight (kg)}} = \underline{\hspace{1.5cm}} \text{ kg}$$

Mayhew et al. (32)

$$\underline{\hspace{2cm}}_{\text{Rep weight (kg)}} / (0.522 + 0.419e^{-0.055 \times \underline{\hspace{1cm}}_{\text{number of reps}}}) = \underline{\hspace{1.5cm}} \text{ kg}$$

O'Conner et al. (37)

$$0.025 \times (\underline{\hspace{2cm}}_{\text{number of reps}} \times \underline{\hspace{2cm}}_{\text{rep weight (kg)}}) + \underline{\hspace{2cm}}_{\text{rep weight (kg)}} = \underline{\hspace{1.5cm}} \text{ kg}$$

Laboratory Activity 11.7

Anaerobic Fitness Measurements

Objectives

1. Explain the different anaerobic power and capacity tests.
2. Gain exposure to basic procedures used to evaluate anaerobic power and capacity.
3. Describe a method for evaluating the stretch–shortening cycle with the use of jumping tests.
4. Examine the Wingate anaerobic test.
5. Examine published norms for various anaerobic power tests.

Definitions

anaerobic capacity—Mean power output achieved during exercise bouts that last between 10 and 120 s. Tests that assess anaerobic capacity typically stress the ATP-PC and glycolytic energy systems. The term *mean power* is often used synonymously with *anaerobic capacity.*

anaerobic power—Mean or peak power output in exercise lasting 10 s or less. Anaerobic power is typically evaluated with tests that stress the phosphagen or ATP-PC systems and require very high intensity to be applied for a short duration.

eccentric utilization ratio—Representation of the ability to use the stretch–shortening cycle; estimated by dividing countermovement vertical jump results by static vertical jump results (42) and typically calculated with peak power output and vertical displacement.

fatigue rate—Rate of decline in performance; may be considered as the degree of power drop-off from the highest power output to the value at the end of the test (28), thus often calculated as percent change between these two values.

force plate—Device that measures ground reaction forces; often considered the gold standard when evaluating vertical jump performance.

horizontal power—Power typically estimated by sprinting tasks such as a 40 m sprint.

Margaria-Kalamen test—Stair test used to evaluate anaerobic peak power (39).

mean anaerobic power—Average power output achieved during a performance test that stresses the anaerobic energy supply mechanisms.

mean power—Average power output achieved during a specified time interval or during a test.

peak anaerobic power—Highest power output achieved during a test in which the anaerobic system is the primary supplier of energy (these tests typically last less than 10 s).

peak power—Highest power output achieved during a test (24).

power—Rate of doing work (35); calculated in terms of either work divided by time (power = work / time) or force multiplied by velocity (power = force × velocity).

power endurance—Ability to repetitively achieve high power outputs or to maintain a level of power output.

rate of fatigue—Rate also known as the *fatigue index* and is represented as the degree of power drop-off during the test

stretch–shortening cycle (SSC)—Active stretch of a muscle (i.e., eccentric muscle ac-tion) followed by an immediate shortening of the same muscle (i.e., concentric muscle action).

switch mat—Device used to measure flight time, which can then be used to determine jump height and power output.

total work—Product of mean power and time.

Wingate anaerobic test (WAnT)—A test of anaerobic power and capacity performed on a cycle ergometer that typically lasts 30 s and uses a resistance equal to 0.075 kp per kg of body mass.

The assessment of anaerobic fitness is an important part of the overall assessment of physical capacity for both athletes and clinical clients. It is well documented that one's ability to express high **power** outputs is a primary contributor to one's ability to be successful in sporting activities (5, 52). For example, Carlock et al. (9) found that an athlete's weightlifting ability is highly correlated with his or her ability to express high power outputs during jumping tasks. Similarly, among American football players and soccer players, performance capacity is strongly related to power-generating capacity as indicated by jumping performance and sprinting performance (5, 13, 14, 52). Because of these relationships, power output is one of the most commonly tested performance characteristics in sport science and athletic settings (52). Thus it is important that students in sport or exercise science understand the various methods for assessing power.

Anaerobic fitness may also be related to an individual's ability to undertake activities of daily living. Persons whose power-generating capacity is compromised are often hindered in their ability to perform simple tasks of daily life. If, however, they undertake activities to improve their power output, they can markedly improve their performance in these tasks. The importance of power-generating capacity in activities of daily living means that exercise physiology students must understand the types of testing available for evaluating a client's power output—and how these tests are implemented.

Generally, *power* can be defined as the rate of performing work (35), whereas **anaerobic power** is the rate of performing work under conditions in which aerobic metabolism offers little contribution (40). Tests of anaerobic power generally evaluate activities in which the primary energy supplier is the phosphagen or ATP-PC system, whereas tests of anaerobic *capacity* examine activities in which the primary energy supplier is the glycolytic system. Typically, anaerobic power tests last less than 10 s, while **anaerobic capacity** tests last longer than 10 s but not longer than 2 min.

When evaluating the anaerobic capacity tests, you can often calculate a **peak power, mean power,** or **fatigue rate**. Peak power is the highest power output attained during a specific test, whereas mean power is an average collected over the duration of the test. Sometimes mean power is referred to as *anaerobic capacity* (28) or as **power endurance**. Fatigue rate is the degree of power drop-off during the test. It is generally calculated during tests based on power endurance or anaerobic capacity as a percentage of peak power output (28). Depending upon the systems and techniques used, anaerobic capacity and power tests can also involve calculation of other variables, such as the time to peak power, the rate of power development, and the velocity of movement.

Numerous anaerobic power and capacity tests can be conducted. In the following laboratory activities, we discuss six tests, two of which examine **anaerobic peak power** output and three of which examine both peak anaerobic power and anaerobic capacity (table 12.1). We also discuss a test that gives an estimation of **horizontal power**.

Table 12.1 Classification of Anaerobic Power Tests

Test name	Variables quantified by test	
	Anaerobic peak power	**Anaerobic capacity**
Countermovement vertical jump test	Yes	
Static vertical jump test	Yes	
Margaria-Kalamen test	Yes	
Bosco 60 s continuous jump test	Yes	Yes
Wingate anaerobic cycle test	Yes	Yes

Sprinting Performance Tests for Estimating Horizontal Power

Sprinting performance is commonly measured in many athletic settings. Successful sprinting depends largely on the individual's ability to exert vertical forces, the ability of the bioenergetic systems to meet the metabolic demands of the activity, and the individual's muscle-fiber-type profile (52).

Horizontal sprinting activities are often used to evaluate horizontal power-generating capacity, but they do not involve measurement of a power output (1). For sprinting activities, the only knowns are the athlete's body mass, the horizontal distance covered, and the time used to cover that distance. Therefore, these variables are used to estimate horizontal power according to the following formula:

$$\text{Horizontal power } (kg \cdot m \cdot s^{-1}) = \text{force } (kg) \times \text{average velocity } (m \cdot s^{-1})$$

This equation can also be represented as follows:

$$\text{Horizontal power } (N \cdot m \cdot s^{-1}) = \text{force } (N) \times \text{average velocity } (m \cdot s^{-1})$$

In both equations, force is represented by the individual's body mass, and average velocity is calculated by dividing the known distance by the time it took to cover that distance. For example, if a 100 kg athlete covered a distance of 200 m in 22 s, he or she would achieve a horizontal power of 909.1 $kg \cdot m \cdot s^{-1}$. This value could be calculated as follows:

$$\text{Horizontal power } (kg \cdot m \cdot s^{-1}) = 100 \text{ kg} \times (200 \text{ m} / 22 \text{ s})$$

$$= 100 \text{ kg} \times 9.09 \text{ m} \cdot s^{-1}$$

$$= 909.1 \text{ kg} \cdot m \cdot s^{-1}$$

Thus, if two individuals of similar mass cover the same distance, the faster individual is considered to have a higher power output.

Horizontal power is therefore usually tested with sprinting activities such as the 30-yard (27.4 m), 40-yard (36.6 m), and 60-yard (54.9 m) sprints (24). When performing sprint testing, you can use either a stopwatch or an electronic timer to quantify the athlete's sprint time. Stopwatches are generally well correlated with electronic timers ($r = 0.98$, $r^2 = 0.95$), but there is usually a 0.24 s error when using a stopwatch, which generally results in a faster running time. Stopwatch times can generally be converted to electronic-equivalent times by using the following equation:

$$\text{Electronic timer time } (s) = 1.0113 \times \text{handheld stopwatch time } (s) + 0.2252$$

This difference in the results achieved by the two timing methods makes it important to know which method was used when comparing results with norms presented in the literature (24).

Jumping Performance Tests for Determining Vertical Power

Vertical power or jumping performance is one of the most popular field tests used to assess anaerobic power in athletic populations (24). The ability to express high vertical power outputs or jump heights has been correlated with a variety of sporting activities, including competitive weightlifting (9, 16), track cycling sprint performance (51), and sprinting performance (26). Jumping ability has also been shown to differentiate between level of play and playing abilities in American football

(13) and soccer players (14). Because of the overall simplicity of jumping tests, they are commonly included as part of field-based performance test batteries (15, 19, 50).

Vertical jumps used for testing vertical power output can be of two types: countermovement and static. The countermovement vertical jump requires the athlete to start in a standing position, from which he or she drops into a squat position and then immediately jumps as high as possible (figure 12.1). The static vertical jump, or squat jump as it is sometimes called (6), requires the athlete to squat with the top of the thigh parallel to the floor (with a knee angle of approximately 90°) and hold this position during a countdown (3, 2, 1, jump) and then jump as high and as fast as possible (17)

Figure 12.1 Countermovement vertical jump positions: *(a)* start position, *(b)* dip, *(c)* maximum height, and *(d-e)* landing.

(figure 12.2). The static vertical jump is often viewed as the preferred testing method because countermovement jump performance can be affected by an individual's skill level (49). However, both types of jumps are commonly used in testing batteries, in which case the countermovement jump test precedes the static jump test. Regardless of which type of jump you use, the ultimate goal in the testing process is to determine the vertical jump displacement that the individual can achieve. This can be done by means of various methods (34).

Figure 12.2 Static vertical jump positions: *(a)* start position, *(b)* take-off position, *(c)* maximum height, and *(d-e)* landing.

Force Plate

The **force plate** is considered by many to be the gold standard when evaluating vertical jumping and power outputs (25). The vertical displacement achieved during a vertical jump performed on a force plate is determined by quantifying flight time (i.e., the time between take-off from the force plate and landing on it during the jump)—the higher the athlete jumps, the longer his or her flight time. Therefore, flight time and gravity can be used to estimate the vertical displacement by means of the following equation (7):

$$\text{Vertical displacement (m)} = \frac{9.81 (m \cdot s^{-2}) \times \text{flight time (s)} \times \text{flight time (s)} \times}{8}$$

The use of a force plate to determine vertical displacements has been shown to be highly reliable (ICC = 0.94–0.99) (18, 31, 36). In examining the power output generated during a jumping activity, you can use two methods when working with force plates.

The first method involves measuring the force generated during a vertical jump, then deriving the velocity of movement achieved during the jump. The force generated during the vertical jump is directly assessed by measuring the ground reaction forces created during the movement (10). The force at each time point during the movement is then divided by the system mass in order to determine acceleration. The acceleration due to gravity is then subtracted from the calculated acceleration in order to represent the acceleration produced by the individual when calculating the velocity of movement. The velocity of movement is then determined by the product of acceleration and time data at each time point (10). The derived velocity is then multiplied by the quantified force in order to calculate the power output achieved in the jumping activity. Generally, this method for quantifying vertical jump power output has been shown to be highly reliable (ICC = 0.93–0.98) (31, 32).

The second method for estimating power output uses the force plate to measure the flight time achieved during the vertical jump in order to calculate the vertical displacement achieved during the jump (7). The reliability or accuracy of the test can be compromised if the subject tries to extend the flight time by tucking his or her legs when jumping or landing. Therefore, it is important to visually monitor the jump and landing in order to make sure that the landing is performed properly. If the jump is properly performed, the vertical displacement determined can then be placed into one of the power estimation equations presented in the section of this chapter in which Formulas for Estimating Vertical Power are presented.

Switch Mat

Another method for evaluating vertical jump displacement uses a relatively new technology known as the **switch mat** (34). Switch mats (which measure 0.6858 by 0.6858 m) contain embedded micro switches that measure the time interval between take-off and landing and quantify it as flight time. As with the force plate, flight time can then be used to determine vertical displacement. To use a switch mat, the subject stands on the mat and performs either a countermovement or a static vertical jump (figure 12.3). Flight time is then determined and placed into the previously presented equation for determining vertical displacement.

Vertical displacement determined with the use of a switch mat has been reported to be highly reliable (ICC = 0.96–0.98) (9). As with the force plate, you must carefully monitor the jumping and landing technique in order to insure that accurate displacements are estimated. Once the vertical displacement is derived, it is then placed into one of the power estimation equations presented later in this chapter (Formulas for Estimating Vertical Power).

Jump-and-Reach

The jump-and-reach test is the most basic method for determining vertical jump displacement and is generally based on the original Sargent jump test (34). This method basically measures the difference between an individual's standing reach height and the height that he or she can jump and touch. Standing reach is determined by having the subject place his or her feet together with his or her dominant arm near a wall (1). The subject then reaches as high as possible so that the palm of the hand is against the marking scale. The highest point achieved is then considered the standing reach height. Next, the subject jumps as high as possible and reaches for the highest point possible. The vertical jump displacement is then measured as the difference between the standing reach height and the maximal reach height achieved during the jump (34).

Figure 12.3 The countermovement vertical jump test performed on a switch mat: *(a)* start position, *(b)* dip, *(c)* maximum height, and *(d)* landing.

Vertical displacement (cm) = jump height (cm) − reach height (cm)

An alternative jump-and-reach method uses a commercial apparatus such as the Vertec, which allows you to measure the reach height and jumping height without having to be near a wall. The device contains a telescopic metal pole and plastic swivel vanes that are separated by 0.05 in. intervals (34). As with the classic vertical jump method, you first determine reach height—in this case, by having the subject reach as high as possible with the dominant hand and push away plastic swivel vanes while standing flat-footed under the Vertec (figure 12.4*a*). Since the Vertec can be raised to specific known distances and each vane corresponds to 0.05 in. increments, the standing height is easily obtained. Depending upon the athlete's jumping ability, the Vertec can then be raised. At this point, the subject performs a two-footed, nonstep jump in an attempt to displace as many of the horizontal vanes as possible (figure 12.4*b*). The maximal height is determined as the highest displacement achieved. The vertical jump displacement is then calculated by means of the same formula used with the classic jump-and-reach method (34).

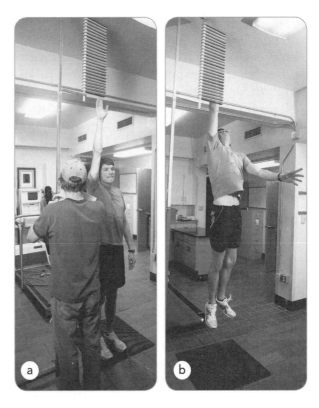

Figure 12.4 Positions for the standing reach-and-jump test using the Vertec: *(a)* establishment of reach height and *(b)* the jump portion of the test.

Though considered very reliable, the jump-and-reach method does have several limitations. Specifically, it gives little insight into the force and velocity aspects of the jumping movements. In addition, recent research suggests that this method underestimates vertical displacement as compared with video analysis and switch mat measurements (37).

Formulas for Estimating Vertical Power

Though the vertical jump test is important, it does have some drawbacks; specifically, it yields only a vertical jump displacement and does not truly yield a power output (24). To compensate for this limitation, numerous methods can be used for estimating the power output achieved during a vertical jump. One oft-used method is the Lewis formula (24):

$$\text{Power (W)} = \sqrt{4.9} \times$$
$$\text{weight (kg)} \times \sqrt{\text{jump height (m)}}$$

However, even though the Lewis formula is popular with coaches, physical educators, and sports scientists, its power estimate does not accurately represent the values achieved for peak or average power output on a force plate (20). In order to address these inaccuracies, Harman et al. (20) developed two formulas to estimate peak and average power output based on force plate data and regression analysis:

$$\text{Peak power (W)} = 61.9 \times \text{jump height (cm)} +$$
$$36 \times \text{body mass (kg)} + 1,822$$

$$\text{Average power (W)} = 21.2 \times$$
$$\text{jump height (cm)} + 23 \times$$
$$\text{body mass (kg)} - 1,393$$

Though these two formulas do improve on the Lewis equation, they were generated with a small sample size ($n = 17$) and a relatively homogenous group. As a result, their validity has been challenged in the scientific literature. Sayer et al. (49) have questioned the validity of the Harman peak and average power equations because they were generated with the use of a static vertical jump test (this test is described later in the chapter). In contrast, a person doing a countermovement vertical jump test engages the stretch reflex and thus can generate significantly more force (3); specifically, the countermovement vertical jump can result in a 10% to 23% higher vertical displacement than a static vertical jump (49). In order to address these issues, Sayer et al. (49) used a force plate system to test both countermovement and static vertical jump performance of a large heterogeneous sample ($n=108$). After performing regression analyses on these data, they generated the following equation:

$$\text{Peak power (W)} = 60.7 \times \text{jump height (cm)} +$$
$$45.3 \times \text{body mass (kg)} - 2,055$$

This equation was found to be a valid method for estimating peak power output from either a static or a countermovement vertical jump, and it has become increasingly popular in research and applied applications. It has also been consistently reported to be highly reliable (ICC = 0.99) (9, 18).

An additional method for evaluation of both peak and average power during a vertical jump has been reported by Johnson and Bahamonde (29). Their equation differs from the Sayer equations in that they include the jumper's height in the equation:

$$\text{Peak power (W)} = 78.5 \times$$
$$\text{jump height (cm)} + 60.6 \times$$
$$\text{body mass (kg)} - 15.3 \times \text{height (cm)} - 1,308$$

$$\text{Average power (W)} = 41.4 \times$$
$$\text{jump height (cm)} + 31.2 \times$$
$$\text{body mass (kg)} - 13.9 \times \text{height (cm)} + 431$$

Bosco Test for Estimating Power Endurance

Power endurance can be defined as the ability to repetitively perform high-power movements such as jumping. One of the most popular methods for estimating power endurance is the Bosco test (7), which is a repetitive jumping test that lasts between 15 and 60 s and requires the athlete to jump as high and as rapidly as possible (8).

Power output is calculated during the Bosco test by determining the number of jumps completed during the work time and the sum of the flight times recorded during that time frame (7). These data can then be placed into the following basic equation in order to calculate mechanical power:

$$\text{Mechanical power (w} \cdot \text{kg}^{-1}) = \frac{g^2 \times T_f \times T_t}{4 \times n \times (T_t - T_f)}$$

where g = 9.81 m $\cdot$ s^{-2}, T_f = total flight time, T_t = total time, and n = number of jumps. The base equation can accommodate a variety of testing lengths, but the most common testing duration is 60 s, in which case the formula would be rewritten as follows:

$$\text{Mechanical power (W} \cdot \text{kg}^{-1}) = \frac{g^2 \times T_f \times 60}{4 \times n \times (60 - T_f)}$$

The mechanical power value calculated by the Bosco test represents the subject's power endurance because it represents an average power output.

Regardless of the test duration chosen, this test should be performed on either a force plate or a switch mat in order to collect the data required for quantifying power endurance.

Determining the Eccentric Utilization Ratio

The ability to use the **stretch–shortening cycle** (SSC) is important for success in numerous sporting activities. The SSC involves the performance of an eccentric muscle action immediately followed by the concentric muscle action. The muscle is activated, stretched, and then immediately shortened, which augments the force and power generated as compared with a concentric-only muscle action. The use of the SSC during a countermovement jump results in a higher power output as compared with a static vertical jump, which involves only a concentric muscle action. Quantifying the effect of training interventions on a person's ability to engage the SSC is of particular importance to sport scientists and strength and conditioning professionals. One method involves quantifying the **eccentric utilization ratio** (EUR), which can be measured to gain insight into the contribution of the SSC to the athlete's performance.

In order to calculate the EUR, the individual must perform a static vertical jump test and a countermovement vertical jump test. The results can be used to calculate the EUR with the following equation:

Eccentric utilization ratio = countermovement
vertical jump / static vertical jump

This basic formula can then be used with countermovement and static vertical jump displacements or power outputs. McGuigan et al. (42) suggest that the EUR is sensitive to training stressors and can be used to monitor training outcomes across a periodized training plan. As a monitoring tool, a higher EUR value indicates greater reliance on the SSC, whereas a lower value indicates less reliance on the SSC. For example, McGuigan et al. (42) present data showing that field hockey players have a higher EUR (1.26) value during the preseason than in the off-season (1.05). These findings seem logical since preseason training generally targets the development of power output, whereas off-season training generally targets strength development. Since higher power outputs are associated with the use of the SSC, it is clear that when training targets power development, improvements in the EUR should be noticed. The influence of training on the EUR is clearly noted by the fact that endurance athletes consistently express the lowest EURs because their training typically does not target the maximization of power output and can negatively affect type II fiber content.

Wingate Anaerobic Test for Determining Anaerobic Cycling Power

The most popular anaerobic cycling power test is the **Wingate anaerobic test (WAnT)**, which was developed at Wingate University in Israel (28). The classic WAnT is performed for a 30 s duration against 0.075 kp per kg of body mass and is designed to test anaerobic performance capacity (28). The WAnT appears ideally suited for this usage because during the test between 60% and 85% of energy supply comes from the ATP-PC and glycolytic energy systems (4, 12, 43). The first 3 to 15 s of the test appear to tax the ATP-PC systems, and the glycolytic systems are used for the remainder of the test (53).

This test can be used to determine the individual's peak anaerobic power, lowest anaerobic power output, mean anaerobic power, total work, and rate of fatigue during an all-out cycling bout performed against a set resistance for a predetermined duration. **Peak anaerobic power** is the highest mechanical power generated during the test. It typically occurs in the first 5 s (53) and is usually calculated as the average power over any 3 or 5 s time period by dividing the amount of work completed and by time:

$$\text{Power} = \frac{\text{work}}{\text{time}} = \frac{\text{force} \times \text{displacement}}{\text{time}}$$

If working with 5 s time periods, the maximum number of revolutions completed in one of the six 5 s intervals is multiplied by the distance covered by the flywheel during one revolution and then divided by 5 s. When using a Monark cycle, the flywheel is 1.62 m in circumference and makes 3.7 rotations per pedal revolution, thus producing a total distance of 6 m per revolution (1).

$$\text{Power (W)} = [\text{force (N)} \times (\text{maximum revolutions} \times 6\ \text{m})] / 5\ \text{s}$$

It is often important to examine relative power outputs, and peak power output can be calculated relative to body mass ($W \cdot kg^{-1}$) with the following equation (28):

$$\text{Power (W} \cdot kg^{-1}) = \text{power (W)} / \text{body mass (kg)}$$

Power can also be calculated relative to lean body mass ($W \cdot kgLBM^{-1}$) with the following equation (28):

$$\text{Power (W} \cdot kg^{-1}) = \text{power (W)} / \text{lean body mass (kg)}$$

The lowest anaerobic power is the lowest power output achieved during any individual 3 or 5 s interval during the 30 s test. This value typically occurs during the last 3 to 5 s of the test and is used in calculating the fatigue index (53).

Total work can be calculated if you know the total number of revolutions completed at the end of the 30 s cycle test and the distance covered by the flywheel per revolution (again, 6 m when using a Monark cycle ergometer) (1).

$$\text{Work (J)} = \text{force setting (N)} \times (\text{rev in 30 s} \times \text{distance covered by flywheel per revolution})$$

Therefore, when using a Monark cycle ergometer the following equation can be can be used to calculate total work:

$$\text{Work (J)} = \text{force setting (N)} \times (\text{rev in 30 s} \times 6\ \text{m})$$

Total work can also be normalized to body mass by dividing the total work by the individual's body mass.

$$\text{Work (J} \cdot kg^{-1}) = \frac{\text{work (J)}}{\text{body mass (kg)}} =$$

$$\frac{\text{force setting (N)} \times (\text{rev in 30 s} \times 6\ \text{m})}{\text{body mass (kg)}}$$

Mean anaerobic power is the average power maintained over the duration of the 30 s test (53). It can be calculated by averaging the values obtained during the ten 3 s or six 5 s segments of the test (28) or by dividing the total amount of work by the duration of the test (1). This value can be calculated with the following equation:

$$\text{Mean power (W)} = \frac{\text{work}}{\text{time}} =$$

$$\frac{\text{force setting (N)} \times (\text{rev in 30 s} \times 6\ \text{m})}{30\ \text{s}}$$

Mean power output can also be normalized to body mass by dividing the total work by the individual's body mass.

$$\text{Mean power (W} \cdot kg^{-1}) = \frac{\text{mean power (W)}}{\text{body mass (kg)}}$$

Rate of fatigue is also known as the *fatigue index* and is represented as the degree of power drop-off during the test (28). It is generally calculated with the following equation:

$$\text{Fatigue index (\%)} = \frac{\text{peak power} - \text{lowest power}}{\text{peak power}} \times 100$$

Typically, a fatigue rate of ≥40% can be found from the first 5 s (peak power) to the last 5 s (lowest power) (1, 53).

When working with the WAnT, two important considerations are the test's duration and the amount of resistance applied. The duration of the WAnT is typically 30 s, while shorter (e.g., 5 s) tests (47) have been used when the investigator wants to examine only peak power output.

Resistance in the WAnT is generally set at 0.075 kp per kg of body mass for children, older adults, and sedentary persons (1, 28). Some researchers have also used this resistance with athletes (53), but it is generally recommended that athletes use between 0.090 and 0.100 kp per kg body mass (table 12.2) (28). Here is the basic

Table 12.2 Optimal Loads for the Wingate Anaerobic Test (WAnT)

Subject		Force load (kp) / body mass (kg)	Reference
Adult male	Sedentary	0.075	Inbar et al. (28)
	Active	0.098	Evans and Quinney (12)
	Athlete	0.098	Evans and Quinney (12)
	PE student	0.087	Dotan and Bar-Or (11)
	Track athlete	0.100	Kirksey et al. (33)
Junior male	Wrestler	0.090	Mirzaei et al. (44)
Adult female	Sedentary	0.075	Inbar et al. (28)
	PE student	0.085	Dotan and Bar-Or (11)
	Track athlete	0.100	Kirksey et al. (33)

method for determining resistance load or force with the WAnT:

$$\text{Force (N)} = \text{body mass (kg)} \times 9.81 \text{ m} \cdot \text{s}^{-2} \times \text{resistance factor}$$

Therefore, when working with a sedentary adult, you would use the following equation to determine the resistance load for the WAnT:

$$\text{Force (N)} = \text{body mass (kg)} \times 9.81 \text{ m} \cdot \text{s}^{-2} \times 0.075$$

When working with an athlete, you would use this equation:

$$\text{Force (N)} = \text{body mass (kg)} \times 9.81 \text{ m} \cdot \text{s}^{-2} \times 0.100$$

Finally, the WAnT has been consistently shown to be highly reliable (with reliability coefficients ranging between 0.89 and 0.96) for both peak and mean anaerobic power (28). This high reliability demonstrates the test's repeatability and ensures that any changes in performance are indeed related to actual changes rather than testing error. In addition, a large body of normative data concerning the results of the WAnT test for various populations is presented in the scientific literature, which allows for comparisons between testing populations; see Inbar et al. (28) for detailed normative data.

Margaria-Kalamen Stair Climb Test for Determining Anaerobic Power

One of the more popular anaerobic stepping tests performed in both athletic and clinical settings is the Margaria stair climb test (39), which was modified by Kalamen (30). The contemporary version of the **Margaria-Kalamen test** is designed to test anaerobic power; specifically, because of its relatively short duration (usually less than 5 s), it tests the contribution of the phosphagen system (39).

In order to perform this test, you need a staircase with a 6 m run-up and at least nine stairs, each which is between 174 and 175 cm in height (50). It is generally recommended that you place an electronic switch mat or photoelectric cell on the third and ninth steps in order to ensure accurate timing (46), but careful timing with a stopwatch can provide data that is somewhat less accurate yet still reasonable (50). The test itself requires the subject to move up the stairs as quickly as possible by taking three steps at a time (i.e., stepping on the third, sixth, and ninth steps). A graphic summary of the general testing set-up is presented in figure 12.5.

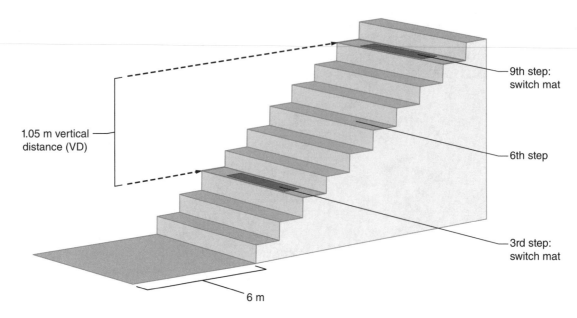

Figure 12.5 Set-up for the Margaria-Kalamen stair climb test.

You can calculate power from this test with the following basic equation:

$$\text{Power} (\text{kg} \cdot \text{m} \cdot \text{s}^{-1}) = \frac{\text{weight (kg)} \times \text{distance (m)}}{\text{time (s)}}$$

where weight is the subject's body mass in kilograms, distance is the vertical height between step 3 and step 9, and time is the number of s it takes the subject to move between step 3 and step 9. Therefore, based on the set-up shown in figure 12.6, the equation could be modified as follows:

$$\text{Power} (\text{kg} \cdot \text{m} \cdot \text{s}^{-1}) = \frac{\text{weight (kg)} \times 1.05\,\text{m}}{\text{time (s)}}$$

If the power value is to be converted to watts, the power value in $\text{kg} \cdot \text{m} \cdot \text{s}^{-1}$ is multiplied by 9.807, which is equivalent to the normal acceleration due to gravity.

The Margaria-Kalamen test is considered to be very reliable, exhibiting a test–retest reliability of $r = 0.85$ and a coefficient of variation of <4% (38). Nonetheless, it should be administered with some caution; specifically, individuals who are less experienced or untrained may have difficulty with the every-third-step foot placement required by the test, thus increasing the risk of injury or inaccurate results.

References

1. Adams GM. *Exercise Physiology Laboratory Manual.* 3rd ed. Boston: McGraw-Hill, 1998.

2. Beam WC and Adams GM. *Exercise Physiology Laboratory Manuel.* 6th ed. Boston: McGraw-Hill, 2011.

3. Bobbert MF, Gerritsen KG, Litjens MC, and Van Soest AJ. Why Is Countermovement Jump Height Greater Than Squat Jump Height? *Med Sci Sports Exer* 28: 1402–1412, 1996.

4. Bogdanis GC, Nevill ME, Boobis LH, and Lakomy HK. Contribution of Phosphocreatine and Aerobic Metabolism to Energy Supply During Repeated Sprint Exercise. *J Appl Physiol* 80: 876–884, 1996.

5. Bompa TO and Haff GG. *Periodization: Theory and Methodology of Training.* 5th ed. Champaign, IL: Human Kinetics, 2009.

6. Bosco C and Komi PV. Mechanical Characteristics and Fiber Composition of Human Leg Extensor Muscles. *Eur J Appl Physiol Occup Physiol* 41: 275–284, 1979.

7. Bosco C, Luhtanen P, and Komi PV. A Simple Method for Measurement of Mechanical Power in Jumping. *Eur J Appl Physiol* 50: 273–282, 1983.

8. Breivik SL. Artistic Gymnastics. In: Winter EM, Jones AM, Davison RCR, Bromley PD, and Mercer TH, eds., *Sport and Exercise Physiology Testing*

Guidelines: Volume I—Sport Testing. London: Routledge, 2007, pp. 220–231.

9. Carlock J, Smith A, Hartman M, Morris R, Ciroslan D, Pierce KC, Newton RU, and Stone MH. The Relationship Between Vertical Jump Power Estimates and Weightlifting Ability: A Field Test Approach. *J Strength Cond Res* 18: 534–539, 2004.

10. Cormie, P., J.M. Mcbride JM, and G.O. Mccaulley GO. Validation of Power Measurement Techniques in Dynamic Lower Body Resistance Exercises. *Journal of Applied Biomechanics* 23: 103–118, 2007.

11. Dotan R and Bar-Or O. Load Optimization for the Wingate Anaerobic Test. *Eur J Appl Physiol Occup Physiol* 51: 409–417, 1983.

12. Evans JA and Quinney HA. Determination of Resistance Settings for Anaerobic Power Testing. *Can J Appl Sport Sci* 6: 53–56, 1981.

13. Fry AC and Kraemer WJ. Physical Performance Characteristics of American Collegiate Football Players. *J Appl Sport Sci Res* 5: 126–138, 1991.

14. Gissis I, Papadopoulos C, Kalapotharakos VI, Sotiropoulos A, Komsis G, and Manolopoulos E. Strength and Speed Characteristics of Elite, Subelite, and Recreational Young Soccer Players. *Res Sports Med* 14: 205–214, 2006.

15. Gore CJ, ed. *Physiological Tests for Elite Athletes.* Champaign, IL: Human Kinetics, 2000, p. 464.

16. Haff GG, Carlock JM, Hartman MJ, Kilgore JL, Kawamori N, Jackson JR, Morris RT, Sands WA, and Stone MH. Force-Time Curve Characteristics of Dynamic and Isometric Muscle Actions of Elite Women Olympic Weightlifters. *J Strength Cond Res* 19: 741–748, 2005.

17. Haff GG, Kirksey KB, Stone MH, Warren BJ, Johnson RL, Stone M, O'bryant HS, and Proulx C. The Effects of Six Weeks of Creatine Monohydrate Supplementation on Dynamic Rate of Force Developlment. *J Strength Cond Res* 14: 426–433, 2000.

18. Haff GG, Stone MH, O'bryant HS, Harman E, Dinan CN, Johnson R, and Han KH. Force-Time Dependent Characteristics of Dynamic and Isometric Muscle Actions. *J Strength Cond Res* 11: 269–272, 1997.

19. Harman E and Garhammer J. Administration, Scoring, and Interpretation of Selected Tests. In: Baechle TR and Earle RW, eds., *Essentials of Strength Training and Conditioning.* 3rd ed. National Strength and Conditioning Association. Champaign, IL: Human Kinetics, 2008, pp. 249–292.

20. Harman EA, Rosenstein MT, Frykman PN, Rosenstein RM, and Kraemer W. Estimation of Human Power Output From Vertical Jump. *J Appl Sport Sci Res* 5: 116–120, 1991.

21. Hawkins SB, Doyle TL, and Mcguigan MR. The Effect of Different Training Programs on Eccentric Energy Utilization in College-Aged Males. *J Strength Cond Res* 23: 1996–2002, 2009.

22. Hespanhol JE, Neto LGS, De Arruda M, and Dini CA. Assessment of Explosive Strength-Endurance in Volleyball Players Through Vertical Jump Testing. *Rev Bras Med Esporte* 13: 160e–163e, 2007.

23. Hoffman JR. *Norms for Fitness, Performance, and Health.* Champaign, IL: Human Kinetics, 2006.

24. Hoffman JR. *Physiological Aspects of Sport Training and Performance.* Champaign, IL: Human Kinetics, 2002, p. 342.

25. Hori N, Newton RU, Andrews WA, Kawamori N, Mcguigan MR, and Nosaka K. Does Performance of Hang Power Clean Differentiate Performance of Jumping, Sprinting, and Changing of Direction? *J Strength Cond Res* 22: 412–418, 2008.

26. Hori N, Newton RU, Nosaka K, and Mcguigan MR. Comparison of Different Methods of Determining Power Output in Weightlifting Exercises. *Strength and Cond J* 28: 34–40, 2006.

27. Housh TJ, Cramer JT, Weir JP, Beck TW, and Johnson GO. *Physical Fitness Laboratories on a Budget.* Scottsdale, AZ: Holcomb-Hathaway, 2009, p. 234.

28. Inbar O, Bar-Or O, and Skinner JS. *The Wingate Anaerobic Test.* Champaign, IL: Human Kinetics, 1996, p. 110.

29. Johnson DL and Bahamonde R. Power Output Estimate in University Athletes. *J Strength Cond Res* 10: 161–166, 1996.

30. Kalamen J. *Measurement of Maximum Muscle Power in Man.* Columbus: Ohio State University, 1968.

31. Kawamori N, Crum AJ, Blumert P, Kulik J, Childers J, Wood J, Stone MH, and Haff GG. Influence of Different Relative Intensities on Power Output During the Hang Power Clean: Identification of the Optimal Load. *J Strength Cond Res* 19: 698–708, 2005.

32. Kawamori N, Rossi SJ, Justice BD, Haff EE, Pistilli EE, O'bryant HS, Stone MH, and Haff GG. Peak Force and Rate of Force Development During Isometric and Dynamic Mid-Thigh Clean Pulls Performed at Various Intensities. *J Strength Cond Res* 20: 483–491, 2006.

33. Kirksey B, Stone MH, Warren BJ, Johnson RL, Stone M, Haff GG, Williams FE, and Proulx C.

The Effects of Six Weeks of Creatine Monohydrate Supplementation on Performance Measures and Body Composition in Collegiate Track and Field Athletes. *Journal of Strength and Conditioning Research* 13: 148–156, 1999.

34. Klavora P. Vertical-Jump Tests: A Critical Review. *Strength & Conditioning Journal* 22: 70, 2000.

35. Komi PV. *Strength and Power in Sport.* 2nd ed. Malden, MA: Blackwell Scientific, 2003, p. 523.

36. Kraska JM, Ramsey MW, Haff GG, Fethke N, Sands WA, Stone ME, and Stone MH. Relationship Between Strength Characteristics and Unweighted and Weighted Vertical Jump Height. *Int J Sports Physiol Perform* 4: 461–473, 2009.

37. Leard JS, Cirillo MA, Katsnelson E, Kimiatek DA, Miller TW, Trebincevic K, and Garbalosa JC. Validity of Two Alternative Systems for Measuring Vertical Jump Height. *J Strength Cond Res* 21: 1296–1299, 2007.

38. Macdougall JD, Wenger HA, and Green HJ. The Purpose of Physiological Testing. In: Macdougall JD, Wenger HA, and Green HJ, eds., *Physiological Testing of the High-Performance Athlete.* Champaign, IL: Human Kinetics, 1991, pp. 1–6.

39. Margaria R, Aghemo P, and Rovelli E. Measurement of Muscular Power (Anaerobic) in Man. *J Appl Physiol* 21: 1662–1664, 1966.

40. Maud PJ, Berning JM, Foster C, Cotter HM, Dodge C, Dekonning JJ, Hettinga F, and Lampen J. Testing for Anaerobic Ability. In: Maud PJ and Foster C, eds., *Physiological Assessments of Human Fitness.* Champaign, IL: Human Kinetics, 2006, pp. 77–92.

41. Maud PJ and Foster C, eds. *Physiological Assessments of Human Fitness.* 2nd ed. Champaign, IL: Human Kinetics, 2006.

42. Mcguigan MR, Doyle TL, Newton M, Edwards DJ, Nimphius S, and Newton RU. Eccentric Utilization Ratio: Effect of Sport and Phase of Training. *J Strength Cond Res* 20: 992–995, 2006.

43. Medbo JI and Tabata I. Anaerobic Energy Release in Working Muscle During 30 seconds to 3 minutes of Exhausting Bicycling. *J Appl Physiol* 75: 1654–1660, 1993.

44. Mirzaei B, Curby DG, Rahmani-Nia F, and Moghadasi M. Physiological Profile of Elite Iranian Junior Freestyle Wrestlers. *J Strength Cond Res* 23: 2339–2344, 2009.

45. Nicklin RC, O'bryant HS, Zehnbauer TM, and Collins A. A Computerized Method for Assessing Anaerobic Power and Work Capacity Using Maximal Cycle Ergometry. *J Appl Sport Sci Res* 4: 135–140, 1990.

46. Plowman SA and Smith DL. *Exercise Physiology for Health, Fitness and Performance.* 3rd ed. Baltimore: Lippincott Williams & Wilkins, 2011.

47. Robinson JM, Stone MH, Johnson RL, Penland CM, Warren BJ, and Lewis RD. Effects of Different Weight Training Exercise/Rest Intervals on Strength, Power, and High Intensity Exercise Endurance. *J Strength and Cond Res* 9: 216–221, 1995.

48. Sands WA, McNeal JR, Ochi MT, Urbanek TL, Jemni M, and Stone MH. Comparison of the Wingate and Bosco Anaerobic Tests. *J Strength Cond Res* 18: 810–815, 2004.

49. Sayers SP, Harackiewicz DV, Harman EA, Frykman PN, and Rosenstein MT. Cross-Validation of Three Jump Power Equations. *Med Sci Sports Exer* 31: 572–577, 1999.

50. Stone MH and O'bryant HO. *Weight Training: A Scientific Approach.* Edina, MN: Burgess, 1987.

51. Stone MH, Sands WA, Carlock J, Callan S, Dickie D, Daigle K, Cotton J, Smith SL, and Hartman M. The Importance of Isometric Maximum Strength and Peak Rate-Of-Force Development in Sprint Cycling. *J Strength Cond Res* 18: 878–884, 2004.

52. Stone MH, Stone ME, and Sands WA. *Principles and Practice of Resistance Training.* Champaign, IL: Human Kinetics, 2007, p. 376.

53. Zupan MF, Arata AW, Dawson LH, Wile AL, Payn TL, and Hannon ME. Wingate Anaerobic Test Peak Power and Anaerobic Capacity Classifications for Men and Women Intercollegiate Athletes. *J Strength Cond Res* 23: 2598–2604, 2009.

SPRINTING PERFORMANCE

EQUIPMENT

- Track or other area affording 80 to 100 m that can be used for sprinting
- Four stopwatches or an electronic timing gate system
- Calculator
- Individual and group data sheets
- Microsoft Excel or equivalent spreadsheet program

> Find the group data sheets for this laboratory online at www.HumanKinetics.com/LaboratoryManualForExercisePhysiology.

TESTING AREA SET-UP

The sprint test requires a track or other area that offers 80 to 100 m in which the following distances can be marked: 9.13 m (10 yd), 36.6 m (40 yd), 45.7 m (50 yd), and 54.9 m (60 yd). Four timers are used in order to time each of the four distances. Position each timer in such a way that he or she can see both the start and finish of the sprint. A summary of the testing area set-up is shown in figure 12.6.

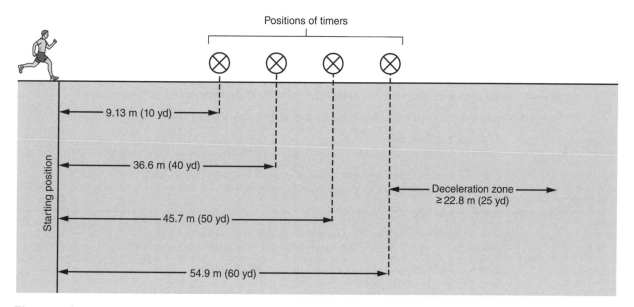

Figure 12.6   Sprint test set-up.

Before beginning a maximal sprint test, have the subject complete a structured warm-up that includes a 5 min period of general warm-up activities (e.g., jogging, cycling, rope jumping) followed by 5 min of dynamic stretching (e.g., leg swings, high knees, walking knee tucks, walking lunges, butt kicks, inchworms, and power skips). Next, have the subject perform moderate-speed runs for the 54.9 m distance; specifically, the subject should perform a series of moderate-speed sprints that get progressively faster. He or she should also practice the start technique. It is recommended that the subject *not* use starting blocks or brace against another person's foot because of the technical skill required to use this technique and the inability to replicate positions consistently. The subject should be in a standing position with a low center of gravity and a slight forward lean (1). Once the warm-up is completed, the testing session can begin.

MAXIMAL SPRINT TEST

In order to effectively perform the sprint test, follow these steps.

Step 1: Set up the sprint area by placing marks at 9.14 m (10 yd), 36.6 m (40 yd), 45.7 m (50 yd), and 54.9 m (60 yd). At the end of the sprint area, include a deceleration zone of at least 22.9 m (25 yd) (figure 12.6).

Step 2: Instruct the subject to remove his or her shoes, then assess his or her height and weight. Record these data on the appropriate location on the individual data sheet for laboratory activity 12.1. The subject should put his or her shoes back on before continuing the test.

Step 3: Have the subject perform a structured warm-up designed to prepare him or her for a maximal sprint bout (see table 12.3).

Step 4: Direct the four timers to stand at the locations noted in figure 12.6. Make sure that they take positions allowing them to see both the start and the finish of the sprint.

Step 5: Instruct the timers to start their stopwatches at the subject's first movement. Tell the subject that the objective of the test is to cover the distance as quickly as possible and that the timers will start their stopwatches upon the subject's first movement.

Step 6: Have the subject move to the start line and assume the starting position.

Step 7: Before you start the test, have the subject and all timers acknowledge that they are ready.

Step 8: Encourage the subject to run as fast as possible through the 9.14 m (10 yd), 36.6 m (40 yd), 45.7 m (50 yd), and 54.9 m (60 yd) lines and then decelerate in the designated zone.

Step 9: Each timer should stop his or her stopwatch when the subject passes the individual finish line at which the timer is standing—9.14 m, 36.6 m, 45.7 m, or 54.9 m. For example, as the sprinter passes the 9.14 m mark, the timer standing at this point should stop his or her watch to mark the sprinter's 9.14 m time.

Step 10: Record the time achieved for each distance to the nearest tenth of a second and note it in the appropriate location on the individual data sheet.

Step 11: Have subject rest and recover for 3 min before initiating the next sprint, then repeat steps 4 through 10.

Table 12.3 Sprint-Specific Warm-Up

Time (min:s)	Activity	
0:00–5:00	General warm-up	Activities such as jogging, cycling, or rope jumping
5:00–10:00	Dynamic stretching warm-up	Activities such as leg swings, high knees, walking knee tucks, walking lunges, butt kicks, inchworms, and power skips
10:00–15:00	Sprint-specific warm-up	Series of moderate-speed sprints that get progressively faster (while practicing correct starting position)
15:00–17:00	Recovery period	Active rest during which the subject prepares for the all-out sprint test

Step 12: After completing the test, have the subject perform a 5 to 10 min cool-down.

Step 13: Calculate the results of the test. Average trials 1 and 2 for each distance covered and record the results in the appropriate locations on the group data sheet.

Step 14: Calculate the velocity of each sprint distance with the following formula:

$$\text{Velocity } (\text{m} \cdot \text{s}^{-1}) = \text{distance (m) / time (s)}$$

Record the results of this calculation on the individual and group data sheets.

Step 15: Calculate the power output for each distance with the following equation:

$$\text{power } (\text{N} \cdot \text{m} \cdot \text{s}^{-1}) = \text{force} \times \text{velocity} = \text{body mass (N)} \times \text{velocity } (\text{m} \cdot \text{s}^{-1})$$

Record the results of this calculation on the individual and group data sheets.

Step 16: Use the following formula to calculate the acceleration between these marks: 0 and 9.14 m, 0 and 36.6 m, 0 and 45.7 m, 0 and 54.9 m, 9.14 m and 36.6 m, 36.6 m and 45.7 m, and 45.7 m and 54.9 m.

$$\text{Acceleration } (\text{m} \cdot \text{s}^{-2}) = \frac{\text{final velocity } (\text{m} \cdot \text{s}^{-1}) - \text{initial velocity } (\text{m} \cdot \text{s}^{-1})}{\text{time during interval (s)}}$$

Record the results of this calculation on the individual and group data sheets. For example, if the sprinter ran 9.14 m in 2 s, then his or her final velocity would be 4.57 m·s⁻¹. Since the subject is starting from a standing position, the initial velocity would be equal to 0 m·s⁻¹. Therefore, the acceleration would be calculated as follows:

$$\text{Acceleration } (\text{m} \cdot \text{s}^{-2}) = \frac{4.57 \text{ m} \cdot \text{s}^{-1} - 0 \text{ m} \cdot \text{s}^{-1}}{2 \text{ s}} = 2.285 \text{ m} \cdot \text{s}^{-2}$$

Thus the individual is accelerating at a rate of 2.285 m·s⁻².

Step 17: Compare the results of this test with those presented in table 12.4.

Step 18: Use Excel or an equivalent spreadsheet program to graph the individual and class velocities achieved at each distance.

Laboratory Activity 12.1

Table 12.4 Percentile Ranks for the 36.6 m Sprint Distance

Percentile	12–13 y		14–15 y		16–18 y	
	Male	Female	Male	Female	Male	Female
90	5.41	5.79	5.02	5.36	4.76	4.93
80	5.63	6.14	5.15	5.68	4.85	5.22
70	5.77	6.49	5.24	6.01	4.90	5.52
60	5.84	6.84	5.32	6.33	4.98	5.82
50	5.97	7.19	5.46	6.65	5.10	6.11
40	6.08	7.54	5.54	6.97	5.13	6.41
30	6.25	7.89	5.78	7.30	5.21	6.71
20	6.32	8.24	6.02	7.62	5.30	7.00
10	6.64	8.59	6.08	7.95	5.46	7.31
Mean	5.99	7.19	5.51	6.65	5.08	6.11

Adapted from Hoffman (24) and Housh et al. (27).

QUESTION SET 12.1

1. Based on your time, what energy systems contributed to your sprinting performance?

2. If the sprint were extended to 400 m, how would the contribution of the various energy systems change?

3. At what interval (0–9.14 m, 9.14–36.6 m, 36.6–45.7 m, or 45.7–54.9 m) was the highest velocity achieved? How does your individual data compare with the class data?

4. At what point during the sprint was the highest acceleration achieved? How does your individual data compare with the class data?

5. How does the class data for the 36.6 m distance compare with the norm table (table 12.4) for sprint performance?

6. Were there any power output differences between intervals? If so, what may have contributed to them?

7. Are there any sex differences in sprint times and power outputs? Physiologically, how could these differences be explained?

8. Explain in detail the relationship between muscle fiber type and sprinting performance. (Hint: What fiber type would be important to have for sprinting—and why?)

Find the case studies for this laboratory online at www.HumanKinetics.com/ LaboratoryManualForExercisePhysiology.

Laboratory Activity 12.1

Name or ID number: _____   Tester: _____

Date: _____ Time: _____

Sex: M / F (circle one)   Age: _____ y

Height: _____ in.   _____ cm

Weight: _____ lb   _____ kg

Temperature: _____ °F   _____ °C

Relative humidity: _____ %

Barometric pressure: _____ mmHg

Location of Testing

❑ Outdoor field

❑ Indoor field

❑ Indoor track

❑ Outdoor track

❑ Gym

❑ Other

Footwear

❑ Jogging shoe

❑ Walking shoe

❑ Tennis shoe

❑ Basketball shoe

❑ Running shoe

❑ Cross-trainer

❑ Other _____

Sprint Times

Round the sprint times to the nearest 0.1 s.

9.14 m sprint			
	Distance (m)	Time (s)	Velocity (m · s⁻¹)
Trial 1	9.14		
Trial 2	9.14		
Mean			

36.6 m sprint			
	Distance (m)	Time (s)	Velocity (m · s⁻¹)
Trial 1	36.6		
Trial 2	36.6		
Mean			

45.7 m sprint			
	Distance (m)	Time (s)	Velocity (m · s⁻¹)
Trial 1	45.7		
Trial 2	45.7		
Mean			

54.9 m sprint			
	Distance (m)	Time (s)	Velocity (m · s⁻¹)
Trial 1	54.9		
Trial 2	54.9		
Mean			

Horizontal power calculation:

$$\text{Power } (\text{N} \cdot \text{m} \cdot \text{s}^{-1}) = \text{force} \times \text{velocity} = \text{body mass (N)} \times \text{velocity } (\text{m} \cdot \text{s}^{-1})$$

Distance	Body mass (N)		Velocity (m·s⁻¹)		Power (N·m·s⁻¹)
9.14 m		×		=	
36.6 m		×		=	
45.7 m		×		=	
54.9 m		×		=	

Note: To convert body mass to newtons, multiply body mass in kg by 9.81 N·kg⁻¹.

Calculating acceleration formula:

$$\text{Acceleration } (\text{m} \cdot \text{s}^{-2}) = \frac{\text{final velocity } (\text{m} \cdot \text{s}^{-1}) - \text{initial velocity } (\text{m} \cdot \text{s}^{-1})}{\text{time during interval (s)}}$$

Interval	Acceleration calculation	Acceleration (m·s⁻²)
Acceleration from 0–9.14 m	_____ (m·s⁻¹) − _____ (m·s⁻¹) / _____ (s)	
Acceleration from 9.14–36.6 m	_____ (m·s⁻¹) − _____ (m·s⁻¹) / _____ (s)	
Acceleration from 36.6 m–45.7 m	_____ (m·s⁻¹) − _____ (m·s⁻¹) / _____ (s)	
Acceleration from 45.7–54.9 m	_____ (m·s⁻¹) − _____ (m·s⁻¹) / _____ (s)	
Acceleration from 0–36.6 m	_____ (m·s⁻¹) − _____ (m·s⁻¹) / _____ (s)	
Acceleration from 0–45.7 m	_____ (m·s⁻¹) − _____ (m·s⁻¹) / _____ (s)	
Acceleration from 0–54.9 m	_____ (m·s⁻¹) − _____ (m·s⁻¹) / _____ (s)	

Laboratory Activity 12.1

JUMPING PERFORMANCE

EQUIPMENT

- Area in which to perform a dynamic warm-up
- Vertec or wall-mounted board for marking vertical displacements
- Calculator
- Individual and group data sheets
- Microsoft Excel or equivalent spreadsheet program

 Find the group data sheets for this laboratory online at www.HumanKinetics.com/ LaboratoryManualForExercisePhysiology.

WARM-UP

This test requires an area with sufficient ceiling clearance and with space for jumpers to perform an appropriate warm-up. As with any physical test, it is essential to have the subject perform a proper warm-up in order to maximize performance. Begin the general warm-up with activities such as jogging, cycling, or rope jumping, then have the subject move on to a dynamic stretching routine consisting of activities that target the lower body (e.g., body-weight squats, squat thrusts, walking lunges, butt kicks, high knees, walking knee tucks). Next, have the subject complete a warm-up tailored to the vertical jump that allows him or her to practice both the countermovement (figure 12.1, p. 308) and the static (figure 12.2, p. 309) vertical jump tests. A summary of this warm-up procedure can be found in table 12.5.

Table 12.5 Specific Warm-Up for Vertical Jump

Time (min:s)	Activity	
0:00–5:00	General warm-up	Activities such as jogging, cycling, or rope jumping
5:00–10:00	Dynamic stretching warm-up	Activities such as leg swings, high knees, walking knee tucks, walking lunges, butt kicks, inchworms, and power skips
10:00–15:00	Warm-up specific to the vertical jump	Series of static and countermovement jumps that allow the jumper to practice jumping positions and motions
15:00–17:00	Recovery period	Period of active rest in which the subject prepares for the all-out jumping tasks

COUNTERMOVEMENT VERTICAL JUMP TEST

Step 1: Set up the testing area by positioning the Vertec with plenty of room around it. If you are using a wall-mounted jump board, make sure that it is clearly marked and easily accessible.

Step 2: Have the subject remove his or her shoes, then determine his or her height and weight. Record these data on the individual data sheet for laboratory activity 12.2. Have the subject put his or her shoes back on.

Step 3: Direct the subject through a structured warm-up that prepares him or her for the vertical jump test (table 12.5).

Step 4: Explain to the subject the position used to establish reach height. Remember that reach height is measured by having the subject stand with the feet together and the dominant side near the wall or Vertec. In this position, have the subject reach as high as possible with his or her dominant hand while the feet remain flat on the floor. If using a wall-mounted measuring device, have the subject keep the palm of his or her hand flat against the wall. Record the highest reach point. If using a Vertec the subject should reach as high as he or she can, moving as many vanes as possible until the highest displacement is achieved. Record the highest displacement as the reach height in the appropriate location on the individual data sheet.

Step 5: After establishing the reach height, explain the procedures associated with the countermovement vertical jump test. Demonstrate the basic jumping process.

Step 6: Position the subject so that his or her feet are approximately shoulder-width apart and instruct the subject to remain in this position (figure 12.1a) during the jump. After achieving this position, instruct the subject to perform a countermovement in which he or she dips the hips and legs while swinging the arms prior to pushing off the ground (figure 12.1b). While the subject is in the air, he or she should reach for the highest possible point on the wall or hit the highest possible vanes of the Vertec. Upon landing, the subject should bend his or her knees to absorb landing forces (figure 12.1c).

Step 7: Have the subject perform a total of three countermovement vertical jumps separated by 2 min each. Record the vertical displacement achieved during each jump on the individual data sheet.

STATIC VERTICAL JUMP TEST

Step 1: After completing the countermovement vertical jump test, explain and demonstrate the static vertical jump test to the subject. As with the countermovement test, have the subject assume a position in which the feet are approximately shoulder-width apart (figure 12.2a). After achieving this position, instruct the subject to squat down so that the top of the thighs are parallel to the floor (figure 12.2b). Have the subject hold this squat position for a count of three, then encourage the subject to rapidly extend his or her knees, hips, and ankles while swinging the arms upward in order to touch the highest possible point on the wall or move the highest possible vane on the Vertec (figure 12.2*c*). Encourage the subject to bend the knees upon landing to absorb landing forces (figure 12.2*d*).

Step 2: Have the subject perform three static vertical jumps separated by 2 min each. Record the vertical displacement achieved during each jump on the individual data sheet.

Step 3: Subtract the reach height from the highest vertical displacement, then convert this measure to cm (if it is initially calculated in inches) and average the results in the appropriate location on the individual data sheet.

CALCULATIONS

Step 1: Calculate the power outputs for both the countermovement and the static vertical jump tests by using the equations presented on the individual data sheet.

Step 2: Calculate the eccentric utilization ratio by dividing the vertical displacements and power outputs calculated from the countermovement vertical jump test by the results of the static jump test.

QUESTION SET 12.2

1. Was there any difference between the highest vertical jump displacements for the countermovement and static vertical jump tests? If so, what neuromuscular characteristics might explain these findings?

2. When you examine the calculated power outputs, how does the Lewis equation compare with the Johnson average and the Harman average power equations? With the Johnson and Harman peak power equations? The Sayer equation? If differences exist between the results for the different formulas, explain what may contribute to these differences.

3. How do the Johnson and Harman equation results compare?

4. Were there any sex differences in either the static or the countermovement vertical jump power outputs? If so, what neuromuscular characteristics might explain these differences?

5. When you look at the eccentric utilization ratio, how do your results compare with those for the class for vertical displacement? For power output?

6. Explain in detail what neuromuscular characteristics can contribute to a high eccentric utilization ratio—and how these characteristics might be affected by training.

 Find the case studies for this laboratory online at www.HumanKinetics.com/ LaboratoryManualForExercisePhysiology.

Name or ID number: _____ Tester: _____

Date: _____ Time: _____

Sex: M / F (circle one)   Age: _____ y

Height: _____ in. _____ cm

Weight: _____ lb _____ kg

Temperature: _____ °F _____ °C

Relative humidity: _____ %

Barometric pressure: _____ mmHg

Location of Testing

❑ Outdoor field

❑ Indoor field

❑ Indoor track

❑ Outdoor track

❑ Gym

❑ Other

Footwear

❑ Jogging shoe

❑ Walking shoe

❑ Tennis shoe

❑ Basketball shoe

❑ Running shoe

❑ Cross-trainer

❑ Other _____

Countermovement Jump Test Results

Reach height = _____ in.   Reach height =_____ in. / 0.3937 = _____ cm

Trial	Jump height (in.)	Vertical displacement (in.)				
		Jump height (in.)	–	Reach height (in.)	=	Displacement (in.)
1			–		=	
2			–		=	
3			–		=	
Mean					=	

Trial	Jump height (cm)	Vertical displacement (cm)				
		Jump height (cm)	–	Reach height (cm)	=	Displacement (cm)
1			–		=	
2			–		=	
3			–		=	
Mean					=	

When working with the power equations, insert the mean displacements into the following equations:

Lewis power (W):

$$\text{Power(W)} = \sqrt{4.9} \times \underbrace{\text{_____}}_{\text{weight (kg)}} \times \sqrt{\underbrace{\text{_____}}_{\text{jump height (cm)}}} = \underline{\hspace{1.5cm}}\ \text{W}$$

Sayer equation (W):

$$\text{Peak power(W)} = 60.7 \times \underbrace{\text{_____}}_{\text{jump height (cm)}} + 45.3 \times \underbrace{\text{_____}}_{\text{body mass (kg)}} - 2{,}055 = \underline{\hspace{1.5cm}}\ \text{W}$$

Harman equations (W):

$$\text{Peak power(W)} = 61.9 \times \underbrace{\text{_____}}_{\text{jump height (cm)}} + 36 \times \underbrace{\text{_____}}_{\text{body mass (kg)}} + 1{,}822 = \underline{\hspace{1.5cm}}\ \text{W}$$

$$\text{Average power(W)} = 21.2 \times \underbrace{\text{_____}}_{\text{jump height (cm)}} + 23 \times \underbrace{\text{_____}}_{\text{body mass (kg)}} - 1{,}393 = \underline{\hspace{1.5cm}}\ \text{W}$$

Johnson equations (W):

Peak power $(W) = 78.5 \times \underline{\hspace{2cm}}_{\text{jump height (cm)}} + 60.6 \times \underline{\hspace{2cm}}_{\text{body mass (kg)}} - 15.3 \times \underline{\hspace{1.5cm}}_{\text{height (cm)}} - 1{,}308 = \underline{\hspace{1.5cm}}$ W

Average power $(W) = 41.4 \times \underline{\hspace{2cm}}_{\text{jump height (cm)}} + 31.2 \times \underline{\hspace{2cm}}_{\text{body mass (kg)}} - 13.9 \times \underline{\hspace{1.5cm}}_{\text{height (cm)}} - 1{,}393 = \underline{\hspace{1.5cm}}$ W

Power calculation summary:

Lewis equation = _____ W

Harman peak power equation = _____ W

Harman average power equation = _____ W

Sayer peak power equation = _____ W

Johnson peak power equation = _____ W

Johnson average power equation = _____ W

Static jump test results:

Reach height = _____ in.   Reach height = _____ in. / 0.3937 = _____ cm

Trial	Jump height (in.)	Vertical displacement (in.)				
		Jump height (in.)	–	Reach height (in.)	=	Displacement (in.)
1			–		=	
2			–		=	
3			–		=	
Mean					=	

Trial	Jump height (cm)	Vertical displacement (cm)				
		Jump height (cm)	–	Reach height (cm)	=	Displacement (cm)
1			–		=	
2			–		=	
3			–		=	
Mean					=	

When working with the power equations, insert the mean displacements into the following equations:

Lewis power (W):

$$\text{Power}(W) = \sqrt{4.9} \times \underline{\hspace{2cm}}_{\text{weight (kg)}} \times \sqrt{\underline{\hspace{2cm}}_{\text{jump height (cm)}}} = \underline{\hspace{2cm}} \text{ W}$$

Sayer equation (W):

$$\text{Peak power}(W) = 60.7 \times \underline{\hspace{2cm}}_{\text{jump height (cm)}} + 45.3 \times \underline{\hspace{2cm}}_{\text{body mass (kg)}} - 2{,}055 = \underline{\hspace{2cm}} \text{ W}$$

Harman equations (W):

$$\text{Peak power}(W) = 61.9 \times \underline{\hspace{2cm}}_{\text{jump height (cm)}} + 36 \times \underline{\hspace{2cm}}_{\text{body mass (kg)}} + 1{,}822 = \underline{\hspace{2cm}} \text{ W}$$

$$\text{Average power}(W) = 21.2 \times \underline{\hspace{2cm}}_{\text{jump height (cm)}} + 23 \times \underline{\hspace{2cm}}_{\text{body mass (kg)}} - 1{,}393 = \underline{\hspace{2cm}} \text{ W}$$

Johnson equations (W):

$$\text{Peak power (W)} = 78.5 \times \underline{\hspace{3cm}}_{\text{jump height (cm)}} + 60.6 \times \underline{\hspace{3cm}}_{\text{body mass (kg)}} - 15.3 \times \underline{\hspace{2cm}}_{\text{height (cm)}} - 1{,}308 = \underline{\hspace{1.5cm}} \text{ W}$$

$$\text{Average power (W)} = 41.4 \times \underline{\hspace{3cm}}_{\text{jump height (cm)}} + 31.2 \times \underline{\hspace{3cm}}_{\text{body mass (kg)}} - 13.9 \times \underline{\hspace{2cm}}_{\text{height (cm)}} - 1{,}393 = \underline{\hspace{1.5cm}} \text{ W}$$

Power calculation summary:

Lewis equation = _____ W

Harman peak power equation = _____ W

Harman average power equation = _____ W

Sayer peak power equation = _____ W

Johnson peak power equation = _____ W

Johnson average power equation = _____ W

To calculate the eccentric utilization ratio, use the following:

Vertical displacement	_____ / Countermovement jump (cm)	_____ static jump (cm)	= _____
Lewis equation	_____ / Countermovement jump (cm)	_____ static jump (cm)	= _____
Harman peak power equation	_____ / Countermovement jump (cm)	_____ static jump (cm)	= _____
Harman average power equation	_____ / Countermovement jump (cm)	_____ static jump (cm)	= _____
Sayer peak power equation	_____ / Countermovement jump (cm)	_____ static jump (cm)	= _____
Johnson peak power equation	_____ / Countermovement jump (cm)	_____ static jump (cm)	= _____
Johnson average power equation	_____ / Countermovement jump (cm)	_____ static jump (cm)	= _____

Laboratory Activity 12.2

JUMPING PERFORMANCE WITH A SWITCH MAT

EQUIPMENT

- Area in which to perform a dynamic warm-up
- Switch mat
- Calculator
- Individual and group data sheets
- Microsoft Excel or equivalent spreadsheet program

Find the group data sheets for this laboratory online at www.HumanKinetics.com/ LaboratoryManualForExercisePhysiology.

WARM-UP

This test requires an area with sufficient ceiling clearance and with space for the subject to perform an appropriate warm-up. As with any physical test, it is essential to have the subject perform a proper warm-up in order to maximize performance. Begin the general warm-up with activities such as jogging, cycling, or rope jumping, then have the subject move on to a dynamic stretching routine consisting of activities that target the lower body (e.g., body-weight squats, squat thrusts, walking lunges, butt kicks, high knees, walking knee tucks). Next, have the subject complete a warm-up tailored to the vertical jump that allows him or her to practice both the countermovement (figure 12.3, p. 311) and static vertical jump tests on the switch mat. A summary of this warm-up procedure can be found in table 12.5.

COUNTERMOVEMENT VERTICAL JUMP TEST

Step 1: Prepare the testing area by positioning the switch mat with plenty of room around it and appropriate ceiling clearance above it.

Step 2: Have the subject remove his or her shoes, then determine his or her height and weight. Record these data in the appropriate location on the individual data sheet for laboratory activity 12.3. Have the subject put his or her shoes back on.

Step 3: Direct the subject to perform a structured warm-up that prepares him or her for the vertical jump test (table 12.5).

Step 4: Explain the countermovement jump testing protocol and demonstrate the technique used in performing this type of jump.

Step 5: Position the subject so that his or her feet are approximately shoulder-width apart (figure 12.3a). Once the subject achieves this position, instruct him or her to dip the hips and legs while keeping the hands at the hips and then rapidly extend the legs and hips to push off of the ground (figure 12.3b). Encourage the subject to extend his or her body as much as possible while in the air (figure 12.3c). Instruct the subject to bend his or her knees when landing to absorb landing forces (figure 12.3d).

Step 6: Have the subject perform a total of three countermovement vertical jumps separated by 2 min each. Record the flight time achieved during each jump on the individual data sheet.

STATIC VERTICAL JUMP TEST

Step 1: Explain the protocol for the static vertical jump and demonstrate the appropriate technique for performing this type of jump.

Step 2: Have the subject assume a position in which the feet are shoulder-width apart and the hands are on the hips (figure 12.2a).

Step 3: Instruct the subject to squat down to a position in which the top of the thighs are parallel to the ground (figure 12.2b) while keeping the hands on the hips.

Step 4: Give a countdown: "3, 2, 1, jump". Encourage the subject to rapidly extend his or her knees, hips, and ankles in order to jump off of the switch mat (figure 12.2c). Encourage the subject to bend his or her knees upon landing to absorb landing forces (figure 12.2d).

Step 5: Have the subject perform a total of three static vertical jumps separated by 2 min each. Record the flight time achieved during each jump on the individual data sheet.

CALCULATIONS

Step 1: For each jump performed in this lab, convert the flight time to vertical displacement with the following equation:

$$\text{Vertical displacement (m)} = \frac{9.81\,(\text{m} \cdot \text{s}^{-2}) \times \text{flight time (s)} \times \text{flight time (s)}}{8}$$

Step 2: Use the various formulas presented on the individual data sheet for this lab to calculate the countermovement and static vertical jump power outputs. Record these values in the appropriate locations on the data sheet.

Step 3: Calculate the eccentric utilization ratio for the vertical displacement and the power outputs achieved during the countermovement and static vertical jumps and record it in the appropriate location on the individual data sheet and record the results.

Step 4: Compare the eccentric utilization ratio with the norms presented in table 12.6.

Table 12.6 Eccentric Utilization Ratio Norms

Sport	Men	Women
Field hockey		1.02 ± 0.13
Rugby Union	1.13 ± 0.14	
Soccer	1.14 ± 0.15	1.17 ± 0.16
Softball	1.03 ± 0.09	1.04 ± 0.13
Athletics: Throwing	1.11 ± 0.10	1.05 ± 0.06
Athletics: Jumping and sprinting	1.11 ± 0.10	1.11 ± 0.10
Untrained	1.17 ± 0.17	

Data from McGuigan et al. (42), Hawkins et al. (21), and unpublished data.

QUESTION SET 12.3

1. Was there any difference between the highest vertical jump displacement for the countermovement and static vertical jump tests? Is this expected? Why or why not? If there is a difference, can it be explained in neuromuscular terms? How?

2. When you examine the calculated power outputs, how does the Lewis equation compare with the Johnson average and the Harman average power equations? With the Johnson and Harman peak power equations? The Sayer equation? What might explain these differences?

3. How do the Johnson and Harman equation results compare?

4. Were there any sex differences in either the static or the countermovement vertical jump power outputs?

5. When you look at the eccentric utilization ratio, how do your results compare with those for the class for vertical displacement? For power output?

6. Explain procedural issues that can increase the reliability of testing on a switch mat. Be specific.

7. What are some pros and cons of using switch mats in a testing situation?

Find the case studies for this laboratory online at www.HumanKinetics.com/ LaboratoryManualForExercisePhysiology.

Laboratory Activity 12.3

Name or ID number: _____ Tester: _____

Date: _____ Time: _____

Sex: M / F (circle one)   Age: _____ y

Height: _____ in.   _____ cm

Weight: _____ lb   _____ kg

Temperature: _____ °F   _____ °C

Relative humidity: _____ %

Barometric pressure: _____ mmHg

Location of Testing

❑ Outdoor field

❑ Indoor field

❑ Indoor track

❑ Outdoor track

❑ Gym

❑ Other

Footwear

❑ Jogging shoe

❑ Walking shoe

❑ Tennis shoe

❑ Basketball shoe

❑ Running shoe

❑ Cross-trainer

❑ Other _____

Countermovement Jump Test Results

Trial	Flight time (s)	Calculation of vertical displacement	=	Displacement (m)
1			=	
2			=	
3			=	
Mean			=	

When working with the power equations, insert the mean displacements into the following equations:

Lewis power (W):

$$\text{Power (W)} = \sqrt{4.9} \times \underset{\text{weight (kg)}}{\underline{\hspace{2cm}}} \times \sqrt{\underset{\text{jump height (cm)}}{\underline{\hspace{2cm}}}} = \underline{\hspace{1.5cm}} \text{ W}$$

Sayer equation (W):

$$\text{Peak power (W)} = 60.7 \times \underset{\text{jump height (cm)}}{\underline{\hspace{1.5cm}}} + 45.3 \times \underset{\text{body mass (kg)}}{\underline{\hspace{1.5cm}}} - 2{,}055 = \underline{\hspace{1.5cm}} \text{ W}$$

Harman equations (W):

$$\text{Peak power (W)} = 61.9 \times \underset{\text{jump height (cm)}}{\underline{\hspace{1.5cm}}} + 36 \times \underset{\text{body mass (kg)}}{\underline{\hspace{1.5cm}}} + 1{,}822 = \underline{\hspace{1.5cm}} \text{ W}$$

$$\text{Average power (W)} = 21.2 \times \underset{\text{jump height (cm)}}{\underline{\hspace{1.5cm}}} + 23 \times \underset{\text{body mass (kg)}}{\underline{\hspace{1.5cm}}} - 1{,}393 = \underline{\hspace{1.5cm}} \text{ W}$$

Johnson equations (W):

$$\text{Peak power (W)} = 78.5 \times \underset{\text{jump height (cm)}}{\underline{\hspace{1.5cm}}} + 60.6 \times \underset{\text{body mass (kg)}}{\underline{\hspace{1.5cm}}} - 15.3 \times \underset{\text{height (cm)}}{\underline{\hspace{1.5cm}}} - 1{,}308 = \underline{\hspace{1.5cm}} \text{ W}$$

$$\text{Average power (W)} = 41.4 \times \underset{\text{jump height (cm)}}{\underline{\hspace{1.5cm}}} + 31.2 \times \underset{\text{body mass (kg)}}{\underline{\hspace{1.5cm}}} - 13.9 \times \underset{\text{height (cm)}}{\underline{\hspace{1.5cm}}} - 1{,}393 = \underline{\hspace{1.5cm}} \text{ W}$$

Power calculation summary:

Lewis equation = _____ W

Harman peak power equation = _____ W

Harman average power equation = _____ W

Sayer peak power equation = _____ W

Johnson peak power equation = _____ W

Johnson average power equation = _____ W

Static Jump Test Results

Trial	Flight time (s)	Vertical displacement calculation	=	Displacement (m)
1			=	
2			=	
3			=	
Mean			=	

Lewis power (W):

$$\text{Power (W)} = \sqrt{4.9} \times \underset{\text{weight (kg)}}{\rule{2cm}{0.4pt}} \times \sqrt{\underset{\text{jump height (cm)}}{\rule{2cm}{0.4pt}}} = \underline{\hspace{1.5cm}} \text{ W}$$

Sayer equation (W):

$$\text{Peak power (W)} = 60.7 \times \underset{\text{jump height (cm)}}{\rule{2cm}{0.4pt}} + 45.3 \times \underset{\text{body mass (kg)}}{\rule{2cm}{0.4pt}} - 2,055 = \underline{\hspace{1.5cm}} \text{ W}$$

Harman equations (W):

$$\text{Peak power (W)} = 61.9 \times \underset{\text{jump height (cm)}}{\rule{2.5cm}{0.4pt}} + 36 \times \underset{\text{body mass (kg)}}{\rule{2.5cm}{0.4pt}} + 1,822 = \underline{\hspace{1.5cm}} \text{ W}$$

$$\text{Average power (W)} = 21.2 \times \underset{\text{jump height (cm)}}{\rule{2.5cm}{0.4pt}} + 23 \times \underset{\text{body mass (kg)}}{\rule{2.5cm}{0.4pt}} - 1,393 = \underline{\hspace{1.5cm}} \text{ W}$$

Johnson equations (W):

$$\text{Peak power (W)} = 78.5 \times \underset{\text{jump height (cm)}}{\rule{2.5cm}{0.4pt}} + 60.6 \times \underset{\text{body mass (kg)}}{\rule{2.5cm}{0.4pt}} - 15.3 \times \underset{\text{height (cm)}}{\rule{1.5cm}{0.4pt}} - 1,308 = \underline{\hspace{1.5cm}} \text{ W}$$

$$\text{Average power (W)} = 41.4 \times \underset{\text{jump height (cm)}}{\rule{2.5cm}{0.4pt}} + 31.2 \times \underset{\text{body mass (kg)}}{\rule{2.5cm}{0.4pt}} - 13.9 \times \underset{\text{height (cm)}}{\rule{1.5cm}{0.4pt}} - 1,393 = \underline{\hspace{1.5cm}} \text{ W}$$

Power calculation summary:

Lewis equation = _____ W

Harman peak power equation = _____ W

Harman average power equation = _____ W

Sayer peak power equation = _____ W

Johnson peak power equation = _____ W

Johnson average power equation = _____ W

To calculate the eccentric utilization ratio, use the following:

Vertical displacement	$\dfrac{\rule{4cm}{0.4pt}}{\text{Countermovement jump (cm)}} \ / \ \dfrac{\rule{4cm}{0.4pt}}{\text{static jump (cm)}} = \rule{3cm}{0.4pt}$	
Lewis equation	$\dfrac{\rule{4cm}{0.4pt}}{\text{Countermovement jump (cm)}} \ / \ \dfrac{\rule{4cm}{0.4pt}}{\text{static jump (cm)}} = \rule{3cm}{0.4pt}$	
Harman peak power equation	$\dfrac{\rule{4cm}{0.4pt}}{\text{Countermovement jump (cm)}} \ / \ \dfrac{\rule{4cm}{0.4pt}}{\text{static jump (cm)}} = \rule{3cm}{0.4pt}$	
Harman average power equation	$\dfrac{\rule{4cm}{0.4pt}}{\text{Countermovement jump (cm)}} \ / \ \dfrac{\rule{4cm}{0.4pt}}{\text{static jump (cm)}} = \rule{3cm}{0.4pt}$	
Sayer peak power equation	$\dfrac{\rule{4cm}{0.4pt}}{\text{Countermovement jump (cm)}} \ / \ \dfrac{\rule{4cm}{0.4pt}}{\text{static jump (cm)}} = \rule{3cm}{0.4pt}$	
Johnson peak power equation	$\dfrac{\rule{4cm}{0.4pt}}{\text{Countermovement jump (cm)}} \ / \ \dfrac{\rule{4cm}{0.4pt}}{\text{static jump (cm)}} = \rule{3cm}{0.4pt}$	
Johnson average power equation	$\dfrac{\rule{4cm}{0.4pt}}{\text{Countermovement jump (cm)}} \ / \ \dfrac{\rule{4cm}{0.4pt}}{\text{static jump (cm)}} = \rule{3cm}{0.4pt}$	

POWER ENDURANCE

EQUIPMENT

- Area in which to perform a dynamic warm-up
- Switch mat
- Calculator
- Individual and group data sheets
- Microsoft Excel or equivalent spreadsheet program

Find the group data sheets for this laboratory online at www.HumanKinetics.com/ LaboratoryManualForExercisePhysiology.

WARM-UP

This test can be performed only on a force platform or switch mat system. As with other vertical jump tests, this test requires an area with sufficient space and ceiling clearance to provide for safety. In addition, it is essential to have the subject perform a jump-specific warm-up that prepares him or her for maximal repetitive jumping in order to ensure the accuracy of the testing protocol. Begin with a general warm-up consisting of activities such as jogging, cycling, or rope jumping, then have the subject move on to a dynamic stretching routine consisting of activities that target the lower body (e.g., body-weight squats, squat thrusts, walking lunges, butt kicks, high knees, walking knee tucks). Next, have the subject complete a warm-up tailored to the vertical jump that allows him or her to practice a series of countermovement vertical jumps (figure 12.1, p. 308) with hands on hips. A summary of this warm-up procedure can be found in table 12.7.

Table 12.7 Specific Warm-Up for Multiple Vertical Jump

Time (min:s)	Activity	
0:00–5:00	General warm-up	Activities such as jogging, cycling, or rope jumping
5:00–10:00	Dynamic stretching warm-up	Activities such as leg swings, high knees, walking knee tucks, walking lunges, butt kicks, inchworms, and power skips
10:00–15:00	Warm-up specific to vertical jump	Series of countermovement jumps that allow the jumper to practice jumping positions and motions
15:00–17:00	Recovery period	Period of active rest in which the subject prepares for the all-out jumping tasks

POWER ENDURANCE JUMP TEST

Step 1: Prep the test area by setting up the switch mat or force plate according to the manufacturer's standards.

Step 2: Have the subject remove his or her shoes, then measure his or her height and weight. Record these values in the appropriate locations on the individual data sheet. Have the subject put his or her shoes back on.

Step 3: Direct the subject through a structured warm-up that prepares him or her for multiple jumping tasks (table 12.7).

Step 4: Explain the power endurance test to the subject. Make sure to tell the subject that he or she is to perform as many countermovement vertical jumps as possible in the 1-minute time frame. Make it clear that the objective also includes jumping as high as possible on each jump.

Step 5: Explain that the subject should bend his or her knees to about 90° during each contact phase of the jumping.

Step 6: Initiate the test by having the subject stand on the force platform or switch mat with his or her hands on hips. Countdown with a "3, 2, 1, jump" sequence. When you say "jump", the subject should jump as high as possible, then continue doing so until 1 min has elapsed.

Step 7: Count the number of jumps completed by the subject in 15 s, 30 s, and 60 s and record these values in the appropriate locations on the individual data sheet.

Step 8: Summate and record the flight times achieved during the 60-second test.

Step 9: Calculate the individual's mechanical power with the following equation:

$$\text{Mechanical power } (W \cdot kg^{-1}) = \frac{g^2 \times T_f \times 60}{4 \times n \times (60 - T_f)}$$

where $g = 9.81$ m $\cdot$ s^{-2}, T_f = total flight time, and n = number of jumps. Record the resulting information on the individual data sheet.

QUESTION SET 12.4

1. How does your mechanical power output compare with that of your classmates? With the norms presented in tables 12.8 and 12.9?

2. How does your mechanical power output for the 15 s, 30 s, and 60 s intervals compare with each other? Is this what you would expect to happen? Give a physiological explanation for this finding.

3. Using Excel, graph the mechanical power outputs for yourself and for the class at 15 s, 30 s, and 60 s. (Hint: Use time as the x-axis and power as the y-axis.)

4. What bioenergetic systems are targeted in this test?

Find the case studies for this laboratory online at www.HumanKinetics.com/ LaboratoryManualForExercisePhysiology.

Table 12.8 Norms for Power Endurance Jump Tests Lasting 60 s

Classification	Number of jumps in 60 s	Power output ($W \cdot kg^{-1}$)
Male basketball players	56.8 ± 4.3	19.8 ± 2.2
Male volleyball players	50.8 ± 2.7	19.6 ± 2.6
Adolescent males	63.2 ± 5.8	22.2 ± 1.8
Female athletes		12.2 ± 2.4
Male athletes		17.8 ± 2.7

Data from Bosco et al. (7), Hespanhol et al. (22), and Sands et al. (48).

Table 12.9 Percentile Ranked Norms for Power Endurance Jump Tests Lasting 60 s

Percentile rank	Male		Female	
	Absolute (W)	Relative ($W \cdot kg^{-1}$)	Absolute (W)	Relative ($W \cdot kg^{-1}$)
95	2,385	29.85	961	15.32
90	1,556	19.90	885	13.46
85	1,481	18.80	848	13.34
80	1,464	17.80	810	13.26
75	1,395	17.35	746	12.80
70	1,367	17.30	740	12.52
65	1,309	16.35	730	11.94
60	1,267	16.10	723	11.80
55	1,249	16.05	705	11.60
50	1,223	15.90	703	11.60
45	1,203	15.55	698	11.42
40	1,172	15.30	667	11.04
35	1,140	15.30	639	10.76
30	1,120	15.10	623	10.08
25	1,101	14.70	619	9.70
20	1,083	14.70	594	9.52
15	1,060	14.45	583	9.40
10	986	14.10	547	9.20
5	922	12.50	470	8.48
Mean	1,289	16.69	700	11.47

Reprinted, by permission, from P. Maud and C. Foster, 2005, *Physiological assessment of human fitness*, 2nd ed. (Champaign, IL: Human Kinetics), 229.

Laboratory Activity 12.4

Name or ID number: _____   Tester: _____

Date: _____ Time: _____

Sex: M / F (circle one)   Age: _____ y

Height: _____ in.   _____ cm

Weight: _____ lb   _____ kg

Temperature: _____ °F   _____ °C

Relative humidity: _____ %

Barometric pressure: _____ mmHg

Location of Testing
- ❏ Outdoor field
- ❏ Indoor field
- ❏ Indoor track
- ❏ Outdoor track
- ❏ Gym
- ❏ Other

Footwear
- ❏ Jogging shoe
- ❏ Walking shoe
- ❏ Tennis shoe
- ❏ Basketball shoe
- ❏ Running shoe
- ❏ Cross-trainer
- ❏ Other _____

Calculation of Mechanical Power

	60 s time interval	30 s time interval	15 s time interval
Number of jumps completed (n)			
Total flight time (T_f)			

Using the following equations, calculate the subject's mechanical power.

Calculation of mechanical power over 60 s:

$$\text{Mechanical power } (\text{W} \cdot \text{kg}^{-1}) = \frac{g^2 \times T_f \times 60}{4 \times n \times (60 - T_f)}$$

$$\frac{9.81 \times \underset{T_f}{\underline{\hspace{1cm}}} \times 60}{4 \times \underset{n}{\underline{\hspace{1cm}}} \times (60 - \underset{T_f}{\underline{\hspace{1cm}}})} = \underset{\text{mechanical power} (\text{W} \cdot \text{kg}^{-1})}{\underline{\hspace{2cm}}}$$

Calculation of mechanical power over 30 s:

$$\text{Mechanical power } (W \cdot kg^{-1}) = \frac{g^2 \times T_f \times 30}{4 \times n \times (30 - T_f)}$$

$$\left. 9.81 \times \underline{}_{T_f} \times 30 \middle/ 4 \times \underline{}_{n} \times (30 - \underline{}_{T_f}) \right) = \underline{}_{\text{mechanical power} (W \cdot kg^{-1})}$$

Calculation of mechanical power over 15 s:

$$\text{Mechanical power } (W \cdot kg^{-1}) = \frac{g^2 \times T_f \times 15}{4 \times n \times (15 - T_f)}$$

$$\left. 9.81 \times \underline{}_{T_f} \times 15 \middle/ 4 \times \underline{}_{n} \times (15 - \underline{}_{T_f}) \right) = \underline{}_{\text{mechanical power} (W \cdot kg^{-1})}$$

ANAEROBIC CYCLING POWER

EQUIPMENT

- Monark cycle or equivalent ergometer equipped to perform a WAnT (e.g., Monark 814E, 824E, and 834E, all of which contain a peg or basket upon which the external load can be applied.
- Revolutions counter that calculates the number of pedal revolutions performed in the test
- Timing apparatus (e.g., laboratory set-timer, stopwatch, or timer attached to the cycle ergometer)
- Calibrated platform-beam scale or electronic digital scale
- Calculator
- Individual and group data sheets
- Microsoft Excel or equivalent spreadsheet program

 Find the group data sheets for this laboratory online at www.HumanKinetics.com/ LaboratoryManualForExercisePhysiology.

LABORATORY ORIENTATION

The WAnT should be performed with a cycle ergometer that can immediately apply an accurate constant force throughout the duration of the test. It is essential to select an appropriate cycle ergometer in order to maximize the accuracy of the test. Typically, testers choose a Monark cycle, or an equivalent, in which a basket device is used to apply the resistance load (figure 12.7). The test should be performed in a well-ventilated area. The basic testing protocol is presented in table 12.10, which outlines four basic phases: warm-up, recovery, 30 s WAnT, and cool-down.

Figure 12.7 Monark cycle ergometer.

Table 12.10 Basic Wingate Procedure

Step		Duration	Activity
1	Warm-up	4–5 min	Cycle at a comfortable pace (about 60–70 rev · min⁻¹) against a resistance equal to 20% of the calculated load for the subsequent test. At the end of each minute, perform a 4- to 6-second sprint; resistance should increase with each sprint until the last sprint is performed against the prescribed load.
2	Recovery	2–5 min	Rest or cycle slowly against a minimal resistance.
3	Wingate test	5–10 s	Pedal against 1/3 of the prescribed resistance at 20–50 rev · min⁻¹.
		2–5 s	Increase pedaling rate to maximum; once maximal pedaling rate is achieved, the prescribed resistance should be applied.
		30 s	Cycle at the highest possible rev · min⁻¹ against the prescribed resistance.
4	Cool-down	2–5 min	Pedal at a low to moderate power level (25–100 W) for 2–5 minutes. If repeated tests are performed, the cool-down should be extended.

Adapted with permission from *Research Quarterly for Exercise and Sport*, Vol. 60, No. 2, 144-151, Copyright 1989 by the American Alliance for Health, Physical Education, Recreation and Dance, 1900 Association Drive, Reston, VA 20191.

- *Warm-up*: As with all testing protocols, a proper warm-up is essential to minimizing injury risk and maximizing performance. The WAnT includes a standardized warm-up in which the subject cycles at about 60 to 70 rev · min⁻¹ against 20% of the calculated resistance load for a total of 4 to 5 min. At the end of each minute, the subject performs an all-out sprint lasting 4 to 6 s. The load increases with each minute until the last 4 to 5 s sprint is performed with the target load.

- *Recovery*: After completing the warm-up, the subject should be given 2 to 5 min to recover. This time is generally spent cycling against minimal resistance.

- *30-second WAnT*: The lead-in to the WAnT consists of two phases that allow the subject to progressively increase his or her cycling rev · min⁻¹ . The first phase, lasting 5 to 10 s, requires the subject to pedal at 20 to 50 rev · min⁻¹ against 1/3 of the prescribed resistance. The second phase requires the subject to increase the pedaling rate to maximum, which generally takes between 2 and 5 s. The prescribed load is then applied while the subject cycles at the highest possible rev · min⁻¹ until the completion of the 30 s time period.

- *Cool-down*: After completing the WAnT, it is essential that the subject perform a cool-down period lasting 2 to 5 min. During this time, the subject should pedal at 25 to 100 W in order to facilitate recovery from the testing bout.

Thus the basic structure of the WAnT is standardized, but it can be modified depending upon the individual needs of the subject. Some typical modifications include lengthening the duration to 60 s or reducing the test duration to 5, 10, or 15 s. One can also modify the resistance used during the test.

If the test is not performed with a computerized cycle ergometer, it requires four testers: timer, force setter, counter, and recorder. The timer initiates the test by yelling "go," maintains control of the testing clock, shouts out the time at 5 s intervals during the test, and yells "stop" when the 30 s time limit is reached (2). The resistance setter establishes the prescribed load, makes sure the load is applied at the right time and is maintained, and lowers the load while the subject performs the cool-down.

The counter's main job is to quantify the number of complete pedal revolutions during each 5 s test interval. A complete pedal revolution involves full rotation of the pedal from its originating position. The counter relays the number of pedal revolutions for each period to the recorder, who writes the number of revolutions down on an individual data sheet throughout the duration of the test. Though this manual process is functional, computers make the testing process much simpler (45), more accurate (thanks to automatic counts of the number of revolutions), and doable with fewer testers.

30-SECOND WAnT

Step 1: Set up the cycle ergometer and check to see if it is in working order.

Step 2: Measure the subject's height and weight without shoes.

Step 3: Calculate the prescribed load based on the subject's training status (see table 12.2). Record this load on the individual data sheet for laboratory activity 12.5.

Step 4: Fit the cycle ergometer to the subject so that while the subject is seated on the bike his or her extended leg has a slight bend (5–15°) (see figures 4.3a [p. 128] for leg angle and 12.8 for basic body position).

Figure 12.8 Body position used in the WAnT.

Step 5: Completely explain the test protocol to the subject. Emphasize the fact that this is an all-out effort lasting 30 s.

Step 6: Have the subject perform the standardized warm-up and recovery periods outlined in table 12.10.

Step 7: Initiate the test with the subject pedaling against 1/3 of the prescribed load at 20 to 50 rev · min⁻¹ for 5 to 10 s.

Step 8: Instruct the subject to increase his or her pedaling rate to maximum. Instruct the force setter to apply the load and the timer to start the timer and simultaneously yell "go" when the subject reaches a maximal pedaling rate.

Step 9: Upon hearing the "go" command, the counter should begin counting pedal revolutions (this step is not performed if you are using a computerized system).

Step 10: The counter should tell the recorder the number of revolutions completed at the end of each 5 s time interval during the test (i.e., at 5 s, then at 10, 15, 20, 25, and 30 s).

Step 11: At the 30 s mark, the timer yells "stop," and the force setter reduces the load so that the subject can pedal at 25 to 100 W for 2 to 5 min or until he or she has recovered. At this time, the counter tells the recorder the number of revolutions completed for the final 5 s interval.

Step 12: Convert the prescribed load to newtons by multiplying the load by 9.80665. Calculate the power output from the total number of revolutions, the distance the flywheel travels per rotation (6 m if using a Monark Cycle Ergometer), and the duration of the interval. Calculate the power output for each time interval and record it in the appropriate location on the individual data sheet.

Step 13: Divide the power output calculated for each time interval by body weight in order to determine a relative time interval. Record these values on the individual data sheet.

Step 14: Calculate the absolute and relative work accomplished during the 30 s time interval and record these values on the individual data sheet.

Step 15: Calculated the absolute and relative mean power outputs accomplished during the 30 s test and record these data on the individual data sheet.

Step 16: Calculate the fatigue index by using the equation supplied on the individual data sheet.

QUESTION SET 12.5

1. What energy system contributes to your ability to perform the WAnT? Bioenergetically, what limits your performance during the first 10 s, in the period from 0 to 20 s, and in the period from 0 to 30 s?

2. What is the typical force setting for the WAnT for an untrained person? For an athlete? Would different force values impact the results of the test? If so, how?

3. How does one calculate the fatigue index, and what does it indicate?

4. During which time interval was your highest power output achieved?

5. How does your absolute peak power output compare with that of the class average and with the norms presented in table 12.11?

6. Using Excel, graph the power outputs achieved in each of the 5 s time intervals.

7. What was your fatigue index, and how does it compare with the class results?

8. Are there differences in performance results on the WAnT between males and females? If so, are these differences related to the peak power, mean power, or fatigue rate? Explain your answer.

9. How would the performance of someone with a high percentage of type I fibers compare with that of a person with a high percentage of type II fibers? What would you expect to see in terms of performance?

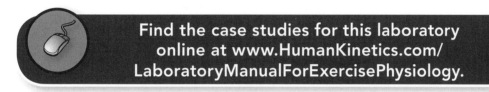

Find the case studies for this laboratory online at www.HumanKinetics.com/ LaboratoryManualForExercisePhysiology.

Table 12.11 Percentile Ranks for College-Age Males and Females for the Wingate Anaerobic Test

Percentile rank	College-age males (18–28 y; n = 60)		College-age females (18–28 y; n = 69)	
	W	W·kg⁻¹	W	W·kg⁻¹
90	662	8.2	470	7.3
80	618	8.0	419	7.0
70	600	7.9	410	6.8
60	577	7.6	391	6.6
50	565	7.4	381	6.4
40	548	7.1	367	6.1
30	530	7.0	353	6.0
20	496	6.6	337	5.7
10	471	6.0	306	5.3

Norm data for WAnT test performed on a Monark cycle ergometer with a resistance of 0.075 kp/ kg body mass. $W \cdot kg^{-1}$ = watts /kilogram body mass

Reprinted, by permission, from J.R. Hoffman, 2006, *Norms for fitness, performance, and health* (Champaign, IL: Human Kinetics), 54. Adapted with permission from *Research Quarterly for Exercise and Sport*, Vol. 60, No. 2, 144-151, Copyright 1989 by the American Alliance for Health, Physical Education, Recreation and Dance, 1900 Association Drive, Reston, VA 20191.

Name or ID number: _____   Date: _____

Tester: _____   Time: _____

Sex: M / F (circle one)   Age: _____ y   Height: _____ in. _____ cm

Temperature: _____ °F _____ °C   Weight: _____ lb _____ kg

Barometric pressure: _____ mmHg   Relative humidity: _____ %

Determination of Prescribed Load

Prescribed load = _____ × _____ = _____ kp
                     body weight (kg)   kp per kg body mass

Note: In most instances, the multiplier should be 0.075 kp/kg body mass.

Force (N) = _____ × 9.80665 = _____ (N)
              force setting in kp

Quantification of Pedal Revolutions

Time interval	Revolutions
0–5 s	
5–10 s	
10–15 s	
15–20 s	
20–25 s	
25–30 s	

Power

Time interval	Peak power
0–5 s	Peak power (W) = (_____ × _____ × 6 m) / 5 s = _____ W force (N) revolutions
5–10 s	Peak power (W) = (_____ × _____ × 6 m) / 5 s = _____ W force (N) revolutions
10–15 s	Peak power (W) = (_____ × _____ × 6 m) / 5 s = _____ W force (N) revolutions
15–20 s	Peak power (W) = (_____ × _____ × 6 m) / 5 s = _____ W force (N) revolutions
20–25 s	Peak power (W) = (_____ × _____ × 6 m) / 5 s = _____ W force (N) revolutions
25–30 s	Peak power (W) = (_____ × _____ × 6 m) / 5 s = _____ W force (N) revolutions

Time interval	Relative power (W · kg⁻¹)
0–5 s	Relative peak power = _____ / _____ = _____ W · kg⁻¹ peak power (W) body mass (kg)
5–10 s	Relative peak power = _____ / _____ = _____ W · kg⁻¹ peak power (W) body mass (kg)
10–15 s	Relative peak power = _____ / _____ = _____ W · kg⁻¹ peak power (W) body mass (kg)
15–20 s	Relative peak power = _____ / _____ = _____ W · kg⁻¹ peak power (W) body mass (kg)
20–25 s	Relative peak power = _____ / _____ = _____ W · kg⁻¹ peak power (W) body mass (kg)
25–30 s	Relative peak power = _____ / _____ = _____ W · kg⁻¹ peak power (W) body mass (kg)

Absolute peak power output   = _____ W

Relative peak power output   = _____ W · kg⁻¹

Absolute lowest power output = _____ W

Work

Time Interval	Work
0–30 s	Total work = (_____ × _____ × 6 m) = _____ J force (N) revolutions
0–30 s	Relative work = (_____ / _____ × 6 m) = _____ J · kg⁻¹ total work (J) body mass (kg)

Mean Power Output

Time interval	Mean power output
0–30s	Mean power output = _____ / 30 s = _____ W total work (J)
0–30 s	Relative mean power output = _____ / _____ = _____ W · kg⁻¹ mean power (W) body mass (kg)

Fatigue Index

Lowest power output   = _____ W

Highest power output = _____ W

Calculating the Fatigue Index

Fatigue index = [(_____ – _____) / _____] × 100 = _____ %

 highest power (W)   lowest power (W)   highest power (W)

THE MARGARIA-KALAMEN STAIR CLIMB TEST

EQUIPMENT

- Staircase with a 6 m run-up and a minimum of 9 stairs that are each 174 to 175 cm in height
- Two contact mats interfaced with a computer
- Calibrated beam scale or electronic digital scale
- Tape to mark start line
- Measuring tape
- Stopwatch
- Calculator
- Individual and group data sheets
- Microsoft Excel or equivalent spreadsheet program

Find the group data sheets for this laboratory online at www.HumanKinetics.com/ LaboratoryManualForExercisePhysiology.

LABORATORY ORIENTATION

When performing the Margaria-Kalamen stair test, it is essential to select an appropriate staircase. The stairs must have a 6 m (20 ft) run-up and a minimum of 9 stairs, each of which is 174 to 175 cm in height (figure 12.5, p. 316). The stairway should be well ventilated.

The basic testing protocol involves the following distinct steps: (1) warm-up, (2) practice trials, and (3) testing (figure 12.9). As with other anaerobic testing protocols, the subject should perform a warm-up that begins with 5 min of general warm-up activities (e.g., jogging, cycling, or rope jumping), then moves on to 5 min of dynamic stretching activities (e.g., leg swings, high knees, walking knee tucks, walking lunges, butt kicks, inchworms, power skips). After completing the

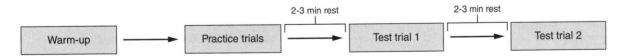

Figure 12.9 Basic procedure for conducting the Margaria-Kalamen stair test.

Table 12.12 Normative Values for the Margaria-Kalamen Stair Sprint Test

Classification	15–20 y		20–30 y		30–40 y		40–50 y		>50 y	
	W	kg·m·s⁻¹	W	kg·m·s⁻¹	W	kg·m·s⁻¹	W	kg·m·s⁻¹	W	kg·m·s⁻¹
Men										
Excellent	>2,197	>224	>2,059	>210	>1,648	>168	>1,226	>125	>961	>98
Good	1,840	188	1,722	176	1,379	141	1,036	106	810	83
Average	1,839	188	1,721	175	1,378	141	1,035	106	809	82
Fair	1,466	149	1,368	139	1,094	112	829	85	642	65
Poor	<1,108	<113	<1,040	<106	<834	<85	<637	<65	<490	<50
Women										
Excellent	>1,789	>182	>1,648	>168	>1,226	>125	>961	>98	>736	>75
Good	1,487	152	1,379	141	1,036	106	810	83	604	62
Average	1,486	152	1,378	141	1,035	106	809	82	603	61
Fair	1,182	121	1,094	112	829	85	642	65	476	49
Poor	<902	<92	<834	<85	<637	<65	<490	<50	<373	<38

Adapted, by permission, from E. Fox, R. Bowers and M. Foss, 1993, *The physiological basis for exercise and sport*, 5th ed. (Dubuque, IA: Wm C. Brown), 676. ©The McGraw-Hill Companies.

general warm-up, the subject should perform several practice trials of the test in order to develop the timing necessary to maximize the accuracy of the test (24). After mastering the technique of running up the stairs 3 steps at a time, the subject is ready to perform the test. Wait 2 to 3 min after the completion of the practice trials, then have the subject perform the stair test twice with a period of 2 or 3 min between the tests.

MARGARIA-KALAMEN STAIR CLIMB TEST

Step 1: Select a staircase that meets the requirements of a 6 m run-up and a minimum of 9 stairs (figure 12.10).

Step 2: Measure the height of each stair and calculate the displacement between the 3rd and 9th stairs.

Step 3: Place the switch mats on steps 3 and 9 (figure 12.6) and put a mark on the floor 6 m (20 ft) from the bottom of the first step. (Note: If a switch mat is unavailable, use a person with a stopwatch to determine this time.)

Step 4: Measure the subject's height and weight without shoes. Record these values on the individual data sheet.

Figure 12.10 Margaria-Kalmen test, running.

Step 5: Direct the subject through a general 10 min warm-up.

Step 6: Have the subject perform several practice runs up the stairs in order to learn how to step on every third step.

Step 7: Initiate the first test 2 to 3 min after completing the last practice run.

Step 8: Instruct the subject to run up the stairs as quickly as possible.

Step 9: The timer will start when the subject steps on the switch mat on step 3 and stop when he or she steps on step 9. (Note: If using a stopwatch, the timer should start the watch when the subject steps on step 3 and stop it when the subject crosses step 9).

Step 10: Record the time on the individual data sheet.

Step 11: Repeat the test 2 to 3 min after completing the first trial. Record the second time value on the data sheet.

Step 12: Calculate power with the following equation:

$$\text{Power}\,(\text{kg} \cdot \text{m} \cdot \text{s}^{-1}) = \frac{\text{weight (kg)} \times \text{distance (m)}}{\text{time (s)}}$$

Step 13: Convert the power value to watts by multiplying by 9.807. Record this value on the individual data sheet.

Step 14: Calculate the relative power by dividing absolute power by body weight. Record this value on the individual data sheet.

QUESTION SET 12.6

1. Explain how power is calculated in the Margaria-Kalamen stair test.

2. How does your power output compare with those presented in the norm tables and the averages for the class?

3. Were there any differences be-tween your individual power outputs for trial 1 and 2? If so, why might this be? If not, why is there no difference?

4. Using Excel, graph the power outputs in W and W·kg⁻¹.

5. Are there any differences between the results for the men and the women in the class for absolute power (W) or relative power (W·kg⁻¹)? What might be the physiological explanation for this?

6. What are some pros and cons of performing this test?

Find the case studies for this laboratory online at www.HumanKinetics.com/ LaboratoryManualForExercisePhysiology.

Laboratory Activity 12.6 Individual Data Sheet

Name or ID number: _____ Date: _____

Tester: _____ Time: _____

Sex: M / F (circle one)   Age: _____ y Height: _____ in. _____ cm

Temperature: _____ °F _____ °C Weight: _____ lb _____ kg

Barometric pressure: _____ mmHg Relative humidity: _____ %

Measurement of Step Height and Time Between 3rd and 9th Steps

Step		Height (m)
1	=	
2	=	
3	=	
4	=	
5	=	
6	=	
Total vertical distance	=	
Trial		**Time (s)**
1	=	
2	=	
Average	=	

Calculation of Power

$$\text{Power}\,(\text{kg} \cdot \text{m} \cdot \text{s}^{-1}) = \frac{\text{weight}\,(\text{kg}) \times \text{distance}\,(\text{m})}{\text{time}\,(\text{s})}$$

Trial 1: Use the time from trial 1 to perform these calculations.

Power (kg·m·s⁻¹) = _____ × _____ / _____ = _____
$\qquad\qquad\qquad\quad$ body weight (kg) $\quad$ distance (m) $\quad$ time (s) $\quad$ power (kg·m·s⁻¹)

Power (W) = _____ × 9.807 = _____
$\qquad\qquad$ power (kg·m·s⁻¹) $\qquad\qquad$ power (W)

Power (W·kg⁻¹) = _____ / _____ = _____
$\qquad\qquad\qquad$ power (W) $\quad$ body weight (kg) $\quad$ power (W·kg⁻¹)

Laboratory Activity 12.6

Trial 2: Use the time from trial 2 to perform these calculations.

Power (kg·m·s⁻¹) = _____ × _____ / _____ = _____
                     body weight (kg)   distance (m)   time (s)   power (kg·m·s⁻¹)

Power (W) = _____ × 9.807 = _____
               power (kgm·s⁻¹)   power (W)

Power (W·kg⁻¹) = _____ / _____ = _____
                    power (W)   body weight (kg)   power (W·kg⁻¹)

Average of trials 1 and 2: Use the average time from trials 1 and 2 to perform these calculations.

Power (kg·m·s⁻¹) = _____ × _____ / _____ = _____
                     body weight (kg)   distance (m)   time (s)   power (kg·m·s⁻¹)

Power (W) = _____ × 9.807 = _____
               power (kg·m·s⁻¹)   power (W)

Power (W·kg⁻¹) = _____ / _____ = _____
                    power (W)   body weight (kg)   power (W·kg⁻¹)

Laboratory Activity 12.6

Pulmonary Function Testing

Objectives

1. Introduce the terms associated with respiratory function.
2. Execute several pulmonary function tests (PFT).
3. Compare your values with normative values for similar individuals (in terms of age and sex).
4. Identify pulmonary disease states and diagnoses.

Definitions

chronic obstructive pulmonary disease (COPD)—Diseases of the lungs characterized by reduced air flow rates.

exercise-induced asthma—Diseases of the lungs characterized by reduced air flow rates during or immediately after exercise.

pulmonary function test (PFT)—Test that measures lung function and encapsulates both volumes and flow rates.

restrictive diseases—Diseases of the lungs characterized by reduced total lung capacities or volumes.

spirometry—Measurement of breath, or lung volume.

tidal volume (V_T)—The amount of air moved per breath.

ventilation ($\dot{V}_E$)—Amount of air moved into the pulmonary system; expressed in $L \cdot min^{-1}$.

Misconceptions abound about our breathing, or **ventilation ($\dot{V}_E$)**. Of course, ventilation is necessary for the exchange of gases between the atmosphere and our metabolism. We use oxygen (O_2) from the atmosphere as the final electron acceptor in the electron transport chain. We must also expire carbon dioxide (CO_2), since this carbon is the result of macronutrient combustion (refer to lab 5). During exercise, O_2 consumption and CO_2 production increase; therefore, we need to increase our ventilation. As you learned in labs 9 and 10, we ventilate much more at higher workloads above the ventilatory threshold. This is the result of greater CO_2 production at higher intensities from the buffering of lactate (see labs 9 and 10), or in other words, respiratory compensation for metabolic acidosis. Thus our lungs act to regulate acid–base balance.

At normal altitudes, our ventilation is more sensitive to the production of CO_2 than it is to the consumption of O_2. Very small increases in PCO_2 results in a proportional increase in $\dot{V}_E$, whereas PO_2 can decrease significantly with little effect on the stimulation of $\dot{V}_E$ (see figure 13.1, *a* and *b*). At high altitudes (>2,000 m or 6,500 ft), the reduced partial pressure of O_2 (which can lead to hypoxia) is capable of stimulating greater amounts of ventilation. Not only do changes in CO_2 and O_2 stimulate us to breathe differently; we also have voluntary control over our breathing.

The increased need for ventilation during exercise as the result of increased CO_2 production and O_2 consumption leads us as exercise physiologists to be interested in its measurement. We move atmospheric gases into our lungs by creating negative pressures in our thoracic cavity.

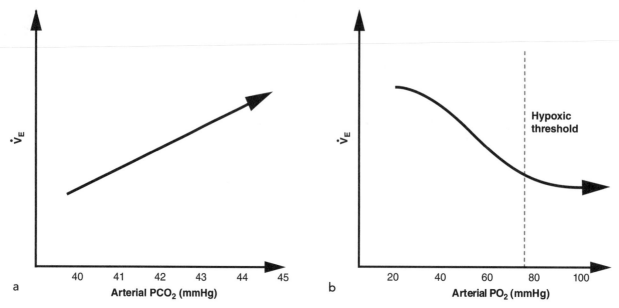

Figure 13.1 The effect of (a) arterial PCO_2 and (b) PO_2 on ventilation.

This is accomplished by the skeletal muscles of inspiration (diaphragm, intercostals) and expiration (intercostals, abdominals), which change the volume within the thoracic cavity—the greater the change in volume, the bigger our breath, or **tidal volume (V_T)**. Ventilation is, therefore, a function of breathing frequency (bf) multiplied by tidal volume (V_T):

$$\dot{V}_E (L \cdot min^{-1}) = bf (b \cdot min^{-1}) \times V_T (L \cdot b^{-1})$$

Tidal volume is one of the measurements made during **spirometry**. Ventilation can be divided into distinct volumes and capacities, as seen in figure 13.2.

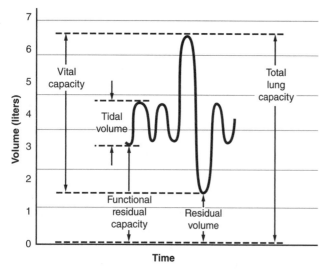

Figure 13.2 Lung volumes from spirometry.

Reprinted, by permission, from NSCA, 2012, *Essentials of personal training*, 2nd ed. (Champaign, IL: Human Kinetics), 25.

Pulmonary Function Testing

A **pulmonary function test (PFT)** can measure the amount (volume) and speed (flow rate) of inspiration and expiration. PFTs can be used to identify the health and capacity of the pulmonary system. Most PFTs can measure tidal volume (V_T), as well as the following:

- **forced vital capacity (FVC)**—Maximum volume forcibly expired after maximum inspiration (in liters).

- **forced expiratory volume (FEV$_{1.0}$)**—Volume of air exhaled in the first second after maximal inhalation; used as a diagnostic tool for limitations in flow rates (expressed in liters but by the definition $L \cdot s^{-1}$).

- **FEV$_{1.0}$/FVC ratio**—Ratio of FEV$_{1.0}$ to FVC, which is a common measure of pulmonary disease (see later in this chapter); in healthy individuals, about 70% to 85% (ratio reduced in COPD patients but may be normal in restrictive diseases).

- **peak expiratory flow (PEF)**—Maximum expiratory flow during a forced expiration from the point of maximum inspiration (total lung capacity); expressed in $L \cdot min^{-1}$ or $L \cdot s^{-1}$ and used to provide a measure of airway caliber (diameter) and airflow (yet is dependent not only on airway caliber but also on lung elastic recoil, patient effort, and patient cooperation); generally less specific than FEV$_{1.0}$ as a diagnostic measure.

- **maximum voluntary ventilation (MVV)**—Maximal amount of air expired in one minute ($L \cdot min^{-1}$).
- **maximum exercise ventilation ($\dot{V}_E max$)**—Maximal volume expired during maximal exercise ($L \cdot min^{-1}$).
- **residual volume (RV)**—Amount of air remaining in the lungs following a maximal expiration (in liters).
- **total lung capacity (TLC)**—Vital capacity and residual volume combined for the total volume of the lungs (in liters).

Lung volumes are largely a function of age, height, and sex. Physical fitness does not significantly affect lung size per se, but it may improve measurement of flow rates, though this idea may be challenged by some evidence concerning swimmers (4, 22).

Volumes for PFT results are usually expressed in terms of BTPS (body temperature and pressure, saturated). You may recall that volumes of gases are dependent on the temperature, barometric pressure, and level of humidity, or water saturation. As a result, expression of gas volumes needs to have a systematic unit of expression. Since the air in our lungs is at body temperature, ambient pressure, and nearly 100% humidity, the units of BTPS are common for PFT results. However, as you exhale air and it cools to the temperature of the room and loses humidity or condenses, its volume changes. Thus, it may be necessary to convert from ambient (or room) temperature, ambient pressure, saturated (ATPS). Your laboratory instructor can indicate whether this is necessary. The equation for converting ATPS to BTPS is as follows (where V_{ATPS} = volume of gas at ATPS, T_A = temperature in your lab, P_B = barometric pressure in your lab, and P_{H2O} = water vapor pressure at the T_A):

$$V_{BTPS} = V_{ATPS} \times [310 / (273 + T_A)] \times [(P_B - P_{H2O}) / (P_B - 47)]$$

PFT as a Tool for Diagnosing Pulmonary Disease

PFT results can be used in diagnoses of respiratory diseases, which are categorized into two different types (based on PFT results): obstructive and restrictive.

An obstructive disease is characterized by an acute or chronic obstruction in the bronchi leading to the alveoli. These obstructions are often referred to as **chronic obstructive pulmonary diseases (COPDs)**. Individuals with these diseases usually have normal lung volumes but restricted flow rates. COPDs can be diagnosed by testing the $FEV_{1.0}$, $FEV_{1.0}$/FVC ratio, PEF, or other flow rate measures. Examples of COPDs include asthma, chronic bronchitis, and emphysema. Asthma differs from the other two in that it can be temporary and reversible. Exercise-induced asthma is a temporary inflammatory response during exercise that obstructs ventilatory flow rates. Exercise-induced asthmatics may have normal flow rates at rest but produce a positive test (i.e., a 15% decrease from pre-exercise measured $FEV_{1.0}$) following 6 to 8 min of vigorous exercise at a target intensity of 85% to 90% of maximum heart rate.

A **restrictive disease** is characterized by reduced total lung capacities or volumes, such as TLC or vital capacity (VC), but the individual may have normal flow rates. Restrictive diseases can be diagnosed by measuring FVC or a slow VC. Examples of restrictive diseases include pulmonary fibrosis, scar tissue, and tumors.

Pulmonary Disease Classifications

For both obstructive and restrictive pulmonary diseases, PFT results are classified by the following comparisons with predicted results (for restrictive diseases, FVC is comparison; for COPDs, % predicted is for $FEV_{1.0}$) (23):

Good: >100% of predicted
Normal: 100% to 80% of predicted
Mild: 80% to 65% of predicted
Moderate: 65% to 50% of predicted
Moderately severe: 50% to 35% of predicted
Severe: <35% of predicted

Individuals with pulmonary disease have increased work of breathing and, if severe enough, limited O_2 delivery to working tissues, including the brain. Under these circumstances, exercise is extremely uncomfortable, and supplemental oxygen may be necessary for activities of daily living. Despite the discomfort of exercise for these individuals, it can alleviate many of their symptoms. For further information on pulmonary rehabilitation, refer to reference sources 10, 11, 27, as well as the following sources from Human Kinetics: 1, 3, 18, 19.

Respiratory Limitations on Exercise

In healthy individuals without respiratory disease classifications, lungs rarely limit function during rest, activities of daily living, or exercise performance. Many people believe that breathing hard during exercise (hyperpnea) means that our lungs are limiting performance. In reality, our lungs are extremely "overbuilt" (7, 8, 14). We know this because their job is to remove CO_2 and maintain O_2 delivery. We know that the lungs are succeeding since the pressure of O_2 in our arteries (PaO_2), the pressure of CO_2 in our arteries ($PaCO_2$), and hemoglobin saturation are maintained even during high intensities in the majority of individuals (24). Although exercise-induced hypoxemia (decrease in PaO_2 and hemoglobin saturation) can occur in some elite athletes (25, 26) and in healthy trained women (15, 16), it is a relatively rare circumstance. In order to diagnose exercise-induced hypoxemia, we also need to be able to measure blood gases and hemoglobin saturation.

A potential noninvasive way to estimate ventilatory limitation is to calculate the $\dot{V}_E$ max/MVV ratio $\times$ 100, which in normal healthy individuals falls somewhere between 50% and 70% (21). In fact, $\dot{V}_E$ max can be predicted by taking 72% of MVV. In addition, breathing reserve can be defined as 100% – ($\dot{V}_E$ max / MVV $\times$ 100) or, more simply, MVV – $\dot{V}_E$, and it can be used to indicate ventilatory limitation (21).

Although lungs typically succeed in maintaining PaO_2 and $PaCO_2$ in most individuals during exercise, respiratory musculature may contribute to fatigue in other ways. Respiratory muscles (diaphragm, intercostals, abdominals) are skeletal muscles subject to the same requirements of our exercising muscles—that is O_2 and fuel delivery—and to fatigue. It is possible that under certain conditions respiratory muscle failure or blood flow redirected toward respiratory muscle work, and thus away from exercising muscles, may contribute to fatigue or limit performance (3, 8, 13, 14, 17, 28, 30).

Though possible, it is generally believed that lung capacities themselves do not respond to exercise training; respiratory musculature, however, responds similarly to the working of skeletal muscles, which may allow a trained athlete to reach greater flow rates ($FEV_{1.0}$ and PEF) and be more resistant to fatigue. Respiratory muscles such as the diaphragm have tremendous oxidative capacities, and while they are not immune to fatigue it is more likely that another mechanism of fatigue would precede the failure of the respiratory musculature (3, 7, 9, 14).

References

1. American Association of Cardiovascular and Pulmonary Rehabilitation. *Guidelines for Pulmonary Rehabilitation Programs.* 4th ed. Champaign, IL: Human Kinetics, 2011.

2. American College of Sports Medicine. *ACSM's Exercise Management for Persons with Chronic Diseases and Disabilities.* 3rd ed. Champaign, IL: Human Kinetics, 2009.

3. Babcock MA, Pegelow DF, Harms CA, and Dempsey JA. Effects of Respiratory Muscle Unloading on Exercise-Induced Diaphragm Fatigue. *J Appl Physiol* 93: 201–206, 2002.

4. Black LF, Offord K, and Hyatt RE. Variability in the Maximal Expiratory Flow Volume Curve in Asymptomatic Smokers and in Nonsmokers. *Am Rev Respir Dis* 110: 282–292, 1974.

5. Clanton TL, Dixon GF, Drake J, and Gadek JE. Effects of Swim Training on Lung Volumes and Inspiratory Muscle Conditioning. *J Appl Physiol* 62: 39–46, 1987.

6. Crapo RO, Morris AH, and Gardner RM. Reference Spirometric Values Using Techniques and Equipment That Meet ATS Recommendations. *Am Rev Respir Dis* 123: 659–664, 1981.

7. Dempsey JA. Is the lung built for exercise? (JB Wolffe Memorial Lecture.) *Med Sci Sports Exer* 18: 143–155, 1986.

8. Dempsey JA, Harms CA, and Ainsworth DM. Respiratory Muscle Perfusion and Energetics During Exercise. *Med Sci Sports Exer* 28: 1123–1128, 1996.

9. Dempsey JA, Sheel AW, Haverkamp HC, Babcock MA, and Harms CA. Pulmonary system limitations to exercise in health. (John Sutton Memorial Lecture: CSEP, 2002.) *Can J Appl Physiol* 28 Suppl: S2–S24, 2003.

10. Ferreira SA, Guimaraes M, and Taveira N. Pulmonary Rehabilitation in COPD: From Exercise Training to "Real Life." *J Bras Pneumol* 35: 1112–1115, 2009.

11. Ghanem M, Elaal EA, Mehany M, and Tolba K. Home-Based Pulmonary Rehabilitation Program: Effect on Exercise Tolerance and Quality of Life in Chronic Obstructive Pulmonary Disease Patients. *Ann Thorac Med* 5: 18–25, 2010.

12. Hankinson JL, Crapo RO, and Jensen RL. Spirometric Reference Values for the 6-s FVC Maneuver. *Chest* 124: 1805–1811, 2003.

13. Harms CA, Babcock MA, McClaran SR, Pegelow DF, Nickele GA, Nelson WB, and Dempsey JA. Respiratory Muscle Work Compromises Leg Blood Flow During Maximal Exercise. *J Appl Physiol* 82: 1573–1583, 1997.

14. Harms CA and Dempsey JA. Does Ventilation Ever Limit Human Performance? In: Ward SA, ed., *The physiology and Pathophysiology of Exercise Tolerance*. New York: Plenum Press, 1996, pp. 91–96.

15. Harms CA, McClaran SR, Nickele GA, Pegelow DF, Nelson WB, and Dempsey JA. Effect of Exercise-Induced Arterial O_2 Desaturation on $\dot{V}O_2$max in Women. *Med Sci Sports Exer* 32: 1101–1108, 2000.

16. Harms CA, McClaran SR, Nickele GA, Pegelow DF, Nelson WB, and Dempsey JA. Exercise-Induced Arterial Hypoxaemiain Healthy Young Women. *J Physiol* 507 (Pt 2): 619–628, 1998.

17. Harms CA, Wetter TJ, St Croix CM, Pegelow DF, and Dempsey JA. Effects of Respiratory Muscle Work on Exercise Performance. *J Appl Physiol* 89: 131–138, 2000.

18. Jobin J, Maltais F, LeBlanc P, and Simard C. *Advances in Cardiopulmonary Rehabilitation*. Champaign, IL: Human Kinetics, 2000.

19. Jobin J, Maltais F, Poirier P, LeBlanc PJ, and Simard C. *Advancing the Frontiers of Cardiopulmonary Rehabilitation*. Champaign, IL: Human Kinetics, 2002.

20. Kory RC, Callahan R, Boren HG, and Syner JC. The Veterans Administration–Army Cooperative Study of Pulmonary Function. I. Clinical Spirometry in Normal Men. *Am J Med* 30: 243–258, 1961.

21. Melissant CF, Lammers JW, and Demedts M. Relationship Between External Resistances, Lung Function Changes and Maximal Exercise Capacity. *Eur Respir J* 11: 1369–1375, 1998.

22. Mickleborough TD, Stager JM, Chatham K, Lindley MR, and Ionescu AA. Pulmonary Adaptations to Swim and Inspiratory Muscle Training. *Eur J Appl Physiol* 103: 635–646, 2008.

23. Morris JF. Spirometry in the Evaluation of Pulmonary Function. *West J Med* 125: 110–118, 1976.

24. Powers SK, Dodd S, Criswell DD, Lawler J, Martin D, and Grinton S. Evidence for an Alveolar-Arterial PO2 Gradient Threshold During Incremental Exercise. *Int J Sports Med* 12: 313–318, 1991.

25. Powers SK, Lawler J, Dempsey JA, Dodd S, and Landry G. Effects of Incomplete Pulmonary Gas Exchange on $\dot{V}O_2$max. *J Appl Physiol* 66: 2491–2495, 1989.

26. Powers SK, Martin D, and Dodd S. Exercise-Induced Hypoxaemia in Elite Endurance Athletes. Incidence, Causes and Impact on $\dot{V}O_2$max. *Sports Med* 16: 14–22, 1993.

27. Riario-Sforza GG, Incorvaia C, Paterniti F, Pessina L, Caligiuri R, Pravettoni C, Di Marco F, and Centanni S. Effects of Pulmonary Rehabilitation on Exercise Capacity in Patients With COPD: A Number Needed to Treat Study. *Int J Chron Obstruct Pulmon Dis* 4: 315–319, 2009.

28. St Croix CM, Morgan BJ, Wetter TJ, and Dempsey JA. Fatiguing Inspiratory Muscle Work Causes Reflex Sympathetic Activation in Humans. *J Physiol* 529 (Pt 2): 493–504, 2000.

29. Stocks J and Quanjer PH. Reference Values for Residual Volume, Functional Residual Capacity and Total Lung Capacity. ATS Workshop on Lung Volume Measurements. Official Statement of The European Respiratory Society. *Eur Respir J* 8: 492–506, 1995.

30. Wetter TJ, Harms CA, Nelson WB, Pegelow DF, and Dempsey JA. Influence of Respiratory Muscle Work on $\dot{V}O_2$ and Leg Blood Flow During Submaximal Exercise. *J Appl Physiol* 87: 643–651, 1999.

LUNG VOLUMES AND CAPACITIES

EQUIPMENT

- Calculator
- Individual and group data sheets

 Find the group data sheets for this laboratory online at www.HumanKinetics.com/ LaboratoryManualForExercisePhysiology.

LUNG VOLUME PREDICTION FORMULAS

Lung volumes are typically predicted using height, weight, and sex. These predictions are usually quite close; in fact, variants are used in clinical diagnoses. The following equations, or ones like them, are built into electronic spirometers to allow percent predicted lung volumes as part of the output data.

Step 1: Obtain height (cm) and weight (kg) from each lab member.

Step 2: Each lab member should calculate his or her own lung volumes given the equations in table 13.1 and use the resulting values to fill out the individual and group data sheets.

Step 3: Norms for lung volumes can be found in table 13.2 (12).

Table 13.1 Equations for Predicting Lung Volumes

Equation	Reference
Women	
FVC (L) = $0.0491 \times H - 0.0216 \times A - 3.59$	(6)
$FEV_{1.0}$ (L) = $0.0342 \times H - 0.0255 \times A - 1.578$	(6)
RV_A (L) = $0.01812 \times H + 0.016 \times A - 2.003$	(29)
RV_B (L) = $0.023 \times H + 0.021 \times A - 2.978$	(4)
RV_C (L) = $0.28 \times FVC$	
Men	
FVC (L) = $0.060 \times H - 0.0214 \times A - 4.65$	(6)
$FEV_{1.0}$ (L) = $0.0414 \times H - 0.0244 \times A - 2.19$	(6)
RV_A (L) = $0.0131 \times H + 0.022 \times A - 1.232$	(29)
RV_B (L) = $0.019 \times H + 0.0115 \times A - 2.24$	(20)
RV_C (L) = $0.24 \times FVC$	

A = age in y; H = height in cm.

Table 13.2 Normal Values for Lung Volumes

Lung volume	<20 y male	20–40 y male	>40 y male	<20 y female	20–40 y female	>40 y female
FVC (L)	3.22	4.93	4.34	2.82	3.53	2.99
$FEV_{1.0}$ (L)	2.77	4.10	3.37	2.51	3.02	2.36
$FEV_{1.0}$/FVC (%)	86.02	83.16	77.65	89.01	85.55	78.93

QUESTION SET 13.1

1. What are the main components that predict lung volumes? Flow rates?

2. How do your lung volumes compare with the age-appropriate norms?

3. What is your RV using the equations provided? What is the percent difference between these three estimations? (You will use these values during the body composition lab on page 409 in chapter 14.)

4. In examining the group data sheet, what do you notice about the predicted lung volumes for men and women? For bigger and smaller people?

Find the case studies for this laboratory online at www.HumanKinetics.com/ LaboratoryManualForExercisePhysiology.

Name or ID number: _____   Date: _____

Tester: _____   Time: _____

Sex: M / F (circle one)   Age: _____ y   Height: _____ in. _____ cm

Temperature: _____ °F _____ °C   Weight: _____ lb _____ kg

Barometric pressure: _____ mmHg   Relative humidity: _____ %

Women

$0.0491 \times$ _____ $-.0216 \times$ _____ $-3.59 =$ _____
    height (cm)   age (y)   FVC (L)

$0.0342 \times$ _____ $-.0255 \times$ _____ $-1.578 =$ _____
    height (cm)   age (y) $FEV_{1.0}$

$0.01812 \times$ _____ $+.016 \times$ _____ $-2.003 =$ _____
    height (cm)   age (y) RV_A (L)

$0.023 \times$ _____ $+.021 \times$ _____ $-2.978 =$ _____
    height (cm)   age (y) RV_B (L)

$0.28 \times$ _____ $=$ _____
    FVC RV_C (L)

Men

$0.060 \times$ _____ $-.0214 \times$ _____ $-4.65 =$ _____
    height (cm)   age (y)   FVC (L)

$0.0414 \times$ _____ $-.0244 \times$ _____ $-2.19 =$ _____
    height (cm)   age (y) $FEV_{1.0}$

$0.0131 \times$ _____ $+.022 \times$ _____ $-1.232 =$ _____
    height (cm)   age (y) RV_A (L)

$0.019 \times$ _____ $+.0115 \times$ _____ $-2.24 =$ _____
    height (cm)   age (y) RV_B (L)

$0.24 \times$ _____ $=$ _____
    FVC RV_C (L)

PULMONARY FUNCTION

EQUIPMENT

- Spirometer
- Noseclips
- Individual and group data sheets

 Find the group data sheets for this laboratory online at www.HumanKinetics.com/ LaboratoryManualForExercisePhysiology.

PULMONARY FUNCTION SPIROMETER TEST

Lab spirometers (figure 13.3) that measure pulmonary function may be electronic or use a bell that floats in water. They also differ in the measures they are capable of making; most, however, collect FVC, $FEV_{1.0}$, and PEF. Your laboratory instructor will demonstrate the use of the particular spirometer in your lab. Here are some pretest instructions that you should follow when testing pulmonary function: avoid exercise, eating, and smoking for 2 hours prior to the test; dress appropriately in unrestrictive clothing; and abstain from alcohol for 4 hours beforehand.

 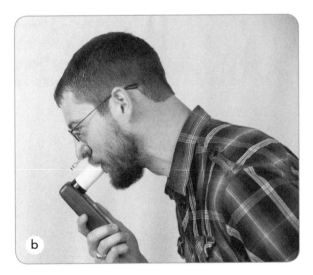

Figure 13.3 An electronic spirometer (a), and in use with nose clips (b).

Step 1: Use your lab spirometer to measure FVC, $FEV_{1.0}$, and PEF. Spirometers typically provide both your absolute value and the % of norm. Since flow rates are being measured, it is important to expire as quickly as possible during these tests.

Step 2: Record the results for your absolute values and % of norm on the individual data sheet (include units). Space is available for additional pulmonary measures that your particular spirometer records.

Step 3: Repeat the trial twice more.

Step 4: Calculate the average for the three attempts.

Step 5: Record data from each trial by entering the information on the individual data sheet.

Step 6: Record your values on the group data sheet.

QUESTION SET 13.2

1. How does your $FEV_{1.0}$ compare with the norm?
2. How might you explain a deviation from the norm?
3. What factors could negatively affect peak expiratory flow?
4. What factors could positively affect forced vital capacity?
5. How are COPDs typically diagnosed using PFT? What are the parameters, and what level of change is required for diagnosis?
6. How does one perform an exercise-induced asthma test? What constitutes a positive test?
7. Based on the group data sheet, was the flow limited for any of your labmates?

Find the case studies for this laboratory online at www.HumanKinetics.com/ LaboratoryManualForExercisePhysiology.

Name or ID number: _____   Date: _____

Tester: _____   Time: _____

Sex: M / F (circle one)   Age: _____ y   Height: _____ in. _____ cm

Temperature: _____ °F _____ °C   Weight: _____ lb _____ kg

Barometric pressure: _____ mmHg   Relative humidity: _____ %

Measure	Trial 1		Trial 2		Trial 3		Average	
	Your value	% of norm	Your value	% of norm	Your value	% of norm	Your value	% of norm
FVC								
$FEV_{1.0}$								
PEF								

EXERCISE-INDUCED VENTILATORY LIMITATIONS

EQUIPMENT

- Spirometer
- Noseclips
- Data from prior $\dot{V}O_2$max tests (lab 9 or 10)
- Individual and group data sheets

 Find the group data sheets for this laboratory online at www.HumanKinetics.com/ LaboratoryManualForExercisePhysiology.

MAXIMUM VOLUNTARY VENTILATION TEST

One indirect measure of ventilatory limitation involves comparing maximum voluntary ventilation (MVV) with maximum ventilation during an incremental test to exhaustion ($\dot{V}_E$). MVV is the maximum amount of air moved ($L \cdot min^{-1}$) at rest. It should be measured for 10 to 15 seconds while the subject is seated, and it is most effective when the subject takes modestly deep breaths at a very high rate (>30 breaths $\cdot min^{-1}$). The $\dot{V}_E$/MVV ratio can be used to predict ventilatory limitations.

Step 1: Select one to three volunteer subjects who have completed a maximal exercise test in a previous lab (lab 9 or 10).

Step 2: Measure MVV on your lab spirometer. It is important for the subject to be seated, since he or she may become dizzy during and following the test. The subject should breathe deeply and as rapidly as possible for the entire duration of the test (10–15 s).

Step 3: Record the results for the subject in the table. Repeat the test after giving the subject a few minutes to recover.

Step 4: Calculate the average for the two attempts.

Step 5: Record $\dot{V}_E$ from the maximal test reports in lab 9 or 10.

Step 6: If you are not one of the subjects with max data, measure and record your own MVV.

Step 7: Record your values on the group data sheet.

QUESTION SET 13.3

1. Compare the MVVs from the class with the maximum $\dot{V}_E$ attained during the $\dot{V}O_2$max tests. Do you think anyone had ventilatory limitation?

2. Why might you see greater $\dot{V}_E$ than MVV yet remain unsure that ventilation is a limitation?

3. What would you measure to determine whether the lungs truly limit exercise performance?

4. Why might dizziness or syncope accompany the measurement of MVV?

 Find the case studies for this laboratory online at www.HumanKinetics.com/ LaboratoryManualForExercisePhysiology.

Subject A

Name or ID number: _____ Date: _____

Tester: _____ Time: _____

Sex: M / F (circle one)   Age: _____ y   Height: _____ in. _____ cm

Temperature: _____ °F _____ °C   Weight: _____ lb _____ kg

Barometric pressure: _____ mmHg   Relative humidity: _____ %

Subject B

Name or ID number: _____ Date: _____

Tester: _____ Time: _____

Sex: M / F (circle one)   Age: _____ y   Height: _____ in. _____ cm

Temperature: _____ °F _____ °C   Weight: _____ lb _____ kg

Barometric pressure: _____ mmHg   Relative humidity: _____ %

Subject C

Name or ID number: _____ Date: _____

Tester: _____ Time: _____

Sex: M / F (circle one)   Age: _____ y   Height: _____ in. _____ cm

Temperature: _____ °F _____ °C   Weight: _____ lb _____ kg

Barometric pressure: _____ mmHg   Relative humidity: _____ %

Subject	Trial 1		Trial 2		Average	Max $\dot{V}_E$	$\dot{V}_E$ / MVV ratio
	MVV	% pred	MVV	% pred			
A							
B							
C							

EXERCISE-INDUCED ASTHMA

EQUIPMENT

- Spirometer
- Noseclips
- Treadmill
- Heart rate monitor
- Individual data sheets

PULMONARY FUNCTION TEST FOR EXERCISE-INDUCED ASTHMA

One of the more useful applications of pulmonary function testing is for diagnostic purposes. **Exercise-induced asthma (EIA)** is a relatively common ailment resulting from a maladaptation that results in bronchoconstriction. In normally functioning individuals, the sympathetic nervous system drive that occurs during exercise results in dilation of the bronchi. However, for several possible reasons, it can result in bronchoconstriction in some individuals, and this effect can be exacerbated in cold or dry air. As detailed earlier, individuals with EIA may have normal flow rates at rest but yield a positive test following 6 to 8 min of vigorous exercise above a certain intensity level.

Step 1: Select one subject with normal lung function and (if possible) one with suspected exercise-induced asthma or other lung dysfunction. (If no one in the class has suspected EIA, you can still perform this test for practice and discuss the results.)

Step 2: Use your lab spirometer to measure FVC, $FEV_{1.0}$, and PEF. Since flow rates are being measured, it is important for the subject to expire as quickly as possible during these tests.

Step 3: Record the subjects' absolute values at rest on the individual data sheet (including units).

Step 4: Have the subjects run on the treadmill for 6 to 8 min at about 85% to 90% of his or her age-predicted max heart rate.

Step 5: Perform a PFT as quickly as possible after the cessation of exercise.

Step 6: Record the test results on the individual data sheet.

Step 7: Calculate the % change in lung parameters from pre-exercise to post-exercise with the following formula:

$$\% \text{ difference} = (\text{pre} - \text{post}) / \text{pre}$$

QUESTION SET 13.4

1. What change is required from pre-exercise to post-exercise to produce a positive test for exercise-induced asthma?

2. Did your subject have exercise-induced asthma? Justify your answer.

3. Why might ventilatory flow rates *increase* following exercise?

 Find the case studies for this laboratory online at www.HumanKinetics.com/ LaboratoryManualForExercisePhysiology.

Subject A

Name or ID number: _____ Date: _____

Tester: _____ Time: _____

Sex: M / F (circle one)   Age: _____ y   Height: _____ in. _____ cm

Temperature: _____ °F _____ °C   Weight: _____ lb _____ kg

Barometric pressure: _____ mmHg   Relative humidity: _____ %

Subject B

Name or ID number: _____ Date: _____

Tester: _____ Time: _____

Sex: M / F (circle one)   Age: _____ y   Height: _____ in. _____ cm

Temperature: _____ °F _____ °C   Weight: _____ lb _____ kg

Barometric pressure: _____ mmHg   Relative humidity: _____ %

	Pre-exercise		Post-exercise		% Δ from pre to post	
	Sub A	Sub B	Sub A	Sub B	Sub A	Sub B
FVC (L)						
$FEV_{1.0}$ (L)						
PEF ($L \cdot s^{-1}$)						

Sub A = subject A (with asthma); sub B = subject B (without asthma).

Body Composition Assessments

Objectives

1. Know the range of values for percent body fat generally considered average or typical for college-age men and women.
2. Know the values for body mass index (BMI) considered indicative of obesity for men and women.
3. Explain the principle of underwater weighing (UWW; also known as *hydrostatic weighing* or *densitometry*) for determining body density and percent body fat.
4. Competently perform BMI, circumference, skinfold, and underwater weighing measurements.
5. Understand the importance of estimating or measuring residual volume for UWW validity.
6. Calculate body density and percent body fat based on the two-component body composition system.
7. Describe common sources of error in body composition measurements.
8. Describe the advantages and disadvantages of various body composition techniques.

Definitions

body mass index (BMI)—Ratio of an individual's weight (kg) to height squared (m^2).

densitometry—Measurement of body density ($D_{body} = mass_{body} / volume_{body}$).

fat-free mass—All body tissue that is not fat, including bone, muscle, organ, and connective tissue.

relative body fat—Amount of body fat expressed as a percent of total body weight.

reliability—Reproducibility of a test, with high reliability meaning that if the same measure is tested several times for the same person, the same answer will result.

residual volume—Amount of air remaining in the lungs after a maximal exhalation.

skinfold calipers—Instrument used to measure the thickness of pinched skinfolds; must provide a standard pinch pressure in order to provide accurate values.

skinfold thickness—Body composition measurement using the thickness of skinfolds, and thus subcutaneous fat, as an estimate of body density; use requires skill for correct skinfold placement.

two-component body composition—Commonly used system in which the body is assumed to have two major components: fat and fat-free (or lean) tissue.

underwater weighing (UWW)—Body composition measurement using the displacement of water to estimate body density.

validity—Accuracy of a measure—that is, how close it is to the "true" value. (In other words, are you *really* measuring what you say you are measuring—in this case, relative body fat?)

vital capacity—Volume of air that can be moved into or out of the lungs in one maximal breath.

Body composition refers to the components that make up the body. Usually, we are interested in the percentage of the body that is adipose (or fatty) tissue versus the percentage that is fat free. The primary reason for this interest is that obesity—extreme overfatness—is closely related to a number of health risks prevalent in Western societies (e.g., cardiovascular disease, peripheral vascular disease, hypertension, and diabetes). There are also less critical reasons for interest in body composition; for example, athletes wishing to maximize their performance want to optimize their amount of muscle and fat, and many individuals are interested in body composition due to their society's perception that extreme leanness is aesthetically desirable. In fact, losing weight (and thus fat) is probably the main reason that Americans exercise.

Thus it is important to know not only body weight but also how much of the weight is fat and how much is lean. Changes resulting from exercise may not be noticeable by a scale weight, but many techniques are available for estimating body composition.

We aim here to expose you to some of the more common techniques of evaluating disease risk based on body composition, including circumferences, underwater weighing (UWW), skinfold thickness (SF), body mass index (BMI), waist and hip circumference, and waist-to-hip ratio (W/H). See table 14.1 for more on methods for determining body composition.

Table 14.1 Methods for Determining Body Composition Using the Two (2)- or Three (3)-Component Model

Method	Description
Anthropometry (2)	Measures body segment girths to predict body fat.
Bioelectrical impedance analysis (BIA) (2)	Measures resistance to electric current to predict body water content, lean body mass, and body fat.
Body plethysmography (2)	Whole-body plethysmography measures air displacement and calculates body density (comparable to water displacement protocol used in underwater weighing).
Computed tomography (CT) (3)	X-ray scanning images body tissues; used to determine subcutaneous and deep fat to predict body fat percentage and to calculate bone mass.
Dual energy X-ray absorptiometry (DEXA, DXA) (3)	X-ray technique used at two energy levels to image body fat; used to calculate bone mass.
Dual photon absorptiometry (DPA) (3)	Beam of photons passes through tissues, differentiating soft tissues from bone tissues; used to predict body fat and calculate bone mass.
Infrared interactance (2)	Infrared light passes through tissues, and interaction with tissue components is used to predict body fat.
Magnetic resonance imaging (MRI) (3)	Magnetic-field and radio-frequency waves are used to image body tissues (similar to CT scan and very useful for imaging deep abdominal fat).
Neutron activation analysis (3)	Beam of neutrons passes through tissues, permitting analysis of nitrogen and other mineral content in the body; used to predict lean body mass.
Skinfold thickness (2)	Measures subcutaneous fat folds to predict body fat content and lean body mass.
Total body potassium (2)	Measures total body potassium, the main intracellular ion, to predict lean body mass and body fat.
Total body water (hydrometry) (2)	Measures total body water by isotope dilution techniques to predict lean body mass and body fat.
Ultrasound (2)	High-frequency ultrasound waves pass through tissues to image subcutaneous fat and predict body fat content.
Underwater weighing (hydrodensitometry)	Underwater weighing technique based on Archimedes' principle predicts body density, body fat, and lean body mass.

Reprinted, by permission, from M. Williams, 2009, *Nutrition for health, fitness, and sport.* 9th ed. New York: McGraw-Hill.

These techniques differ by sophistication of equipment, validity, reliability, general assumptions, and ease of use. Some (BMI, circumferences, waist-to-hip ratio) do not estimate body composition at all but do correlate strongly with obesity and other lifestyle diseases.

Knowledge of **relative body fat** (%BF) can be useful in pursuing goals of weight gain or loss. It is possible to calculate a desired body weight from a target relative body fat by means of a simple equation:

$$\text{Desired body weight} = \frac{BW - (BW \times \frac{\%BF}{100})}{1 - (\frac{\text{desired }\%BF}{100})}$$

This equation can be rewritten as follows (LBM indicates lean body mass):

$$\text{Desired body weight (BW)} = \frac{LBM}{\text{desired }\%LBM}$$

These goal weight predictions assume that the person is capable of reducing body fat without changes in lean body mass or **fat-free mass**. Whether or not this is true may depend on the person's methods chosen of weight loss or gain.

Of course, some fat in our bodies is necessary, such as for cellular membranes, vitamins, hormones and surrounding our nerves (Schwann cells). Storage fat in adipose tissue is available for use as a fuel; nonetheless, it is often called "nonessential" or "storage fat." It is this fat that people are interested in reducing for reasons of health, athletic performance, or vanity. Essential fat is thought to make up about 3% of the body by weight in men and about 12% in women (6). The remaining fat is that which is considered nonessential or storage fat despite its usefulness as a dense fuel source. Different classifications exist for relative body fat (combined essential and nonessential) dependent on sex and age. They differ by percentile ranks, at-risk cutoffs, and target recommendations. Tables 14.2 and 14.3 list some examples of these classifications.

Table 14.2 Body Composition (Normative References)

%	20–29 y	30–39 y	40–49 y	50–59 y	60–69 y	70–79 y
Male						
90	7.9	11.9	14.9	16.7	17.6	17.8
80	10.5	14.5	17.4	19.1	19.7	20.4
70	12.7	16.5	19.1	20.7	21.3	21.6
60	14.8	18.2	20.6	22.1	22.6	23.1
50	16.6	19.7	21.9	23.2	23.7	24.1
40	18.6	21.3	23.4	24.6	25.2	24.8
30	20.6	23.0	24.8	26.0	25.4	26.0
20	23.1	24.9	26.6	27.8	28.4	27.6
10	26.3	27.8	29.2	30.3	30.9	30.4
Female						
90	14.8	15.6	17.2	19.4	19.8	20.3
80	16.5	17.4	19.8	22.5	23.2	24.0
70	18.0	19.1	21.9	25.1	25.9	26.2
60	19.4	20.8	23.8	27.0	27.9	28.6
50	21.0	22.6	25.6	28.8	29.8	30.4
40	22.7	24.6	27.6	30.4	31.3	31.8
30	24.5	26.7	29.6	32.5	33.3	33.9
20	27.1	29.1	31.9	34.5	35.4	36.0
10	31.4	33.0	35.4	36.7	37.3	38.2

Adapted, by permission, from Cooper Institute, *Physical fitness assessments and norms for adults and law enforcement* (Dallas, TX: The Cooper Institute). 52, 53. For more information: www.cooperinstitute.org.

Table 14.3 Percent Body Fat Standards for Adults, Children, and Physically Active Adults

	NR*	Low	Mid	High	Obesity
Recommended %BF levels for adults and children					
Male					
6–17 y	<5	5–10	11–25	26–31	>31
18–34 y	<8	8	13	22	>22
35–55 y	<10	10	18	25	>25
>55 y	<10	10	16	23	>23
Female					
6–17 y	<12	12–15	16–30	31–36	>36
18–34 y	<20	20	28	35	>35
35–55 y	<25	25	32	38	>38
>55 y	<25	25	30	35	>35

	Low	Mid	Upper
Recommended %BF levels for physically active adults			
Male			
18–34 y	5	10	15
35–55 y	7	11	18
>55 y	9	12	18
Female			
18–34 y	16	23	28
35–55 y	20	27	33
>55 y	20	27	33

*NR = not recommended; %BF = percent body fat.

Reprinted, by permission, from V. Heyward, 2010, *Advanced fitness assessment and exercise prescription,* 6th ed. (Champaign, IL: Human Kinetics), 190.

Body Composition Testing Models

Several models exist for describing differences in the composition of our bodies See figure 14.1 for more on methods for determining body composition. The **two-component body composition** model (such as UWW and SF) splits the body into two parts—fat mass (FM) and fat-free mass (FFM). FM is self-evident, whereas FFM is made up of protein, water, and bone. Other multicomponent models separate one or more of the components within the FFM, such as bone density (by dual energy X-ray absorptiometry [DEXA] or magnetic resonance imaging [MRI]) or total body water (by stable radioisotopes).

The multicomponent models are considered more accurate, or valid, since they remove some assumptions made in the two-component models. FM and FFM in the two-component model are assumed to have consistent densities within a certain population or age group. For FM, that density is 0.90 $g \cdot ml^{-1}$, and for FFM the density is 1.1 $g \cdot ml^{-1}$. Siri used these values to derive the equation used to estimate percent body fat in adult Caucasians (11):

$$\% \text{ body fat} = (495 / \text{density}) - 450$$

Because of slight variances in lean mass (FFM) densities, different equations exist based on age group, sex, and race (see table 14.4).

The specificity of these equations increases the accuracy of the body composition measurement.

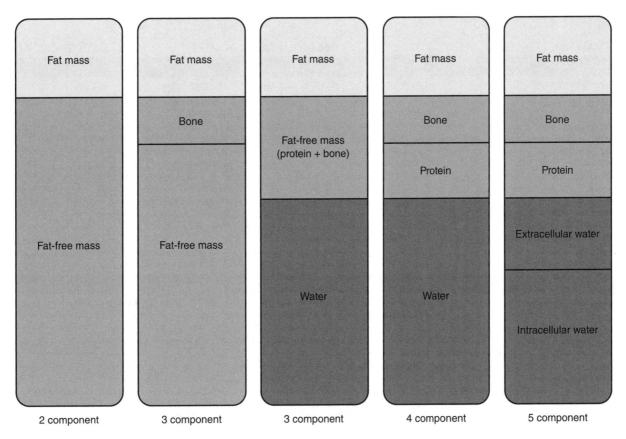

Figure 14.1 Body composition models.

Table 14.4 Population-Specific Two-Component Model Formulas for Converting Body Density to Percent Body Fat

Population	Age (y)	Sex	%BF[a]	FFB_d (g · ml⁻¹)*
Race or ethnicity				
African American	9–17	Female	(5.24 / Db) – 4.82	1.088
	19–45	Male	(4.86 / Db) – 4.39	1.106
	24–79	Female	(4.85 / Db) – 4.39	1.106
American Indian	18–62	Male	(4.97 / Db) – 4.52	1.099
	18–60	Female	(4.81 / Db) – 4.34	1.108
Japanese native	18–48	Male	(4.97 / Db) – 4.52	1.099
	18–48	Female	(4.76 / Db) – 4.28	1.111
	61–78	Male	(4.87 / Db) – 4.41	1.105
	61–78	Female	(4.95 / Db) – 4.50	1.100
Singaporean (Chinese, Indian, Malay)		Male	(4.94 / Db) – 4.48	1.102
		Female	(4.84 / Db) – 4.37	1.107

(continued)

Table 14.4 *(continued)*

Population	Age (y)	Sex	%BF[a]	FFB$_d$ (g · ml⁻¹)*
Race or ethnicity *(continued)*				
White	8–12	Male	(5.27 / Db) – 4.85	1.086
	8–12	Female	(5.27 / Db) – 4.85	1.086
	13–17	Male	(5.27 / Db) – 4.85	1.092
	13–17	Female	(5.27 / Db) – 4.85	1.090
	18–59	Male	(4.95 / Db) – 4.50	1.100
	18–59	Female	(4.96 / Db) – 4.51	1.101
	60–90	Male	(4.97 / Db) – 4.52	1.099
		Female	(5.02 / Db) – 4.57	1.098
Hispanic		Male	NA	NA
	20–40	Female	(4.87 / Db) – 4.41	1.105
Athletes				
Resistance-trained	24 ± 4	Male	(5.21 / Db) – 4.78	1.089
	35 ± 6	Female	(4.97 / Db) – 4.52	1.099
Endurance-trained	21 ± 2	Male	(5.03 / Db) – 4.59	1.097
	21 ± 4	Female	(4.95 / Db) – 4.50	1.100
All sports	18–22	Male	(5.12 / Db) – 4.68	1.093
	18–22	Female	(4.97 / Db) – 4.52	1.099
Clinical populations				
Anorexia nervosa	15–44	Female	(4.96 / Db) – 4.51	1.101
Obesity	17–62	Female	(4.95 / Db) – 4.50	1.100
Spinal cord injury (paraplegic or quadriplegic)	18–73	Male	(4.67 / Db) – 4.18	1.116
		Female	(4.70 / Db) – 4.22	1.114

FFB$_d$ = fat-free body density; Db = body density; %BF = percent body fat; NA = no data available for this population subgroup.

[a]Multiply value by 100 to calculate %BF.

*FFB$_d$ is based on average values reported in selected research articles.

Reprinted, by permission, from V. Heyward and D. Wagner, 2004, *Applied body composition assessment*, 2nd ed. (Champaign, IL: Human Kinetics), 9.

However, the term *accuracy* is often misused. We assume that if a body composition measure is accurate it is close to the "real value," whereas *precision* refers to the repeatability of the measurement. In body composition, the terms **validity** and **reliability** are often used to replace the terms accuracy and precision, respectively. A body composition technique is considered valid if it measures what it says it measures—relative body fat. To be reliable, the body composition measurement must generate reproducible results. For example, weighing someone on a scale is a simple, reliable measure, but in reference to relative body fat it is not particularly valid. UWW, on the other hand, estimates relative body fat and is one of our most valid measurements, but by comparison to body weight measurement it is not as reliable. Consider using these terms in relation to body composition techniques to reduce confusion.

Body Mass Index for Categorizing Body Composition

Body mass index (BMI) is a very simple and reliable measure that is commonly used in clinical situations and epidemiological research to catego-

Table 14.5   Classification of Weight by Body Mass Index (BMI), Waist Circumference, and Associated Disease Risks

BMI	Classification	Risk level by waist circumference*			
		Men		Women	
		≤102 cm	>102 cm	≤88 cm	>88 cm
<18.5	Underweight				
18.5-24.9	Normal				
25.0-29.9	Overweight	Increased	High	Increased	High
30.0-34.9	Class 1 obesity	High	Very high	High	Very high
35.0-39.9	Class 2 obesity	Very high	Very high	Very high	Very high
≥40.0	Class 3 obesity	Extremely high	Extremely high	Extremely high	Extremely high

* Disease risk for type 2 diabetes, hypertension, and coronary heart disease.

Adapted from the NHLBI (8).

rize individuals with regard to obesity (1). BMI is calculated by the following equation:

$$BMI = Wt \ (kg) \ / \ Ht^2 \ (m)$$

The theory behind this method is that weight-to-height ratios across the general population have a positive relationship with percent body fat. The National Institutes of Health (NIH) have used BMI to create definitions of overweight and obesity (8) (table 14.5). Obesity is associated with a three-fold greater risk of diabetes and hypertension and a twofold greater risk of hypercholesterolemia (high blood cholesterol). Individuals classified as obese warrant an intervention such as calorie restriction (8). The popularity of BMI is based in part on its ease of measurement and the potential database of BMI data from doctor's office visits. However, since BMI does not separate how much weight is fat or lean body mass, a muscular person could be falsely described as obese. BMI is therefore a very reliable measure but has questionable validity as a measure of body composition.

Circumference Measurements and Health Risk

Another simple measurement associated with increased health risk is the waist-to-hip ratio (circumference of the waist divided by circumference of the hips). As W/H ratio increases, the risk increases twofold for heart attack, stroke, hyper-

tension, diabetes mellitus, gallbladder disease, and death (1). Waist circumference is the smallest circumference below the ribcage and above the umbilicus while the client is standing with the abdominal muscles relaxed. Hip measurement is made at the level of the greatest circumference of the buttocks in a level horizontal plane. Measure the waist while facing the client. Measure the hips from the side. Record the results to the nearest 0.5 cm. Repeated measures should yield results within 0.5 cm. If possible, use a spring-loaded tape to normalize tension. Ratios above 0.90 for men and 0.80 for women relate to sharp increases in disease risk for hypertension, high cholesterol, cardiovascular disease, and diabetes (1) (see table 14.8 on page 397).

Skinfold Thickness as a Measure of Body Fat

Estimation of body fat based on **skinfold thickness** has become quite common, largely because of the relative ease of measurement with minimal equipment. Some **skinfold calipers** are available for very a low price, but many of these devices are grossly inaccurate. Instruments such as the ones that you will likely use in this lab can cost several hundred dollars. As you will learn in this lab, the skill of the person making the measurements is quite important for obtaining meaningful measurements.

The rationale for using skinfold thickness is that an age-dependent proportion of body fat is deposited subcutaneously. Thus by measuring the amount of adipose tissue that can be pinched, it is possible to gain some indication of the amount of overall body fat (see figure 14.2).

The process of turning this information (i.e., skinfold thickness at several sites) into a number representing body density is based on research results. This type of research compares the sum of skinfolds from a number of anatomical sites to a "gold standard" measure of body density, such as hydrostatic (or underwater) weighing (4). It has been shown that curvilinear or quadratic relationships better predict body density from the sum of skinfolds compared with linear regression (4), which is why the equations require squaring the sum of several skinfolds.

The relationship between subcutaneous fat and total body fat varies with race, age, and sex. Consequently, most equations that have been developed to predict body fatness are population specific and should not be used for subjects of a different race, age, or sex (see table 14.6). In general, estimation of percent body fat from skinfold measures has an error of about 3.5% (9).

The different equations require different skinfold sites to be measured. It is important to understand the technique of measuring skinfold sites and which equations require particular sites. Although it may seem intuitive that body fat prediction from the sum of skinfolds is improved by using more sites, this is not necessarily the case. Validity is not improved by using more than three sites (9). This lab demonstrates the skinfold sites used in the more popular equations.

Unless otherwise appropriate, we choose the commonly used equations developed by Jackson and Pollock for men (4) and Jackson, Pollock, and Ward for women (5), as shown in the accompanying highlight box. These equations are considered generalized equations because the development populations included subjects across a wide age span. Consequently, these equations can be used across various age groups and have been validated for both athletic and nonathletic populations. Separate equations are used for men and women to estimate body density. Once a body density has been determined, body composition is estimated from the Siri equation mentioned earlier (11):

$$\% \text{ body fat} = (495 \,/\, \text{density}) - 450$$

This holds true unless another population-specific equation is more appropriate (see population-specific equations in table 14.6).

Underwater Weighing Method

Underwater weighing (UWW), also called *hydrostatic weighing* or *hydrodensitometry,* is often considered the most valid method for estimating relative body fat (figure 14.3). The technique is called **densitometry** because it is based on the fact that fat and lean tissues have different densities. Underwater weighing depends on the fact that water has a density of approximately $1 \text{ g} \cdot \text{ml}^{-1}$ (see table 14.7 for the exact density of water at different temperatures) and that fat floats because its density is approximately $0.9 \text{ g} \cdot \text{ml}^{-1}$. On the other hand, lean tissue (that is, all other

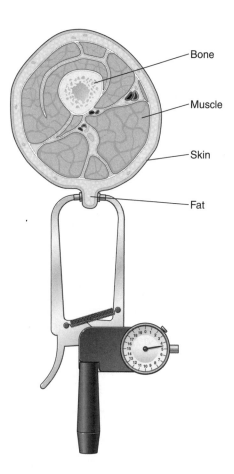

Figure 14.2 Subcutaneous skinfold measurement.

Table 14.6 Skinfold Prediction Equations

SKF sites	Population subgroups	Equation
Σ7SKF (chest + abdomen + thigh + triceps + subscapular + suprailiac + midaxilla	Black or Hispanic women, 18–55 y	Db (g · cc^{-1})[a] = 1.0970 − 0.00046971(Σ7SKF) + 0.00000056(Σ7SKF)^2 − 0.00012828(age)
	Black men or male athletes, 18–61 y	Db (g · cc^{-1})[a] = 1.1120 − 0.00043499(Σ7SKF) + 0.00000055(Σ7SKF)^2 − 0.00028826(age)
Σ4SKF (triceps + anterior suprailiac + abdomen + thigh)	Female athletes, 18–29 y	Db (g · cc^{-1})[a] = 1.096095 − 0.0006952(Σ4SKF) + 0.0000011(Σ4SKF)^2 − 0.0000714(age)
Σ3SKF (triceps + suprailiac + thigh)	White or anorexic women, 18–55 y	Db (g · cc^{-1})[a] = 1.0994921 − 0.0009929(Σ3SKF) + 0.0000023(Σ3SKF)^2 − 0.0001392(age)
Σ3SKF (chest + abdomen + thigh)	White men, 18–61 y	Db (g · cc^{-1})[a] = 1.109380 − 0.0008267(Σ3SKF) + 0.0000016(Σ3SKF)^2 − 0.0002574(age)
Σ3SKF (abdomen + thigh + triceps)	Black or white collegiate male and female athletes, 18–34 y	%BF = 8.997 + 0.2468(Σ3SKF) − 6.343(gender[b]) − 1.998(race[c])
Σ2SKF (triceps + calf)	Black or white boys, 6–17 y Black or white girls, 6–17 y	%BF = 0.735(Σ2SKF) + 1.2 %BF = 0.610(Σ2SKF) + 5.1

ΣSKF=sum of skinfolds (mm).

[a]Use population-specific conversion formulas to calculate %BF (percent body fat) from Db (body density).

[b]Male athletes = 1; female athletes = 0.

[c]Black athletes = 1; white athletes = 0.

Reprinted, by permission, from V. Heyward, 2010, *Advanced fitness assessment and exercise prescription*, 6th ed. (Champaign, IL: Human Kinetics), 204.

Generalized Body Density Formulas

Body Density Equations for Men (4)

$$D_b = 1.10938 − (0.0008267 × ΣSKF3) + (0.0000016 × ΣSKF3^2) − (0.0002574 × age)$$

(Sum of 3 skinfolds [mm] = chest + abdomen + thigh)

$$D_b = 1.1120 − (0.00043499 × ΣSKF7) + (0.00000055 × ΣSKF7^2) − (0.00028826 × age)$$

(Sum of 7 skinfolds [mm] = chest + midaxillary + subscapular + triceps + abdomen + suprailium + thigh)

Body Density Equations for Women (5)

$$D_b = 1.0994921 − (0.0009929 × ΣSKF3) + (0.0000023 × ΣSKF3^2) − (0.0001392 × age)$$

(Sum of 3 skinfolds [mm] = triceps + suprailium + thigh)

$$D_b = 1.0970 − (0.00046971 × ΣSKF7) + (0.00000056 × ΣSKF7^2) − (0.00012828 × age)$$

(Sum of 7 skinfolds [mm] = chest + midaxillary + subscapular + triceps + abdomen + suprailium + thigh)

Figure 14.3 Underwater weighing method.

tissues excluding fat) sinks because its density is greater than that of water (about 1.1 g·ml⁻¹). The underwater weighing technique, then, is based on the two-component body composition model in which the body is simply divided into fat and fat-free (or lean) tissue. Since every person is neither total FM nor total FFM, all people have a body density between 0.9 and 1.1 g·ml⁻¹.

When estimating percent body fat by underwater weighing, it is necessary to determine body density, which, by definition, equals mass divided by volume: $D_B = M / V$. Body volume is calculated using Archimedes' principle, which states that when an object is placed in water it is buoyed up by a counterforce equal to the water it displaces. The volume of water displaced would equal the loss of weight while the object is totally submerged. By subtracting the weight measured in water (M_W) from that measured in air (M_A), we obtain the weight of water displaced ($M_A - M_W$). This weight value is converted to a volume by dividing it by the density of water (D_W). Therefore,

$$\text{Density} = \frac{M}{V} = \frac{M_A}{(M_A - M_W)/D_W}$$

The determined volume must be corrected for the amount of gas in the body, both in the lungs and in the gastrointestinal tract. The gas in the gastrointestinal (V_{GI}) system is assumed to be 100 ml for all subjects. The subject will remove as much air from the respiratory system as possible with a maximal expiration. However, even after a maximal expiration, a certain volume of air remains in the lungs, and this volume is called the **residual volume (RV)**. If RV has not been measured, we can estimate it from **vital capacity**. In fact, in lab 13 you estimated RV with three different equations (RV_A, RV_B, and RV_C). Thus,

Table 14.7 Water Density at Various Temperatures.

Water (°C)	Temp (°F)	D_w (g·ml⁻¹)	Water (°C)	Temp (°F)	D_w (g·ml⁻¹)
0	32	0.999	30	86	0.9957
4	39	1.000	31	88	0.9954
22	72	0.9978	32	89.5	0.9950
23	73	0.9975	33	91	0.9947
24	75	0.9973	34	93	0.9944
25	77	0.9971	35	95	0.9941
26	79	0.9968	36	97	0.9937
27	81	0.9965	37	99	0.9934
28	82	0.9963	38	100	0.9930
29	84	0.9960	39	102	0.9926

the body density equation can be rewritten and expressed as follows:

$$Density = \frac{M}{V} = \frac{M_A}{[(M_A - M_W)/D_W] - RV - V_{GI}}$$

It is important to use consistent density units, such as grams per ml ($g \cdot ml^{-1}$) or kilograms per liter ($kg \cdot L^{-1}$). Once the body density has been determined, body composition is estimated from the Siri equation mentioned earlier (11).

$$\% \text{ body fat} = (495 / density) - 450$$

This holds true unless another population-specific equation is more appropriate (see population-specific equations in table 14.4).

Underwater weighing involves several potential sources of error. Recall that this technique provides an *estimate* of percent body fat based on body density. The equations used depend on the assumed densities for fat and nonfat tissue, which were originally developed on the basis of relatively few samples. Any error in these assumptions can affect the estimates. We know that the densities of different tissues included in the nonfat component vary between some individuals. For example, bone density varies depending on age, disease, and ethnicity. Older individuals and children have lower bone densities than do young adults, and bone density in blacks is greater than in whites. Another source of error lies in the correction for residual volume and gastrointestinal gas volume. The former can cause error when residual volume is simply estimated rather than measured by nitrogen washout or helium dilution rebreathing (7). Finally, as you will see during the lab, skill is required both of the subject and of the technician in obtaining accurate readings for underwater weight. Many subjects are uncomfortable submerging themselves after a maximal exhalation. However, when the measurements are performed by skilled and experienced personnel under well-controlled conditions, the error has been estimated to be ± 3% (7). Furthermore, when the measurements are made before and after an intervention (e.g., initiating a diet or exercise program), the technique can provide individuals with useful information about changes in percent body fat.

While these labs focus on some of the more common and accessible body composition techniques, you should also be aware of others, including bioelectrical impedance, dual energy X-ray absorptiometry, and air displacement plethysmography.

Bioelectrical Impedance Analysis

Bioelectrical impedance analysis (BIA) is based on the principle that lean tissue conducts current better than fat tissue does. A low-level electrical current is passed through the body, and impedance is determined by the BIA analyzer. BIA allows an estimation of total body water that is proportional to the amount of lean tissue. Many assumptions and guidelines are needed with this body composition analysis, such as the following: no eating or drinking for 4 h prior, no exercise for 12 h prior, no alcohol consumption for 48 h prior, no diuretic medications. In addition, results can be affected by sex, the female menstrual cycle, age, fitness, and race. Population-specific equations are available, but the user of a BIA unit is often at the mercy of the code programmed into the unit's software. Typically, electrodes are placed on the wrists and ankles. In many commercially available BIA units, clients stand on electrodes (much like standing a bathroom scale) or hold onto electrodes with their hands. Electrical current moves in the route of least resistance, so units that utilize only two points of contact make assumptions on resistance through the rest of the body.

Dual Energy X-Ray Absorptiometry

Dual energy X-ray absorptiometry (DXA or DEXA) is considered a three-component model, providing estimates of bone, fat, and lean tissue densities. It requires an expensive piece of equipment (costing more than $70,000) that is typically used in research in the evaluation of bone mineral density. It produces a full-body X-ray that is evaluated by the manufacturer's software for the various tissue densities. For this reason, it is difficult to evaluate the technique's validity, since the algorithms are not usually available to the user. Even so, many researchers consider the DEXA to be the gold standard in body composition (2, 10, 12). It offers additional benefits in that it can provide regional body composition assessment, such as relative body fat of the right or left arm or leg.

Air Displacement Plethysmography

Like UWW, air displacement plethysmography (ADP), sometimes referred to by the manufacturer's name, BOD POD, is a method that estimates body density and is thus a two-component model. Instead of using the displacement of water, however, it uses the displacement of air in a sealed compartment as the underlying principle. Again, it is an expensive piece of equipment (costing more than $30,000) but is considered quite valid (3). It benefits from being able to measure elderly persons and members of other populations who are uncomfortable being submerged in water.

References

1. Bray GA and Gray DS. Obesity. Part I—Pathogenesis. *West J Med* 149: 429–441, 1988.

2. Friedl KE, DeLuca JP, Marchitelli LJ, and Vogel JA. Reliability of Body-Fat Estimations From a Four-Compartment Model by Using Density, Body Water, and Bone Mineral Measurements. *Am J Clin Nutr* 55: 764–770, 1992.

3. Heyward VH and Wagner DR. *Applied Body Composition Assessment*. 2nd ed. Champaign, IL: Human Kinetics, 2004.

4. Jackson AS and Pollock ML. Generalized Equations for Predicting Body Density of Men. *Br J Nutr* 40: 497–504, 1978.

5. Jackson AS, Pollock ML, and Ward A. Generalized Equations for Predicting Body Density of Women. *Med Sci Sports Exer* 12: 175–181, 1980.

6. McArdle WD, Katch FI, and Katch VL. *Exercise Physiology: Energy, Nutrition, and Human Performance*. Baltimore: Lippincott Williams & Wilkins, 2007.

7. Morrow JR, Jr., Jackson AS, Bradley PW, and Hartung GH. Accuracy of Measured and Predicted Residual Lung Volume on Body Density Measurement. *Med Sci Sports Exer* 18: 647–652, 1986.

8. National Heart, Lung, and Blood Institute. *Clinical guidelines on the identification, evaluation, and treatment of overweight and obesity in adults: The evidence report*. Bethesda, MD: National Institutes of Health, 1998.

9. Pollock ML and Jackson AS. Research Progress in Validation of Clinical Methods of Assessing Body Composition. *Med Sci Sports Exer* 16: 606–615, 1984.

10. Prior BM, Cureton KJ, Modlesky CM, Evans EM, Sloniger MA, Saunders M, and Lewis RD. In Vivo Validation of Whole Body Composition Estimates From Dual-Energy X-Ray Absorptiometry. *J Appl Physiol* 83: 623–630, 1997.

11. Siri WE. Body Composition From Fluid Spaces and Density: Analysis of Methods. 1961. *Nutrition* 9: 480–491 (discussion 480, 492), 1993.

12. Withers RT, Smith DA, Chatterton BE, Schultz CG, and Gaffney RD. A Comparison of Four Methods of Estimating the Body Composition of Male Endurance Athletes. *Eur J Clin Nutr* 46: 773–784, 1992.

BMI AND CIRCUMFERENCE DATA

EQUIPMENT

- Stadiometer or measuring tape
- Scale (accurate to a tenth of a kilogram)
- Anthropometric tape (preferably with spring)
- Individual and group data sheets

Find the group data sheets for this laboratory online at www.HumanKinetics.com/ LaboratoryManualForExercisePhysiology.

BODY MASS INDEX CALCULATION

Step 1: Each lab subject should be prepared to participate in the appropriate attire (light, nonbulky clothing to avoid mismeasurement of weight and girth).

Step 2: Students should record their own data in their individual data tables and work as evaluators for their labmates.

Step 3: Remove your shoes, jewelry, wallet, keys, and any excess clothing prior to weighing yourself on the scale. Record your weight in kg to the preciseness allowed by your equipment.

Step 4: Measure the subject's height by using a stadiometer or a measuring tape attached to a wall. Have the subject remove his or her shoes and stand with his or her back to the stadiometer or wall with the heels on the floor. Use a lever arm or similar device placed on the top of the head at the highest point. Be sure that the arm or similar device is parallel to the ground, then record the height in centimeters to the nearest millimeter.

Step 5: To demonstrate the reliability of this measure, repeat for a total of three trials. You should measure each member of your group once, then repeat in order to demonstrate true reliability.

Step 6: Record your data on the individual data sheet and collect data for the group data sheet.

CIRCUMFERENCE MEASUREMENTS

Waist and hip circumferences are best measured with a spring-loaded tape measure such as the Gulick measuring tape. This allows correct (4 oz) and reproducible tension to be placed on the tape.

Step 1: To measure the circumference of the waist, wrap the tape measure around the subject from his or her anterior side. Place the tape at the narrowest circumference between the umbilicus and the lowest rib. Resolve any twists in the tape. Cross the tape and switch hands to avoid having your arms cross each other. Confirm that the tape is parallel to the ground all the way around the subject. (Doing this measure in front of a mirror can be helpful.) Have the subject relax his or her abdominal muscles. If you do have a spring-loaded tape, pull the tape to the calibrated tension. Record the measurement in cm to the nearest mm at the end of a normal expiration.

Step 2: To measure the circumference of the hip, wrap the tape measure around the subject from his or her lateral side. Place the tape at the point of the largest circumference of the buttocks. Resolve any twists in the tape. Cross the tape and switch hands to avoid having your arms cross each other. (Doing this measure in front of a mirror can be helpful.) Confirm that the tape is parallel to the ground all the way around the subject. If you do have a spring-loaded tape, pull the tape to the calibrated tension. Record the measurement in cm to the nearest mm.

Step 3: Repeat steps 1 and 2 to complete the individual datasheet.

Step 4: Fill out the body composition group data sheet and complete the calculation of averages, BMI, and W/H ratio.

QUESTION SET 14.1

1. What are potential sources of error for BMI and W/H ratio?

2. What are the risk stratifications for BMI and W/H ratio? (See table 14.8.)

3. What sex-specific differences do you notice for W/H ratio on the group data sheet?

Find the case studies for this laboratory online at www.HumanKinetics.com/ LaboratoryManualForExercisePhysiology.

Table 14.8 Waist-to-Hip Circumference Ratio Norms for Men and Women

Age	Low	Moderate	High	Very high
		Risk		
	Low	Moderate	High	Very high
Men				
20–29	<0.83	0.83–0.88	0.89–0.94	>0.94
30–39	<0.84	0.84–0.91	0.92–0.96	>0.96
40–49	<0.88	0.88–0.95	0.96–1.00	>1.00
50–59	<0.90	0.90–0.96	0.97–1.02	>1.02
60–69	<0.91	0.91–0.98	0.99–1.03	>1.03
Women				
20–29	<0.71	0.71–0.77	0.78–0.82	>0.82
30–39	<0.72	0.72–0.78	0.79–0.84	>0.84
40–49	<0.73	0.73–0.79	0.80–0.87	>0.87
50–59	<0.74	0.74–0.81	0.82–0.88	>0.88
60–69	<0.76	0.76–0.83	0.84–0.90	>0.90

Adapted from Bray and Gray (10).

Laboratory Activity 14.1 Individual Data Sheet

Name or ID number: _____ Date: _____

Tester: _____ Time: _____

Sex: M / F (circle one)   Age: _____ y   Height: _____ in. _____ cm

Temperature: _____ °F _____ °C   Weight: _____ lb _____ kg

Barometric pressure: _____ mmHg   Relative humidity: _____ %

	Weight (kg)	Height (cm)	Waist (cm)	Hip (cm)	Evaluator (initials)
Trial 1					
Trial 2					
Trial 3					
Avg					

Calculation of BMI: Use average weight and height measurements.

BMI = _____ $kg \cdot m^{-2}$

Calculation of W/H: Use average waist and hip measurements.

W/H = _____ cm / _____ cm = _____

TECHNIQUES FOR MEASURING SKINFOLD THICKNESS

EQUIPMENT

- Skinfold calipers (Handle the calipers carefully—they are delicate and expensive!)
- Stadiometer or measuring tape
- Scale (accurate to a tenth of a kilogram)
- Individual and group data sheets

Find the group data sheets for this laboratory online at www.HumanKinetics.com/ LaboratoryManualForExercisePhysiology.

STANDARD SKINFOLD TEST PROCEDURES

Accurate skinfold measurement requires practice. Because of the variability present even in measurements taken by experienced technicians, three separate measures are made at each site, all on the same side of the body (e.g., the right side). Once a technician is experienced, it is common to repeat the measurements of sites only twice (provided they are within 10% of each other). This is an opportunity to practice professionalism. Subjects will remove clothing to allow access to skinfold sites. Always have the subject move clothing himself or herself—never reach under the subject's clothes. Making jokes is rarely useful in relieving tension; instead, foster a direct, professional clinical manner to put subjects at ease. Caution the subject not to try to assist you or lean away, since doing either can change anatomical orientations. It is common to measure and mark the skinfold site with a marker. However, this can affect the learning experience when several evaluators will be measuring the same subject; for this reason, we suggest measuring but not marking the subject in this lab.

Step 1: Subjects should wear loose clothing that facilitates measuring skinfolds at the appropriate sites.

Step 2: Each student should serve as a subject for two evaluators and serve as an evaluator for two subjects (if possible, one male and one female).

Step 3: Avoid making measurements through clothing. If this is not possible to avoid (e.g., as with a woman's sports bra), then subtract the thickness of the clothing from your skinfold measurement.

Step 4: Perform repeated measures at multiple sites by making one measure at each site (following the data sheet order), then repeating this sequence in a circuit. See the data sheet and the following list of anatomical site procedures for specialized instructions for each measurement.

Step 5: Pinch the appropriate skinfold using the thumb and index finger 1 cm (0.4 in.) proximal from where the jaws of the calipers will be placed. Inexperienced technicians often make the mistake of having their fingers too close together when taking a skinfold. Start wide and bring your fingers together to create a fold. It is impossible to make too big a fold, but it is possible to underestimate the fold by pinching too closely.

Step 6: Place the jaws of the calipers half the distance between the base of the skinfold that is not held with the fingers and the crest of the fold (see figure 14.2). The calipers should be perpendicular to the fold. Do not twist the calipers so they are easier to read.

Step 7: Record the reading from the scale on the calipers. Most calipers are accurate to the nearest 0.5 mm. Do not release the skinfold with your fingers until the calipers have been removed. Do not allow the calipers to snap shut!

Step 8: Proceed through the rest of the anatomical skinfold sites (descriptions follow these steps) and record all results for two evaluators at each of the nine sites.

Step 9: Skinfold trial outliers (>30% different from other trials) can be eliminated. Calculate averages for the remaining trials for a given evaluator on the individual data sheet.

Step 10: Each student should fill out the individual data sheet with his or her own skinfold measurements.

Step 11: Complete the sex-specific three- and seven-site equations for body density given earlier. Estimate body fat using the Siri equation. Use these data to answer the lab activity questions and report them in the group data sheet.

ANATOMICAL SITE SKINFOLD TEST PROCEDURES

These specific procedures complement the standard skinfold procedures already given. They provide the necessary details for working with the sites used by the Jackson/Pollock (4) and Jackson/Pollock/Ward (5) equations.

Triceps

- The subject stands, facing away from the technician with the right arm hanging straight down.
- The technician, standing behind the subject, pinches a site 1 cm (0.4 in.) above the midpoint between the shoulder (acromion process of the scapula)

and the tip of the elbow (inferior portion of the olecranon process of the ulna) on the posterior aspect of the triceps.

- Place the jaws of the calipers 1 cm below and perpendicular to the vertical fold. See figure 14.4a.

Suprailium

- Grasp the skinfold just above the iliac crest at the level of the anterior axillary line along the natural cleavage of the skinfold (running diagonally down the crest toward the umbilicus).
- Place the jaws of the calipers 1 cm (0.4 in.) distal and perpendicular to the diagonal fold. See figure 14.4b.

Abdomen

- Grasp a vertical skinfold 2 cm (0.8 in.) to the left of and 1 cm (0.4 in.) above the umbilicus.
- Place the jaws of the calipers 1 cm below and perpendicular to the vertical fold (to the right of the umbilicus). See figure 14.4c.

Chest

- For men, the chest skinfold is the midpoint between the anterior axillary line (the front of the armpit) and the nipple. Grasp the skinfold 1 cm (0.4 in.) above the midpoint in a diagonal line along the angle of the pectoralis major muscle. Place the jaws of the calipers 1 cm below and perpendicular to the diagonal fold.
- For women, take a diagonal fold as high as possible on the anterior axillary fold.
- Place the jaws of the calipers 1 cm below and perpendicular to the diagonal fold. See figure 14.4d.

Thigh

- Have the subject stand with all of his or her weight on the left leg and a slight bend at the knee in the right leg.
- Have the subject lift up his or her shorts to gain access to the anterior aspect of the thigh. Technicians often take folds too distally because of longer shorts.
- Take a vertical fold on the anterior aspect of the thigh 1 cm (0.4 in.) above the halfway point between the inguinal crease and the proximal border of the patella.
- Place the jaws of the calipers 1 cm below and perpendicular to the vertical fold. See figure 14.4e.

Midaxillary

- Take a vertical fold along the midaxillary line at the level of the xiphoid process of the sternum. The subject's arm is routinely placed on the tester's shoulder to best gain access to the midaxillary line.
- Place the jaws of the calipers 1 cm (0.4 in.) below and perpendicular to the vertical fold. See figure 14.4f.

Laboratory Activity 14.2

Subscapular

- Take a diagonal fold 1 to 2 cm (0.4–0.8 in.) from the inferior angle of the scapula. The subject can fold his or her arm behind the back to expose the scapula; once the spot is located, the arm is returned to the normal position.
- Place the jaws of the calipers 1 cm below and perpendicular to the diagonal fold. See figure 14.4g.

Biceps

- Take a vertical fold on the anterior aspect of the arm over the belly of the biceps muscle 1 cm (0.4 in.) above the level used to mark the triceps. The subject should rotate his or her palm anteriorly.
- Place the jaws of the calipers 1 cm below and perpendicular to the vertical fold. See figure 14.4h.

Calf

- Take a vertical fold on the medial aspect at the level of the maximal calf circumference with the knee and the hip flexed to 90 degrees (rest the foot on a stool or chair).
- Place the jaws of the calipers 1 cm (0.4 in.) below and perpendicular to the vertical fold. See figure 14.4i.

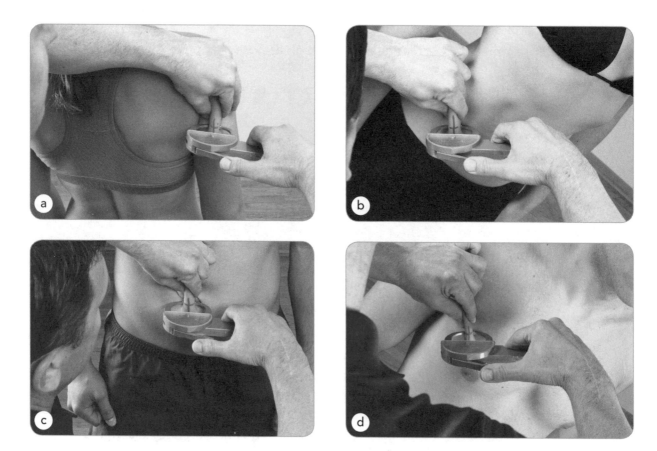

Figure 14.4 Skinfold measurement sites: *(a)* triceps, *(b)* suprailium, *(c)* abdomen, *(d)* chest, . . . *(continued)*

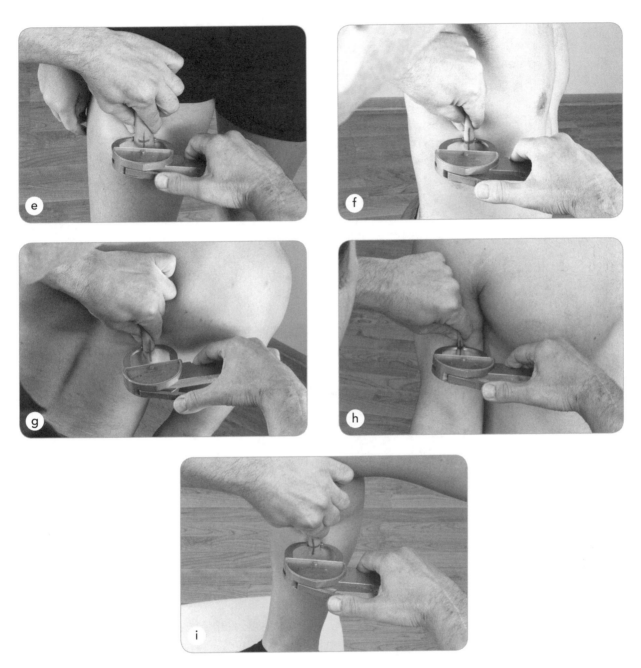

Figure 14.4 *(continued)* . . . *(e)* thigh, *(f)* midaxillary, *(g)* subscapular, *(h)* biceps, and *(i)* calf.

QUESTION SET 14.2

1. Calculate body fat using the numbers from the two evaluators. How did the skinfold measurements compare between the evaluators? How did the reproducibility turn out within each evaluator's measurements? What does this suggest about the reliability of skinfold measures? What does it suggest about the development of skinfold measurement skill?

2. What are potential sources of error for skinfold measurement?

3. Place all of the body composition measurements you performed on a continuum (go from least to most reliable). Do the same for validity. Briefly defend your answers.

4. What explanations might there be for differences in the %BF values between underwater weighing and skinfolds?

5. Briefly describe how you (as an exercise specialist, coach, researcher, athletic trainer, or physical therapist) might be able to use the information obtained from estimating body composition with your student, athlete, subject, or patient.

6. How do your own values for each of the measurements compare with the norms? What is the mean of all the measurements for your %BF? Which one do you believe is the most valid, and why?

7. Calculate your own desired body weight. How do you plan to attain (or maintain) this ideal weight? What assumptions are you making?

Find the case studies for this laboratory online at www.HumanKinetics.com/ LaboratoryManualForExercisePhysiology.

Laboratory Activity 14.2　Individual Data Sheet

Name or ID number: _____ Date: _____

Tester: _____ Time: _____

Sex: M / F (circle one)　Age: _____ y　Height: _____ in. _____ cm

Temperature: _____ °F _____ °C　Weight: _____ lb _____ kg

Barometric pressure: _____ mmHg　Relative humidity: _____ %

Measurement	Trial 1	Trial 2	Trial 3	Mean	Evaluator
Chest					
Midaxillary					
Subscapula					
Triceps					
Biceps					
Abdomen					
Supraillium					
Thigh					
Calf					

*Skinfold measurements are in mm.

Calculation of body density and body fat from Σ3SKF and Σ7SKF equations:

D_b (Σ3SKF) = _____ $g \cdot ml^{-1}$

D_b (Σ7SKF) = _____ $g \cdot ml^{-1}$

%BF (Σ3SKF) = _____ %

%BF (Σ7SKF) = _____ %

Laboratory Activity 14.2

ESTIMATING RELATIVE BODY FAT USING HYDRODENSITOMETRY

EQUIPMENT

- Underwater weighing (UWW) tank (custom tub in your exercise science lab or a swimming pool)
- Scale poised within the tank (suspended balance or load cell) that is able to support a chair or bench on which the subject sits to be suspended in the water
- Scale (accurate to a tenth of a kilogram)
- Stadiometer or measuring tape
- Individual data sheet

SUSPENDED BALANCE UNDERWATER WEIGHING

Step 1: Pretest instructions include a 3 h fast and abstinence from exercise for 24 h.

Step 2: Measure the subject's weight in the air. The subject should wear a swimsuit, ideally one that minimizes air trapped in suit, and remove all jewelry. Weigh the subject while he or she is dry.

Step 3: Have the subject shower to remove oils and lotions that can contaminate the UWW tank.

Step 4: The water temperature should be between 33 °C and 36 °C (for comfort).

Step 5: Calibrate the tank's scale before the subject gets into the tank. This is done simply by hanging known weights on the scale above the water level.

Step 6: The subject should get into the tank and submerge himself or herself to get completely wet, including the hair. Air bubbles should be removed from the bathing suit, hair, and skin, as well as from the swing or any other part of the equipment.

Step 7: Have the subject submerge up to his or her neck in the tank without touching the scale. Note the weight on the scale at this point. This is the "tare weight." The subject displaces water, which will change the weight of the scale. If the subject is wearing a weight during the underwater weighing, the weight should be secured to the swing when the tare weight is determined. Some systems allow you to zero at this tare weight. This process should be performed for every subject who is weighed, since different-size bodies can alter the tare weight.

Step 8: The subject should then sit in the chair or swing and practice the underwater weighing procedure. It may be necessary to have the subject wear a weight (about 2–3 kg), such as a diver's belt, in order to submerge fully. The subject should expel most of the air in the lungs before submerging, then submerge and expel any additional air. It is essential that neither the subject nor the swing touch the side of the tank. The subject should try to be as still as possible. Slight movements cause large oscillations in the scale, thus making it difficult to read. Have the subject attempt to count to 10 slowly and stay submerged during this time if possible. Record the underwater weight on the data sheet.

Step 9: Repeat the procedure with the subject three to five times until at least three stable readings are obtained. A classmate should make the readings of the scale. The scale may oscillate significantly; it should stabilize as the subject becomes suspended. The scale reader should determine the extremes of the needle oscillations when the fluctuations are stabilized and record the *midpoint* between these extremes. The goal is to read the scale to the nearest 20 g, but it may be possible only to read to the nearest 50 g. This level of precision may vary between systems. If your lab has load cells, even greater precision is possible.

Step 10: Fill out the body composition lab table, including your calculations for the subjects' percent body fat. (Use all three predicted RVs. Be sure to retain significant digits.)

QUESTION SET 14.3

1. What are potential sources of error for UWW?

2. What was the difference between the three RVs used for the calculations? What effect did this difference have on the calculated percent body fat? What % difference does this make in body fat?

3. For each of your UWW subjects, calculate the total amount (kg) of fat weight and lean body mass based on your underwater weighing results. Show your calculations.

4. Why might separate equations be developed for resistance trained individuals in the conversion from Db to %BF?

Find the case studies for this laboratory online at www.HumanKinetics.com/ LaboratoryManualForExercisePhysiology.

Name or ID number: _____   Date: _____

Tester: _____   Time: _____

Sex: M / F (circle one)   Age: _____ y   Height: _____ in. _____ cm

Temperature: _____ °F _____ °C   Weight: _____ lb _____ kg

Barometric pressure: _____ mmHg   Relative humidity: _____ %

Mass in air (M_A) _____ kg

RVA = _____ L

RVB = _____ L

RVC = _____ L

Water temp (T_w) _____ °C

D_w (table 14.1): _____

Tare weight: _____ g

Trial	1	2	3	4	5
Mass in water (M_w; g)					

Circle the three highest M_w values and use the average of those three: _____ g

$$\underbrace{\rule{3cm}{0.4pt}}_{\text{Mean } M_w \text{ (g)}} - \underbrace{\rule{3cm}{0.4pt}}_{\text{Tare weight (g)}} = \underbrace{\rule{3cm}{0.4pt}}_{\text{Net } M_w \text{ (g)}}$$

Calculations Using RV_A

$$\cfrac{\overbrace{\rule{5cm}{0pt}}^{M_A\,(g)}}{\cfrac{\underbrace{\rule{2cm}{0pt}}_{M_A\,(g)} - \underbrace{\rule{2cm}{0pt}}_{M_W\,(g)}}{\underbrace{\rule{2cm}{0pt}}_{D_W\,(g\cdot ml^{-1})}} - \underbrace{\rule{2cm}{0pt}}_{RV_A\,(ml)} - 100\ ml^*} = \underbrace{\rule{4cm}{0pt}}_{\text{Body density } (D_b;\, g\cdot ml^{-1})}$$

$$(495 / \underbrace{\rule{3cm}{0pt}}_{D_b\,(g\cdot ml^{-1})}) - 450 = \underbrace{\rule{3cm}{0pt}}_{\%\,fat}$$

Calculations Using RV$_B$

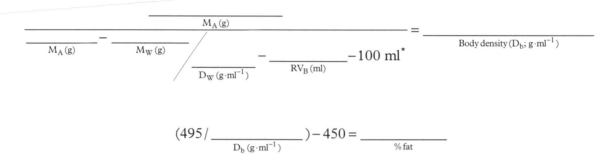

$$\frac{\cfrac{\cfrac{M_A\,(g)}{\underline{\hspace{3cm}}} }{\cfrac{\overline{M_A\,(g)} - \overline{M_W\,(g)}}{D_W\,(g\cdot ml^{-1})}} - \frac{}{RV_B\,(ml)} - 100\ ml^{*}} = \frac{}{\text{Body density}\,(D_b;\,g\cdot ml^{-1})}$$

$$(495 / \underset{D_b\,(g\cdot ml^{-1})}{\underline{\hspace{3cm}}}) - 450 = \underset{\%\,fat}{\underline{\hspace{3cm}}}$$

Calculations Using RV$_C$

$$\frac{\cfrac{\cfrac{M_A\,(g)}{\underline{\hspace{3cm}}} }{\cfrac{\overline{M_A\,(g)} - \overline{M_W\,(g)}}{D_W\,(g\cdot ml^{-1})}} - \frac{}{RV_C\,(ml)} - 100\ ml^{*}} = \frac{}{\text{Body density}\,(D_b;\,g\cdot ml^{-1})}$$

$$(495 / \underset{D_b\,(g\cdot ml^{-1})}{\underline{\hspace{3cm}}}) - 450 = \underset{\%\,fat}{\underline{\hspace{3cm}}}$$

*Note: This 100 ml is the estimated volume of gas in the gastrointestinal tract.

Laboratory Activity 14.3

Electrocardiograph Measurements

Objectives

1. Introduce the terminology associated with electrocardiography.
2. Understand the electrical activity of the heart and the 3-D view of the heart through ECG leads.
3. Learn the procedure for electrode placement for a 12-lead ECG and observe a resting ECG.
4. Learn to determine heart rate and heart axis from an ECG recording.
5. Demonstrate a 12-lead ECG during submaximal exercise.
6. Learn basic interpretation of ECG strips.

Definitions

ectopic pacemaker—Heart cell that is not in the sinoatrial node but originates electrical activity in the heart.

electrocardiograph (ECG or EKG)—Graphical recording of the electrical activity of the heart.

heart axis—Angular position of the heart within the chest cavity.

mean vector—Sum of the electrical vectors of the heart, which determines the heart's placement in the body, or axis.

pacemaker—Heart cell or group of cells that originates the electrical activity for the entire heart (normally, the sinoatrial node).

12-lead ECG—Set of leads originating from 10 electrodes placed on a subject to record electrical activity of the heart.

All cells in our bodies can conduct a current of electricity, and two of the most electrically active tissues are the brain and heart. Currents conducted in these tissues can be recorded through electrodes placed on the skin and connected to an **electrocardiograph (ECG or EKG)**. The resulting three-dimensional graphical recording provides detailed information about heart's rate and rhythm, current performance, and history or risk of injury. Many clinicians spend a lifetime pouring over ECGs to characterize a subject's heart performance. Clinical exercise physiologists often use the stress of exercise to expose individuals who may be at risk for a heart injury. Certifications in this area may allow a student to seek employment in the growing field in clinical exercise physiology and cardiac rehabilitation (1).

Electrical Activity of the Heart

The electrical activity in a normal heart originates in the sinoatrial (SA) node, which serves as the normal **pacemaker** of the heart. However, any heart cell, or myocyte, can initiate pacemaking, and when a cell not in the SA node acts as the pacemaker it is called an **ectopic pacemaker**. From the SA node, electrical activity spreads through the atria, taking advantage of the electrical gap junctions or intercalated discs between myocytes. These specialized junctions allow the spread of depolarization without the need for a chemical synapse. The atria and ventricles of a heart are divided by a thin connective tissue; therefore, they act as separate electrical units. The wave of atrial depolarization is funneled through the atrial–ventricular (AV) node. Upon its reaching the AV node, there is a slight pause before a rapid depolarization is sent down the AV bundle, bundle branches, and Purkinje fibers (3). See figure 15.1.

These conductive fibers in the ventricles are specialized to conduct the electrical signal very rapidly, which is particularly important to the ventricles since they have the responsibility of ejecting blood to our entire bodies, including (against the force of gravity) to our brain. This rapid transmission of the depolarization through the ventricles allows a coordinated contraction to improve the efficiency of blood ejection. The wave of depolarization from cell to cell using intercalated discs travels at about $0.3 \text{ m} \cdot \text{s}^{-1}$, whereas the bundle branches and Purkinje fibers can conduct the electrical current at $>3 \text{ m} \cdot \text{s}^{-1}$. An ECG records this electrical activity and provides information about the quantity and quality of the heart's rhythm.

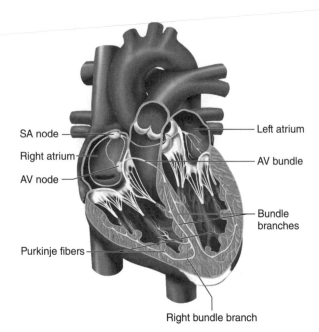

Figure 15.1 Conduction system of the heart.

Reprinted, by permission, from J.H. Wilmore, D.L. Costill, and W. L. Kenney, 2008, *Physiology of sport and exercise*, 4th ed. (Champaign, IL: Human Kinetics), 128.

An ECG readout has three parts: the **P wave**, the **QRS complex**, and the **T wave**. These waves represent the depolarization of the different heart chambers and thus the contraction of these areas (see figure 15.2).

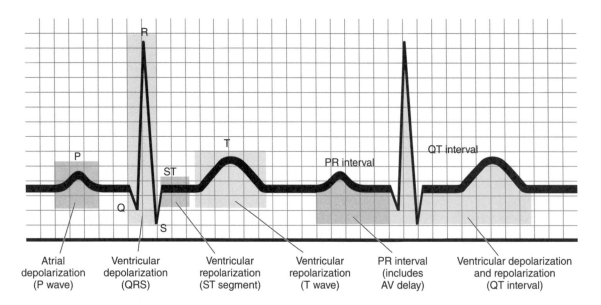

Figure 15.2 Waves of the electrocardiogram.

Reprinted, by permission, from W.L. Kenney, J.H. Wilmore, and D.L. Costill, 2012, *Physiology of sport and exercise*, 5th ed. (Champaign, IL: Human Kinetics), 147.

Electrical impulses travel from the SA node in the right atrium throughout both atria to generate the first wave of the ECG—the P wave, which thus represents atrial depolarization. The impulse then travels to the AV node, where there is a slight pause as the ventricles fill, then travels through the AV Bundle and to the Purkinje fibers located in the walls of the ventricles. As the wave of depolarization spreads from the AV node throughout the ventricles, a wave of deflection is shown on the ECG. Known as the QRS complex, it represents ventricular depolarization. The final wave of the ECG tracing, the T wave, represents repolarization of the ventricles. The atrial repolarization is obscured on the tracing by the QRS complex; therefore, it cannot be seen.

In the heart, the electrical events of depolarization precede the mechanical events of contraction. However, much information about the heart's contraction can be determined from the electrical activity. The axes for an ECG recording are in millivolts (mV) on the y-axis and time on the x-axis, which allows a clinician to determine time delays and quantity of electrical activity. For example, large P waves or QRS complexes indicate greater amounts of muscle in the atria or ventricles, respectively. This type of hypertrophy may indicate valvular defects or even adaptations to exercise. The time for atrial depolarization (P wave) takes longer than ventricular depolarization (QRS). This fact relates to the previously mentioned specialized fibers within the ventricles. In addition, the direction of the electrical signal can be determined. ECG leads have direction, so if a wave is in the direction of an ECG lead, then the wave will be upright, or positive. Conversely, if the wave is in the opposite direction of the lead it will be downward, or negative. As you might then expect, if a wave is equally upward and downward in its inflection, it is considered to be at 90° to the ECG lead.

Placement of ECG Leads

Although an ECG can be done with just one electrode and is commonly done with four, a **12-lead ECG** is completed with 10 carefully placed electrodes (see figure 15.4). Together, the electrodes make up leads, which can be thought of as electrical vectors that have direction from a negative to positive pole. This allows creation of a 3-D picture of the heart and the direction of its electrical activity. There are 6 limb leads from the 4 torso or limb electrodes and 6 chest leads from the 6 chest electrodes. Leads are considered to be the voltage difference between one or more electrodes. The bipolar limb leads (I, II, and III) are called such because they use the combination of positive and negative poles between limb electrodes. For example, lead I is the voltage between the left arm (LA) and right arm (RA) electrodes. The other 3 limb leads are called the augmented limb leads—aVR, aVL, and aVF. These leads are unipolar leads, even though they also have negative and positive poles; the negative pole is a composite of several other electrodes. They are considered augmented because the negative pole is a combination of two electrodes, which augments the signal strength. For example, aVR uses the RA as the positive electrode, and the negative pole is a combination of the LA and left leg (LL) electrodes. Note that the right leg (RL) acts as an electrical ground. See figure 15.3.

The chest or precordial leads (V1–V6) are considered unipolar as well, and they view the heart in a horizontal plane due to their close proximity.

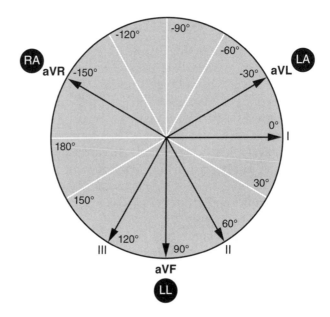

Figure 15.3 ECG lead vectors.

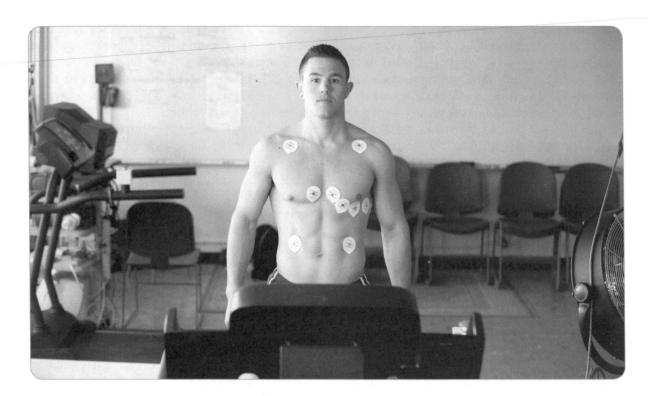

Figure 15.4 12-lead electrode placement.

Their placement across the chest allows additional electrical vectors to aid in the determination of electrical direction, heart rate, or axis.

Interpreting the ECG Recording

Much can be determined from an ECG recording or "strip," including heart rate, heart axis, clinical diagnoses, and performance of the heart contractions. We focus here on heart rate and axis as relatively easy determinations. Normal sinus rhythm occurs when the SA node is acting as the pacemaker of the heart. Sinus bradycardia involves a normal heart rhythm but slow heart rate at rest (below 60 beats $\cdot$ min^{-1}). Sinus tachycardia involves a normal heart rhythm but elevated heart rate at rest (>100 beats $\cdot$ min^{-1}).

Calculating heart rate from an ECG strip involves relatively easy math. Normally, ECG paper speed is 25 mm $\cdot$ s^{-1} or 1,500 mm $\cdot$ min^{-1}. Therefore, in order to calculate heart rate, one simply needs to measure the number of mm between R waves and use this simple equation:

$$\text{Heart rate (beats} \cdot \text{min}^{-1}) = 1500 \,/\, \text{R-to-R distance (mm)}$$

This measurement can be made with a metric ruler or with the knowledge that each small box on the ECG strip is 1 mm (0.04 s) and each bigger box is 5 mm (0.2 s). It is common to measure the distance from R to R across several cardiac cycles to get an average of the R-to-R distances, since heart rate variability is significant in some individuals. For example, imagine we measure 50 mm between the four R waves in a normal sinus rhythm ECG. This would be an average of 16.67 mm $\cdot$ beat^{-1}, since this covers three R-to-R distances. The HR could be calculated as follows: 1500 mm $\cdot$ min^{-1} / 16.67 mm $\cdot$ beat^{-1} = 90 beats $\cdot$ min^{-1}. In figure 15.2, the 18 mm between the R waves would yield an HR of 83.33 beats $\cdot$ min^{-1}. Clearly, this method depends on the assumed paper speed time of 25 mm $\cdot$ s^{-1}, so it is essential to confirm that this holds true for your ECG machine. Other methods exist for calculating heart rate from the ECG strip, such as dividing 300 by the number of larger boxes (5 mm). This approach may save time and give the interpreter a rough estimate of heart rate. Some ECG strips provide tick marks in 1, 3, or 6 s intervals to allow an estimate of heart rate by multiplying the number of R intervals within the tick marks by 60, 20, or 10 respectively. However, these estimates require that the interpreter estimate the number of partial beats between the time intervals, which introduces bias and potential error. Measuring the distance between R waves remains the most accurate, albeit time consuming, method of attaining heart rate from an ECG strip. In reality, ECG machines automatically calculate heart rate; however, we suggest that the lab instructor mask this output in order for students to learn the process of performing this calculation.

The axis of the heart is its angular position within the chest cavity. Heart axis can be affected by body position, heart hypertrophy, pregnancy, and some disease states. Determining **heart axis** from the direction of the electrical vectors is also useful in locating the specific location of damage in a diseased heart. The mean direction of the total electrical activity of the heart gives it its axis, or **mean vector**. Because of the mass of the ventricles, the direction of the QRS complex in the limb leads (I, II, III, aVL, aVR, and aVF), is used to determine axis. Because of the wave of depolarization from the SA node throughout the heart, a normal healthy heart under sinus rhythm has a 60° axis, which causes a positive (upward) QRS deflection in lead II. (Refer back to figure 15.3 for degrees of the heart axis in relation to the electrical vectors.) This axis causes an equally positive and negative deflection (isoelectric) in the leads perpendicular to this axis, in this case the aVL lead, and an almost completely negative deflection in the aVR lead, since it is in a near opposite direction. Normal heart axes range from −30° to 90°, whereas a right heart axis ranges from 90° to 180°, and a left heart axes ranges from −30° to −90°. Thus most hearts have an axis between leads aVL, I, II, and aVF; therefore, a positive QRS deflection in all or most of these leads suggests normal heart axis. The axis is then narrowed to a more specific degree amount by the examination of isoelectric and negative QRS leads. Deviations from normal axis can indicate left or right ventricular hypertrophy, right or left bundle branch blocks, other electrical abnormalities, or a myocardial infarction (heart attack).

ECG as a Tool for Diagnosing Cardiac Abnormalities

This laboratory introduces you to the concepts of the ECG. It is not meant to provide complete training regarding the diagnostic uses of ECG; however, several simple abnormalities can be dis-

cussed in this introductory lab. For a more complete discussion of ECG interpretation see source 2 in this chapter's reference list. The following list presents ECG abnormalities that differ from a normal sinus rhythm.

- **ST segment abnormalities**—The period of time between S and the beginning of the T wave can indicate abnormalities in blood flow to the heart. ST segment elevation indicates an already existing heart attack or some level of necrotic heart tissue. On the other hand, reduced coronary blood flow, or ischemia, puts a person at risk for a future heart attack. Heart ischemia appears as ST segment depression on an ECG. The severity of the discrepancy of the ST segment from the isoelectric line is proportional to the severity of the disease. ST segment depression may be unapparent at rest yet appear during the stress of exercise. See figure 15.5.

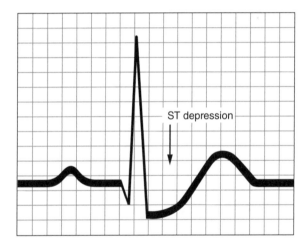

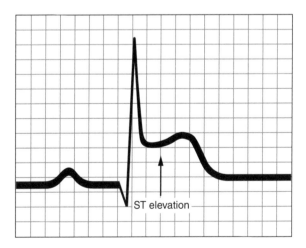

Figure 15.5 ST segment depression and elevation.

• **PVCs (preventricular contractions)**— These contractions occur when a myocyte in the ventricles initiates the heart contraction, thereby acting as an ectopic pacemaker. PVCs are identified by a wide QRS (see figure 15.6). They are relatively common at rest, but if the frequency increases significantly during exercise they may result in poor ventricular performance. A sequence of several PVCs indicates greater severity of this abnormality. The term *bigeminy* refers to an abnormal heartbeat, such as a PVC, occurring every other cardiac cycle (see figure 15.7). *Trigeminy* refers to abnormal beats occurring every third cardiac cycle.

• **Ventricular tachycardia**—Ventricular tachycardia, or V-tach, is a run of consecutive PVCs. If V-tach is sustained, it requires immediate medical attention, typically including CPR (cardiopulmonary resuscitation) or use of an AED (automated external defibrillator) to shock the heart back into sinus rhythm.

• **Ventricular fibrillation**—Ventricular fibrillation, or V-fib, is a continual spasm of the ventricles disallowing appropriate blood ejection.

This abnormality requires immediate medical attention, typically including CPR (cardiopulmonary resuscitation) or use of an AED (automated external defibrillator) to shock the heart back into sinus rhythm. See figure 15.8.

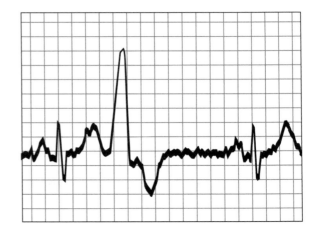

Figure 15.6 Preventricular contraction (PVC).

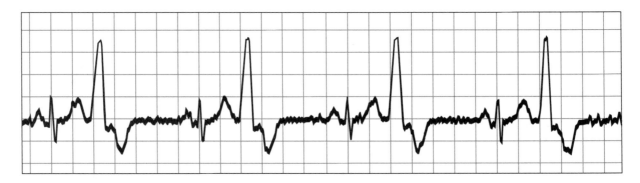

Figure 15.7 Bigeminy.

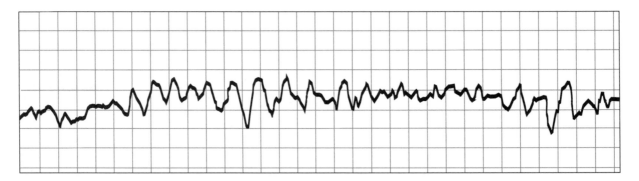

Figure 15.8 Ventricular fibrillation (V-fib).

- **Asystole**—Asystole involves a flatline ECG requiring immediate medical attention, typically including use of an AED (automated external defibrillatory) to shock the heart back into sinus rhythm.

- **PACs (premature atrial contractions)**—PAC's are similar to PVC's, in that they are ectopic contractions of the atria. But because they are in the atria, they are far less worrisome than PVC's. PAC's normally go untreated, however, they may lead to atrial flutter or fibrillation. See figure 15.9.

- **Atrial flutter/fibrillation**—Atrial flutter is an uncoordinated, very rapid (>250 beats · min⁻¹!) contraction of the atria. Though not necessarily dangerous in itself, it jeopardizes complete filling of the ventricles and is normally found in older individuals with other heart disease or chronic obstructive pulmonary disease. It may be temporary but may also lead to atrial fibrillation. Flutter is different from fibrillation in that the atria beat regularly in flutter but irregularly in fibrillation. Atrial flutter/fibrillation may go unnoticed by the patient; however, it may also lead to more serious ventricular problems. Atrial fibrillation is more likely to cause an irregular contraction of the ventricles and may also predispose a patient for stroke. See figure 15.10.

Heart Rate Response to Exercise

As you have likely observed in previous labs, heart rate increases during graded exercise in direct proportion to the intensity of the exercise. This regulation of heart rate during exercise is controlled by the autonomic nervous system (3). Heart rates below 100 beats · min⁻¹ are due to parasympathetic innervations of the SA node. Increasing the heart rate from rest to exercise involves both parasympathetic withdrawal and sympathetic stimulation of the SA node. See figures 15.11 and 7.1 (p. 166).

With each increase in intensity during a graded exercise test, heart rate increases for the first 1 or 2 min, then levels off at steady state as it meets the new metabolic demands—the higher the intensity, the longer it takes the heart rate to reach steady state. At maximum intensity (exhaustion), the heart rate *may* plateau; this is then the maximum heart rate.

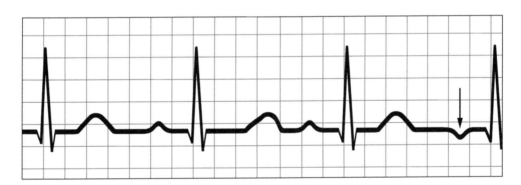

Figure 15.9 Premature atrial contractions (PACs).

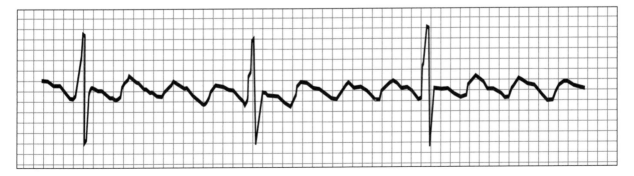

Figure 15.10 Atrial flutter/fibrillation.

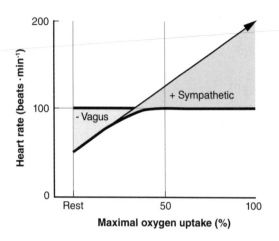

Figure 15.11 Autonomic nervous system control of heart rate.

Reprinted, by permission, from W.L. Kenney, J.H. Wilmore, and D.L. Costill, 2012, *Physiology of sport and exercise,* 5th ed. (Champaign, IL: Human Kinetics), 146.

Aside from changes in heart rate, ECGs may look different during exercise of increasing intensity. Often, a person with occasional PVCs at rest may have a normal sinus rhythm during exercise. Conversely, and more dangerously, the frequency of PVCs or PACs may increase with exercise. In addition, changes may occur in the ST segment; of particular interest is ST segment depression, which indicates myocardial ischemia. The intensity at which ST segment depression appears is referred to as an *ischemic threshold,* and it constitutes reason to stop a test

(1). In your introduction to ECGs, you are very unlikely to observe such changes, which are less common among young, apparently healthy individuals such as most of your labmates. Changes you may observe include increased level of noise from the recruitment of respiratory muscles and the jostling of the electrodes, which make ECG interpretation difficult at higher intensities. This noise can be reduced by keeping the electrode connections as still as possible and having the subject refrain from overly swinging the arms.

Athletes often get misdiagnosed with pathological hearts due to the effect of exercise training on their hearts (2). Common changes in an athlete's heart can include sinus bradycardia, changes in ST segment and T waves (often appearing as ST segment elevation due to the increased size of the T wave), ventricular hypertrophy, bundle branch or AV blocks, and various arrhythmias. These are thought of not as clinically significant but as byproducts of the heart's adaptations to exercise.

References

1. American College of Sports Medicine. *ACSM's Guidelines for Exercise Testing and Prescription.* 8th ed. Philadelphia: Lippincott Williams & Wilkins, 2010.

2. Thaler MS. *The Only EKG Book You'll Ever Need.* Philadelphia: Lippincott Williams & Wilkins, 2007.

3. Kenney WL, Wilmore JH, and Costill DL. *Physiology of Sport and Exercise.* 5th ed. Champaign, IL: Human Kinetics, 2012.

RESTING ECG

EQUIPMENT

- ECG machine
- Disposable ECG electrodes
- Alcohol prep pads, gauze, scrubbing pad, disposable razor
- Individual data sheet

6-LEAD RESTING ECG TEST

Step 1: All students should obtain a resting 6-lead ECG of their heart.

Step 2: ECG electrodes measure electricity of the heart; therefore, any impedance to conductivity impairs the electrical recording. Remove all hair, body oil, dead skin, and lotion from the electrode sites prior to placing the disposable electrode.

Step 3: Prep your labmate, using the four limb electrodes (LA, RA, LL, RL). Locate the correct site for electrode placement (see the following list) and prep the area by shaving (if necessary), scrubbing with an abrasive pad, and cleaning with alcohol. Women should be instructed ahead of time to wear a swimsuit top or sports bra. Use discretion with female subjects and maintain a professional approach at all times. The limb electrodes are often placed on the ankles and wrists; however, since we are exercise physiologists, we will prep the limb electrodes on the torso as if the subject were going to exercise. Follow these anatomical locations (see also figure 15.12):

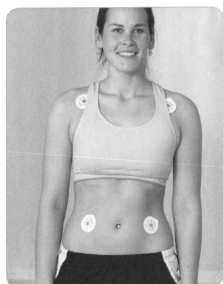

LA (left arm): Just below subjects mid left clavicle above pectoralis muscle, medial to anterior deltoid

RA (right arm): Just below subjects mid right clavicle above pectoralis muscle, medial to anterior deltoid

LL (left leg): Between subjects left obliquus abdominis and rectus abdominis muscle, level with the umbilicus

RL (right leg): Between subjects right obliquus abdominis and rectus abdominis muscle, level with the umbilicus

Figure 15.12 6-lead electrode placement.

Step 4: Connect the labeled ECG wires to the correct ECG electrodes.

Step 5: Obtain an ECG on each person while he or she sits quietly in a chair.

Step 6: Be sure to note the ECG paper speed.

Step 7: Complete the individual data sheet.

QUESTION SET 15.1

1. Using the data from your personal ECG, calculate your heart rate with the 3 methods described (1500 / R-to-R distance; # of R-to-R in tick marks; 300 / # big boxes). Show your work and attach your ECG.

2. Did these three methods result in different heart rates? Why or why not?

3. Describe the cascade of electrical activity through the heart with a normal sinus rhythm. What controls heart rate?

4. Label one cardiac cycle identifying the P wave, QRS complex, T wave, PR interval, and ST segment. Explain what each represents in terms of heart action.

5. What is the purpose of using a 6-lead ECG as opposed to only 1 lead?

6. Determine the axis of your heart. Does your heart axis fall within the normal range?

Find the case studies for this laboratory online at www.HumanKinetics.com/ LaboratoryManualForExercisePhysiology.

Name or ID number: _____ Date: _____

Tester: _____ Time: _____

Sex: M / F (circle one)   Age: _____ y   Height: _____ in. _____ cm

Temperature: _____ °F _____ °C   Weight: _____ lb _____ kg

Barometric pressure: _____ mmHg   Relative humidity: _____ %

Heart Rate

A. Distance between 4 R waves _____ mm / 3 = _____ mm · beats^{-1}

   1,500 mm · min^{-1} / _____ mm · beat^{-1} = _____ beats · min^{-1}

B. 300 big boxes · min^{-1} / _____ big boxes · beats^{-1} = _____ beats · min^{-1}

C. ECG tick mark distance = _____ s

# of R waves · tick^{-1} _____ × _____ ticks · min^{-1} = _____ beats · min^{-1}

Heart Axis

A. Lead with the most positive defection _____ = _____ °

B. Lead with the nearest isoelectric defection _____ = _____ °

C. Lead with the most negative defection _____ = _____ °

D. Estimated heart axis = _____ °

▶ Paste ECG here:

EFFECTS OF BODY POSITION ON THE AXIS OF THE HEART

EQUIPMENT

- ECG machine
- Disposable ECG electrodes
- Alcohol prep pads, gauze, scrubbing pad, disposable razor
- Individual data sheet

6-LEAD BODY POSITION ECG TEST

Step 1: Prep one student in your group for a 6-lead ECG of his or her heart.

Step 2: ECG electrodes measure electricity of the heart; therefore, any impedance to conductivity impairs the electrical recording. Remove hair, body oil, dead skin, and lotion from the electrode sites prior to placing the disposable electrode.

Step 3: Prep your subject, using the four limb electrodes (LA, RA, LL, RL). Locate the correct site for electrode placement (see the following list) and prep the area by shaving (if necessary), scrubbing with an abrasive pad, and cleaning with alcohol. Women should be instructed ahead of time to wear a swimsuit top or sports bra. Use discretion with female subjects and maintain a professional approach at all times. The limb electrodes are often placed on the ankles and wrists; however, since we are exercise physiologists, we will prep the limb electrodes as if the subject were going to exercise. Follow these anatomical locations (see picture of correct limb electrode placement in figure 15.12):

LA (left arm): Just below subjects mid left clavicle above pectoralis muscle, medial to anterior deltoid

RA (right arm): Just below subjects mid right clavicle above pectoralis muscle, medial to anterior deltoid

LL (left leg): Between subjects left obliquus abdominis and rectus abdominis muscle, level with the umbilicus

RL (right leg): Between subjects right obliquus abdominis and rectus abdominis muscle, level with the umbilicus

Step 4: Connect the labeled ECG wires to the correct ECG electrodes.

Step 5: Obtain an ECG on your subject with him or her in the supine position. Allow 2 to 3 min for the subject to relax and come to steady state.

Step 6: Have the subject sit in a chair slumped over. Allow about 2 min in this position, then obtain an ECG recording.

Step 7: Have the subject stand erect. Allow about 2 min in this position, then obtain an ECG recording.

Step 8: Note the ECG paper speed.

Step 9: Complete the individual data sheet.

QUESTION SET 15.2

1. Using the ECG recordings from your subject in the three body positions, calculate heart rate while supine, sitting, and standing. Show the technique that you used and attach the ECG.

2. How does body position affect heart rate? What is occurring physiologically to cause these changes?

3. Using the ECG recordings from your subject in the three body positions, determine heart axis while supine, sitting, and standing.

4. How does body position affect heart axis? What is occurring to cause these changes?

 Find the case studies for this laboratory online at www.HumanKinetics.com/ LaboratoryManualForExercisePhysiology.

Name or ID number: _____ Date: _____

Tester: _____ Time: _____

Sex: M / F (circle one)   Age: _____ y   Height: _____ in. _____ cm

Temperature: _____ °F _____ °C Weight: _____ lb _____ kg

Barometric pressure: _____ mmHg   Relative humidity: _____ %

Heart Rate

A. Supine: Distance between 4 R waves _____ mm /3 = _____ mm · beats^{-1}

  1,500 mm · min^{-1} / _____ mm · beats^{-1} = _____ beats · min^{-1}

B. Sitting: Distance between 4 R waves _____ mm/3 = _____ mm · beat^{-1}

  1,500 mm·min^{-1} / _____ mm · beats^{-1} = _____ beats · min^{-1}

C. Standing: Distance between 4 R waves _____ mm /3 = _____ mm · beat^{-1}

  1,500 mm · min^{-1} / _____ mm · beats^{-1} = _____ beats · min^{-1}

Heart Axis

A. Supine: Lead with the most positive defection _____ = _____ °

  Lead with the nearest isoelectric defection _____ = _____ °

  Lead with the most negative defection _____ = _____ °

  Estimated heart axis = _____ °

B. Sitting: Lead with the most positive defection _____ = _____ °

  Lead with the nearest isoelectric defection _____ = _____ °

  Lead with the most negative defection _____ = _____ °

  Estimated heart axis = _____ °

C. Standing: Lead with the most positive defection _____ = _____ °

  Lead with the nearest isoelectric defection _____ = _____ °

  Lead with the most negative defection _____ = _____ °

  Estimated heart axis = _____ °

▶ Paste ECG here:

SUBMAXIMAL EXERCISE EFFECTS WITH THE 12-LEAD ECG

EQUIPMENT

- ECG machine
- Treadmill
- Disposable ECG electrodes
- Alcohol prep pads, gauze, scrubbing pad, disposable razor
- Individual data sheet

12-LEAD ECG

Step 1: Prep one student in your group for a 12-lead ECG of his or her heart.

Step 2: ECG electrodes measure electricity of the heart; therefore, any impedance to conductivity impairs the electrical recording. Remove all hair, body oil, dead skin, and lotion from the electrode sites prior to placing the disposable electrode.

Step 3: Prep your subject, using 10 electrodes (LA, RA, LL, RL, V1–V6). Locate the correct site for electrode placement (see the following list) and prep the area by shaving (if necessary), scrubbing with an abrasive pad, and cleaning with alcohol. Use discretion with female subjects and maintain a professional approach at all times. Women should be instructed ahead of time to wear a swimsuit top or sports bra. Women may want to find the electrode sites themselves. A curtained area in the lab can provide a convenient space for women to prepare for a 12-lead ECG, perhaps with the aid of a female labmate. The limb electrodes are often placed on the ankles and wrists; however, since we are exercise physiologists, we will prep the limb electrodes for exercise. Follow these anatomical locations (see figure 15.13):

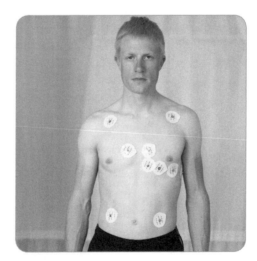

Figure 15.13 12-lead electrode placement.

LA (left arm): Just below subjects mid left clavicle above pectoralis muscle, medial to anterior deltoid

RA (right arm): Just below subjects mid right clavicle above pectoralis muscle, medial to anterior deltoid

LL (left leg): Between subjects left obliquus abdominis and rectus abdominis muscle, level with the umbilicus

RL (right leg): Between subjects right obliquus abdominis and rectus abdominis muscle, level with the umbilicus

For the chest leads, place the electrodes in the order given in the following list. You must palpate for the intercostal spaces between the ribs (you cannot rely on sight). Begin by palpating just below the clavicle left of the sternum. The first space is the first intercostal space. Then count downward to the fourth intercostal space for V2. Left and right refer to the subject's left and right.

V2: On the left sternal border on the fourth intercostal space

V1: On the right sternal border on the fourth intercostal space

V4: On the midclavicular line in the fifth intercostal space

V3: At the midpoint of a straight line between V2 and V4

V5: On the anterior axillary line horizontal to V4

V6: On the midaxillary line horizontal to V4 and V5

Note that V1, V2, and V4 are in the spaces between the ribs, not on the rib bones themselves. This difference is important because bone does not conduct electricity well.

Step 4: Connect the labeled ECG wires to the correct ECG electrodes.

Step 5: Collect the ECG wires and secure them to be safe and keep them out of the subject's way while he or she exercises on the treadmill. This is most frequently done with a soft Velcro strap.

Step 6: Have the subject attain steady state while in the standing position for 2 to 3 min. It is advisable to get a resting ECG in the position in which the subject will be exercising—that is, standing for a treadmill or sitting for a bike. This is useful due to the effects of body position on the ECG as demonstrated in lab activity 15.2. If a treadmill is not available, it is possible to use a bicycle ergometer; however, clinical testing is more frequently done on treadmills. Obtain a 12-lead resting ECG recording for your subject.

Step 7: Have the subject straddle the treadmill. Bring the speed and grade of the treadmill to 1.7 mph and 10% grade. This is the first stage of the Bruce protocol that is often used in exercise stress testing (see lab 9 for details). Allow 3 min for the subject to come to steady state. Obtain a 12-lead exercise ECG recording for your subject. (See figure 15.14.)

Step 8: Increase speed and grade to 2.5 mph and 12% grade. Allow 3 min for the subject to come to steady state. Obtain a 12-lead exercise ECG recording.

Step 9: Increase speed and grade to 3.4 mph and 14% grade. Allow 3 min for the subject to come to steady state. Obtain a 12-lead exercise ECG recording.

Step 10: Stop the treadmill and allow your subject to recover while sitting in a chair. Monitor the 12-lead ECG. Following 3 min of recovery, obtain a 12-lead ECG recording.

Step 11: Fill out the corresponding individual data sheet with data from the test.

QUESTION SET 15.3

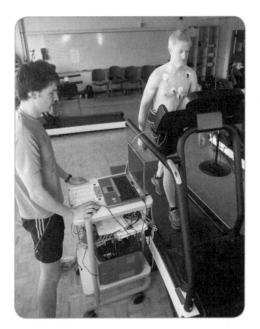

Figure 15.14 Exercising ECG.

1. Using the ECG recordings, calculate the subject's heart rate while at rest, during each stage of exercise, and during recovery. Show the technique that you used and attach the ECGs.

2. How does exercise of increasing intensity change heart rate? What controls this change?

3. Using the ECG recordings, determine the subject's heart axis while at rest, during each stage of exercise, and during recovery.

4. How does exercise of increasing intensity change heart axis?

5. What changes are anticipated in an ECG recording in the transition from rest to exercise? What abnormal or pathological changes may occur in the transition from rest to exercise?

Find the case studies for this laboratory online at www.HumanKinetics.com/ LaboratoryManualForExercisePhysiology.

Name or ID number: _____   Date: _____

Tester: _____   Time: _____

Sex: M / F (circle one)   Age: _____ y   Height: _____ in. _____ cm

Temperature: _____ °F _____ °C   Weight: _____ lb _____ kg

Barometric pressure: _____ mmHg   Relative humidity: _____ %

Time (min)	Speed (mph)	Grade (%)	Power (kg · m · min⁻¹)	Avg R-to-R (mm)	HR (beats · min⁻¹)	Heart axis (°)	RPE
0	0	0	Rest				
3	1.7	10					
6	2.5	12					
9	3.4	14					
12	0	0	Recovery				

▶ Paste ECG here:

Units of Measure and Conversions

Length (Distance)

1 m = 39.370 in. = 3.281 ft = 1.094 yd

1 km = 0.621 mi

1 cm = 0.394 in.

1 in. = 2.540 cm

1 ft = 12 in. = 30.480 cm

1 yd = 3 ft = 0.914 m

1 mi = 5,280 ft = 1,760 yd = 1,609.344 m

Velocity

$1 \text{ mi} \cdot \text{h}^{-1}$ (mph) = $26.822 \text{ m} \cdot \text{min}^{-1}$ = $1.467 \text{ ft} \cdot \text{s}^{-1}$ = $0.447 \text{ m} \cdot \text{s}^{-1}$

$1 \text{ m} \cdot \text{s}^{-1}$ = 2.237 mi/h (mph) = $3.281 \text{ ft} \cdot \text{s}^{-1}$

Force

1 N = 0.225 lb of force = 0.102 kg of force

1 lb force = 4.448 N

1 kg force = 9.81 N

Torque

$1 \text{ N} \cdot \text{m}$ = 0.738 ft-lb

Mass and Weight

1 kg = 2.205 lb

1 kp = 1 kg

1 g = 0.035 oz

1 lb = 16 oz = 0.454 kg

1 oz = 28.350 g

(1 L of water weighs 1 kg)

Energy

$1 \text{ kcal} = 4.186 \text{ kJ} = 426.935 \text{ kg} \cdot \text{m} = 1.163 \text{ W} \cdot \text{h}$

$1 \text{ BTU} = 0.252 \text{ kcal} = 1.055 \text{ kJ} = 107.586 \text{ kg} \cdot \text{m}$

$1 \text{ J} = 1 \text{ N} \cdot \text{m} = 0.102 \text{ kg-m} = 0.239 \text{ cal}$

$1 \text{ kg} \cdot \text{m} = 1 \text{ kp} \cdot \text{m} = 9.807 \text{ J} = 2.342 \text{ cal}$

1 L of oxygen consumed = 5.05 kcal = 21.143 kJ (at an RER of 1.00)

Power

$1 \text{ W} = 1 \text{ J} \cdot \text{s}^{-1} = 6.118 \text{ kg} \cdot \text{m} \cdot \text{min}^{-1} = 0.860 \text{ kcal} \cdot \text{hr}^{-1} = 0.00134 \text{ hp}$

$1 \text{ kg} \cdot \text{m} \cdot \text{min}^{-1} = 1 \text{ kp} \cdot \text{m} \cdot \text{min}^{-1} = 0.163 \text{ W}$

Pressure

1 atm = 760 mmHg = 101.325 kPa = 14.696 psi

1 mmHg = 1 torr = 0.0193 psi = 133.322 Pa = 0.00132 atm

1 kPa = 0.01 mbar

Temperature

$°C = 0.555 \times [(°F) - 32]$

$°F = 1.8 \times [(°C) + 32]$

Volume

1 L = 1.057 qt

1 qt = 0.946 L = 2 pt = 32 oz

1 (U.S.) gal = 4 qt = 128 oz = 3.785 L

1 c = 8 oz = 0.237 L

1 oz = 2 tbsp = 6 tsp = 29.574 ml

1 tbsp = 3 tsp = 14.787 ml

1 tsp = 4.929 ml

Quantity of a Substance

1 mol of a gas = 22.4 L (standard conditions) = 6.022×10^{23} molecules (Avogadro's number)

1 L of gas (standard conditions) = 44.6 mmol

mol = mass (g) / molecular mass

molarity of a solution = mol of substance / L of solvent

Estimation of the O_2 Cost of Walking, Running, and Leg Ergometry

Use the following equations to estimate the O_2 requirement for walking and running.

Treadmill Walking

The oxygen cost of horizontal treadmill walking can be estimated for treadmill speeds of 50 to 100 $m \cdot min^{-1}$ by using a formula based on the linear relationship between treadmill speed and oxygen consumption ($\dot{V}O_2$ in $ml \cdot kg^{-1} \cdot min^{-1}$). The relationship possesses a slope of 0.1 and a y-intercept of 3.5 $ml \cdot kg^{-1} \cdot min^{-1}$ (resting $\dot{V}O_2$). Thus:

$$\dot{V}O_2 \text{ (horizontal component)} =$$
$$\frac{0.1 \, ml \cdot kg^{-1} \cdot min^{-1}}{m \cdot min^{-1}} \times$$
$$\text{speed} (m \cdot min^{-1}) + 3.5 \, ml \cdot kg^{-1} \cdot min^{-1}$$

The oxygen cost of graded treadmill walking is as follows:

$$\dot{V}O_2 \text{ (vertical component)} =$$
$$\frac{1.8 \, ml \cdot kg^{-1} \cdot min^{-1}}{m \cdot min^{-1}} \times$$
$$\text{speed} (m \cdot min^{-1}) \times \text{fractional grade}$$

The $\dot{V}O_2$ required for graded treadmill walking is the sum of the horizontal and vertical O_2 costs. Thus, the O_2 cost of walking at 90 $m \cdot min^{-1}$ up a 7.5% grade would be as follows:

Horizontal $\dot{V}O_2 =$
$[0.1 \, ml \cdot kg^{-1} \cdot min^{-1} \times 90 \, m \cdot min^{-1}] +$
$3.5 \, ml \cdot kg^{-1} \cdot min^{-1} = 12.5 \, ml \cdot kg^{-1} \cdot min^{-1}$

Vertical $\dot{V}O_2 = 1.8 \, ml \cdot kg^{-1} \cdot min^{-1} \times [90$ $m \cdot min^{-1} \times 0.075] = 12.2 \, ml \cdot kg^{-1} \cdot min^{-1}$

Total $\dot{V}O_2 = 12.5 \, ml \cdot kg^{-1} \cdot min^{-1} + 12.2$ $ml \cdot kg^{-1} \cdot min^{-1} = 24.7 \, ml \cdot kg^{-1} \cdot min^{-1}$

Treadmill Running

The oxygen cost for horizontal and graded treadmill running for speeds greater than 134 $m \cdot min^{-1}$ can be calculated in a manner similar to that used for treadmill walking. The $\dot{V}O_2$ ($ml \cdot kg^{-1} \cdot min^{-1}$) for horizontal treadmill running is calculated using the following formula:

$$\dot{V}O_2 \text{ (horizontal component)} =$$
$$\frac{0.2 \, ml \cdot kg^{-1} \cdot min^{-1}}{m \cdot min^{-1}} \times$$
$$\text{speed} (m \cdot min^{-1}) + 3.5 \, ml \cdot kg^{-1} \cdot min^{-1}$$

The $\dot{V}O_2$ for graded treadmill running is given by the following:

$$\dot{V}O_2 \text{ (vertical component)} =$$
$$\frac{0.9 \, ml \cdot kg^{-1} \cdot min^{-1}}{m \cdot min^{-1}} \times$$
$$\text{speed} (m \cdot min^{-1}) \times \text{fractional grade}$$

Total $\dot{V}O_2$ = horizontal $\dot{V}O_2$ + vertical $\dot{V}O_2$

Leg Ergometry

The oxygen cost for leg ergometry, or stationary cycling, can be calculated in a similar manner:

$$\dot{V}O_2 \, (ml \cdot kg^{-1} \cdot min^{-1}) = (10.8 \times \text{workload (W)} \times 1 / \text{body wt (kg)}) + 7 \, ml \cdot kg^{-1} \cdot min^{-1}$$

or

$$\dot{V}O_2 \, (ml \cdot kg^{-1} \cdot min^{-1}) = 1.8 \times \text{workload} (kg \cdot m \cdot min^{-1}) \times \text{body wt (kg)} + 3.5 \, ml \cdot kg^{-1} \cdot min^{-1}$$

Haldane Transformation

Current day metabolic carts often take away from students the knowledge of what is actually being collected and calculated. In reality, they only measure O_2 and CO_2 in the expired breath (F_EO_2 and F_ECO_2) and the volume of air expired ($\dot{V}_E$), while the rest of the metabolic cart output is generated by calculations based on the Haldane transformation. The following calculations assume expired gases from an exercising individual were collected in Douglas bags and analyzed with appropriate equipment for F_EO_2, F_ECO_2, and $\dot{V}_E$.

The Haldane transformation is based on the fact that N_2 is neither produced nor utilized by the body. Thus, the volume of N_2 inhaled must be equivalent to the volume of N_2 exhaled. This fact is expressed by the formula below: (see definitions below)

$$\dot{V}_I \times F_IN_2 = \dot{V}_E \times F_EN_2$$

We can rearrange this formula to solve for V_I as follows:

$$\dot{V}_I = (\dot{V}_E \times F_EN_2) / F_IN_2$$

We know that $F_IN_2 = 0.7904$ and $F_EN_2 = 1.0 - F_EO_2 - F_ECO_2$. Both F_EO_2 and F_ECO_2 can be measured by gas analyzers, while $\dot{V}_E$ can be measured by a tissot or other gas volume analyzer. Thus, $\dot{V}_I$ can be calculated from these measured values and constants.

Constants

$$F_ICO_2 = 0.0003$$

$$F_IO_2 = 0.2093$$

$$F_IN_2 = 0.7904$$

From Douglas bag collection and gas analysis, you will know the other three unknown parameters (fraction of expired CO_2 and O_2 and expired gas volume or $\dot{V}_E$). With these, along with ambient temperature (T_A) and barometric pressure (P_A), you can determine energy expenditure.

Conversion of V_{ATPS} to V_{STPD}

In all metabolic calculations, gas volumes are always expressed at Standard Temperature, Pressure, Dry (STPD). Thus it is necessary to reduce a gas volume to STPD before plugging $\dot{V}_E$ into the $\dot{V}_I$ equation. To do so, use the following formula:

$$\dot{V}_{E\,STPD} = \dot{V}_{E\,ATPS} \times [273 / (273 + T_A)]\,[(P_A - P_{H2O}) / 760]$$

Calculation of Oxygen Consumption and Carbon Dioxide Production

Calculation of oxygen consumption or use ($\dot{V}O_2$) is relatively simple. The amount of oxygen exhaled is subtracted from the amount of oxygen inhaled as follows:

$$\dot{V}O_2 = (\dot{V}_I \times F_IO_2) - (\dot{V}_E \times F_EO_2)$$

Carbon dioxide production ($\dot{V}CO_2$) is the difference between the CO_2 expired and the CO_2 inspired. It can be calculated as follows:

$$\dot{V}CO_2 = (\dot{V}_E \times F_ECO_2) - (\dot{V}_I \times F_ICO_2)$$

Of course, from these results, RER can be easily calculated as $\dot{V}CO_2 / \dot{V}O_2$.

This may seem like a confusing array of equations, and for this reason we suggest the following order of operations:

1. From the expired gas collected, express $\dot{V}_E$ in $L \cdot min^{-1}$ (from tissot or similar device to measure total gas expired and knowledge of collection time).

2. Using the equations presented earlier in this appendix, convert to STPD.

3. Using the relevant equations, calculate $\dot{V}_I$.

4. Using the relevant equations and the results from F_EO_2 and F_ECO_2, calculate $\dot{V}O_2$.

5. Using the relevant equations and the results from F_EO_2 and F_ECO_2, calculate $\dot{V}CO_2$.

6. Using the relevant equations, calculate RER.

Definitions of Terms

$\dot{V}CO_2$—amount of carbon dioxide produced ($L \cdot min^{-1}$)

$\dot{V}O_2$—amount of oxygen consumed ($L \cdot min^{-1}$)

$\dot{V}_I$—volume of gas inspired ($L \cdot min^{-1}$)

$\dot{V}_E$—volume of gas expired ($L \cdot min^{-1}$)

F_IN_2—fraction of inspired nitrogen

F_EN_2—fraction of expired nitrogen

F_ICO_2—fraction of inspired carbon dioxide

F_ECO_2—fraction of expired carbon dioxide

F_IO_2—fraction of inspired oxygen

F_EO_2—fraction of expired oxygen

T_A—ambient temperature (°C)

P_A—atmospheric pressure (mmHg)

P_{H2O}—partial pressure of water vapor (mmHg)

APPENDIX D

Metabolic Cart Information

Several companies provide machines capable of measuring oxygen consumption, carbon dioxide production, and ventilation. They include the following:

- ParvoMedics: www.parvo.com
- COSMED: www.cosmed.it
- New Leaf: www.newleaffitness.com
- AEI Technologies: www.aeitechnologies.com
- VacuMed: www.vacumed.com

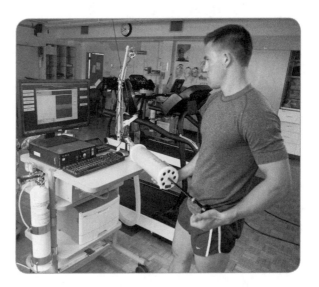

These carts work on the basis of a relatively simple concept. They measure the amount of O_2 and CO_2 in the room air and in expired breath, as well as the volume of expired breath. They do so by means of O_2 and CO_2 analyzers and a flow meter. Knowing these results, along with the atmospheric temperature and pressure, enables you to do simple calculations to determine $\dot{V}O_2$ and $\dot{V}CO_2$ (see appendix C).

It is essential to properly calibrate these carts. To do so, your lab needs to contain reference gases of known concentration (usually around 16.0% O_2 and 4.0% CO_2) and a turbine of known volume (usually 3 L). It may also be necessary to purchase

a portable weather station for measuring ambient barometric pressure, humidity, and temperature. You can then calibrate the cart to the room air concentrations of O_2 and CO_2 (20.93% O_2 and 0.03% CO_2, respectively) and the reference gases. You can also calibrate its flow meter with the known turbine. Your lab instructor will lead you through your lab's metabolic cart calibration.

Calibration of Equipment

The entire field of exercise physiology is based on the measurement of metabolic responses to exercise at a given absolute or relative workloads. Equipment such as bicycle ergometers and treadmills are essential in eliciting measureable workloads. Because of this, it is important to be able to determine if a workload is accurately reflected by what is "dialed in" for a piece of equipment. For example, one must be able to determine whether a treadmill set to 6 mph and 8% grade is actually moving at that speed and grade. This appendix aids students in calibrating their labs' treadmills and bicycle ergometers.

Calibration of a Treadmill

Countless brands of treadmills exist on the market today, and it is not unusual for treadmills to be significantly wrong about speed or grade. Because of this, it is important to calibrate the speed and grade as described here. To do this, you will need some tape, a measuring device, stopwatch and a calculator.

Speed

To check the speed of a treadmill, place a piece of tape on the belt perpendicular to its movement. Measure the belt length by measuring from the front edge of this piece of tape to where the belt loops under the machine. Then, place another piece of tape where this measurement stopped (you may want to label tape pieces as *1,2,3,. . .* or *A, B, C, . . .*). Move the belt along by standing on the treadmill and pushing with your feet, but do not start the treadmill at this point. Once the belt length is measured as accurately as possible, remove all but one piece of tape. Start the treadmill at a relatively slow speed (0.5-2 mph). With a stopwatch, measure the amount of time it takes for the piece of tape to make 20 revolutions to the nearest hundredth of a second. To calculate actual speed of the treadmill, take the belt length multiplied by number of revolutions divided by the time. Or, you can use this formula:

Treadmill speed (distance / time) =
(belt length × 20) / time for 20 revolutions

This process should be repeated a minimum of two times at this speed to reduce human error. Also, repeat this procedure at four different speeds to see if any difference in actual speed is a function of the speed the treadmill is running. When timing revolutions at higher speeds, it is important to use more than 20 revolutions to increase the validity and reliability of the measurement. It may also be advised to repeat this procedure with someone walking on the treadmill, as his or her weight may slow the motor of the treadmill.

If significant differences are found between treadmill speed and actual speed, check the treadmill instruction manual for specific calibration specifications. Your lab may also have a service contract for the treadmill if it is under warranty. Even if the treadmill cannot be fixed, corrections can be made while performing exercise tests. For example if a reading of "4 mph" corresponds to an actual speed of 3.8 mph, then calculations for workload can be adjusted accordingly.

Grade

It is also important to calibrate the grade of your lab treadmill. Most lab treadmills express grade as percent. That is, they express grade as rise over run times 100. This is similar to, but different from degree slope. For example, a 10% grade is equal to a 5.7° slope. Because treadmills and prediction equations are expressed in *% grade,* these same units are used below.

It may go unrecognized that because of different treadmills and different floors, a treadmill reading of 0% grade may not be correct. To check this, use a level on the treadmill. With the treadmill off, and set to 0% grade, the tread should be level to the ground. Surprisingly, this is often not the case, as floors settle and move with time. Another way to check this if you don't have a level is to measure from the ground to the top of the tread in the back of the treadmill and at the front. The measurements should be identical. If the treadmill is not at zero, then it may be possible to shim the treadmill up to zero.

To check the % grade you must first know the *run*. Measure the run from the front edge to the back edge of the treadmill at a convenient clear point, such as the roller. Set the treadmill at 0 mph and at 1-3% grade. Now measure the 'rise' by measuring the difference between the front and back of the treadmill. However, by raising the front of the treadmill, the 'run' has shortened. To account for this, you should remeasure the run using a carpenter's square aligned with the front of the treadmill, ensuring a right angle. Otherwise, you can use the Pythagorean theorem to calculate the run. The pythagorean theorem is

$$\text{rise}^2 + \text{run}^2 = \text{hypotenuse}^2$$

Rewritten to solve for run, it is:

$$\text{run} = \sqrt{\text{rise}^2 + \text{hypotenuse}^2}$$

where the hypotenuse is the original length of the treadmill at zero percent grade. Once rise and run are determined, calculate actual percent grade:

$$\% \text{ grade} = \text{rise} / \text{run} \times 100$$

Repeat this procedure to ensure reliability of the measurement. Repeat this procedure at several different grades to ensure correct calibration.

If the percent grade of your treadmill is off but the zero is correct, your particular treadmill instruction manual may have calibration specifications. You may also have a service contract with your treadmill if it is under warranty. At the very least, on-site corrections can be made while performing exercise tests. For example if 10% grade is really 9.8%, then calculations for workload can be adjusted accordingly.

Calibration of a Bicycle Ergometer

Most bicycle ergometers are based on the premise that flywheel resistance is provided by a strap or rope. Thus under the premise of a fixed pedal cadence, the tighter the strap, the greater resistance to the flywheel, and thus the greater the exercise intensity (see lab 7, p. 182 for directions on how to calculate $kg \cdot m \cdot min^{-1}$ on a bicycle). Some new ergometers are electronically braked, which makes calibration difficult or impossible. However, the following is a description of how to calibrate a bicycle ergometer similar to a Monark, which are the types often used in exercise physiology labs. It involves three steps: zeroing, resistance calibration, and checking the strap condition.

Zeroing

The first thing to check is that when there is zero resistance to a flywheel, the scale reads *0*. Many of the scale boards that allow the pendulum to indicate resistance in kilograms (kg or kp) actually move themselves. There may be a thumbscrew that can be loosened, which will allow you to move the 0 over the pendulum, and then retighten the screw. This is often done with the strap or rope completely removed from the flywheel to ensure zero resistance. Be careful not to allow the strap to fall into the enclosed flywheel, however, since it can be difficult to retrieve. See figure E.1.

Figure E.1 The scale board, pendulum, and flywheel of a bicycle ergometer.

Resistance Calibration

The next step of calibration also requires the strap to be disconnected. From the top strap, hang a weight between 0.25 and 7.0 kg. With this known weight on the strap, the resistance indicator on the bike should display the appropriate amount of weight. Ideally, several weights are hung to ensure calibration between the ranges used in an exercise test. Be sure the weight does not rub or hook on anything while hanging from the strap. If the resistance indicator on the bicycle is not displaying the correct weight, several things can be done. Once several weights across a range have been tested, a linear regression can be completed. This would allow one to calculate the 'real' resistance when a particular weight is indicated on the bike. Many models of bicycle ergometers allow you to adjust the pendulum as well. Adjusting the pendulum might involve using a hex wrench or other equipment to move weight toward or away from the pendulum in order to achieve the correct weight calibration.

Strap Condition

Since resistance on the flywheel is a function of strap friction, the nylon straps can become worn, develop burrs, or even melt over time. Examine your straps regularly to ensure good condition and replace them when needed. Strap life can be a result of a dirty flywheel. Clean the flywheel with a dry cloth during calibration, paying special attention to burrs or other surface irregularities that may affect contact with the strap.

Monark provides an online version of their instruction manual for the popular 828E model that is available at www.fitnesslyceum.com/ProductVideo/monark-828e-exercise-bike-manual.pdf

Certifications in the Field of Exercise Science

Not all certifications are created equal. Many personal trainer exams can be taken online and completed without qualifications. Some of the most reputable certifications are offered by two main governing bodies—the American College of Sports Medicine (ACSM, www.acsm.org) and the National Strength and Conditioning Association (NSCA, http://nsca-lift.org/).

ACSM has three tracks of certification: (1) health and fitness certifications focused on apparently healthy adults, (3) clinical certifications focused on clients with disease, and (3) specialty certifications for working with certain populations. For more information, see ACSM's certification web page: www.acsm.org/AM/Template.cfm?Section=Get_Certified.

ACSM Health Fitness Certifications

Certified Group Exercise Instructor (GEI)

An ACSM-certified group exercise instructor is proficient in developing and implementing various forms of exercise in a group setting, modifying exercises based on individual and group needs, and creating a positive and effective exercise environment. (www.acsm.org)

Certified Personal Trainer (CPT)

As an ACSM-certified personal trainer, you will be qualified to develop and implement exercise programs for apparently healthy individuals and those who have medical clearance to exercise. (www.acsm.org)

Certified Health Fitness Specialist (HFS)

Considered ACSM's advanced personal trainer certification, the certified health fitness specialist certification requires the specialist to have at least an associate's degree and be qualified to work with special populations with medically controlled diseases who have been cleared by their physician for independent exercise. (www.acsm.org)

Clinical Certifications

Certified Clinical Exercise Specialist (CES)

ACSM-certified clinical exercise specialists are health care professionals with a bachelor's degree who typically work in cardiovascular or pulmonary rehabilitation programs, physicians' offices, or medical fitness centers. (www.acsm.org)

Registered Clinical Exercise Physiologist (RCEP)

ACSM-certified clinical exercise specialists are health care professionals with a graduate degree who typically work in cardiovascular or pulmonary rehabilitation programs, physicians' offices, or medical fitness centers. (www.acsm.org)

Specialty Certifications

Certified Cancer Exercise Trainer (CET)

This certification enables trainers to use basic understanding of cancer diagnoses, surgeries, treatments, symptoms, and side effects to assess, develop exercise programs, and train clients who are in any of the various stages of cancer diagnosis or treatment. (www.acsm.org)

Certified Inclusive Fitness Trainer (CIFT)

A person with CIFT training is a fitness professional who assesses, develops, and implements an individualized exercise program for persons who have a physical, sensory, or cognitive disability and who are healthy or have medical clearance to perform independent physical activity. (www.acsm.org)

Physical Activity in Public Health Specialist (PAPHS)

A person with PAPHS training promotes physical activity in public health through engaging and educating key decision makers about the effect of and need for legislation, policies, and programs that promote physical activity. (www. acsm.org)

NSCA Certifications

"NSCA focuses its certifications on working with apparently healthy populations. Its offerings include the certified strength and conditioning specialist (CSCS) and the certified personal trainer (CPT)." (http://nsca-lift.org/)

For more information, see NSCA's certification website: http://nsca-lift.org/certification.shtml.

Certified Strength and Conditioning Specialist (CSCS)

The CSCS program was initiated in 1985 to identify individuals who possess the knowledge and skills to design and implement safe and effective strength and conditioning programs. The program encourages a higher level of competence among practitioners that raises the quality of strength training and conditioning programs provided by those who are CSCS certified. To be eligible to take the exam, you must be CPR certified and hold a BA or BS degree, be enrolled as a college senior at an accredited college or university, or hold a degree in chiropractic medicine. (http://nsca-lift.org/)

Certified Personal Trainer (NSCA-CPT)

The NSCA-CPT credential is designed for professionals who work one-on-one with their clients in a variety of environments, including YMCAs, schools, health and fitness clubs, and clients' homes. The exam thoroughly tests the knowledge and skills necessary to successfully train active and sedentary physically healthy individuals, as well as those who are elderly or obese. Personal trainers with specialized expertise may also be involved in training clients with orthopedic, cardiovascular, and other chronic conditions. NSCA-CPT examination candidates must be CPR certified in order to sit for the exam. (http://nsca-lift.org/)